BIOORGANIC CHEMISTRY

Volume II   Substrate Behavior

CONTRIBUTORS

ARIEH ABRAHAM

NIELS H. ANDERSEN

RONALD E. BARNETT

HARRY P. BROQUIST

H. F. DELUCA

KEIICHIRO FUKUMOTO

ERWIN GLOTTER

F. PETER GUENGERICH

RICHARD K. HILL

MASATAKA IHARA

TETSUJI KAMETANI

ISAAC KIRSON

DANIEL J. KOSMAN

JAMES P. KUTNEY

DAVID LAVIE

PERCY S. MANCHAND

WALTER G. NIEHUAS, JR.

YOSHIMOTO OHTA

RONALD J. PARRY

EUGENE G. SANDER

H. K. SCHNOES

DANIEL D. SYRDAL

JAMES D. WHITE

BIOORGANIC CHEMISTRY

Edited by
E. E. van Tamelen

Department of Chemistry
Stanford University
Stanford, California

Volume II
SUBSTRATE BEHAVIOR

A treatise to supplement Bioorganic Chemistry
An International Journal

Edited by
E. E. van Tamelen

ACADEMIC PRESS   New York   San Francisco   London   1978

A Subsidiary of Harcourt Brace Jovanovich, Publishers

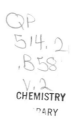

ACADEMIC PRESS, INC.
111 Fifth Avenue, New York, New York 10003

United Kingdom Edition published by
ACADEMIC PRESS, INC. (LONDON) LTD.
24/28 Oval Road. London NW1

Library of Congress Cataloging in Publication Data

Main entry under title:

Bioorganic chemistry.

    Includes bibliographies and index.
    CONTENTS: v. 1.   Enzyme action.–v. 2.   Substrate
behavior.–v. 3.   Macro- and multimolecular systems.
    1.   Biological chemistry.   2.   Chemistry, Organic.
I.   Van Tamelen, Eugene E.,   Date
QP514.2.B58   574.1'92   77-45994
ISBN 0–12–714302–5

Contents

List of Contributors

Numbers in parentheses indicate the pages on which the authors' contributions begin.

ARIEH ABRAHAM (57), Agricultural Research Organization, Volcani Center, Bet Dagan, Israel

NIELS H. ANDERSEN (1), Department of Chemistry, University of Washington, Seattle, Washington

RONALD E. BARNETT* (39), Department of Chemistry, University of Minnesota, Minneapolis, Minnesota

HARRY P. BROQUIST (97), Department of Biochemistry, Vanderbilt University School of Medicine, Nashville, Tennessee

H. F. DELUCA (299), Department of Biochemistry, College of Agricultural and Life Sciences, University of Wisconsin—Madison, Madison, Wisconsin

KEIICHIRO FUKUMOTO (153), Pharmaceutical Institute, Tohoku University, Aboyama, Sendai, Japan

ERWIN GLOTTER† (57), Department of Organic Chemistry, The Weizmann Institute of Science, Rehovot, Israel

F. PETER GUENGERICH (97), Department of Biochemistry, Vanderbilt University School of Medicine, Nashville, Tennessee

*Present address: Department of Biochemistry and Nutrition, Virginia Polytechnic Institute and State University, Blacksburg, Virginia.

†Present address: Faculty of Agriculture, The Hebrew University of Jerusalem, Rehovot, Israel.

RICHARD K. HILL (111), Department of Chemistry, University of Georgia, Atlanta, Georgia

MASATAKA IHARA (153), Pharmaceutical Institute, Tohoku University, Aboyama, Sendai, Japan

TETSUJI KAMETANI (153), Pharmaceutical Institute, Tohoku University, Aboyama, Sendai, Japan

ISAAC KIRSON (57), Department of Organic Chemistry, The Weizmann Institute of Science, Rehovot, Israel

DANIEL J. KOSMAN (175), Bioinorganic Graduate Research Group, Departments of Chemistry and Biochemistry, State University of New York at Buffalo, Buffalo, New York

JAMES P. KUTNEY (197), Department of Chemistry, University of British Columbia, Vancouver, Canada

DAVID LAVIE (57), Department of Organic Chemistry, The Weizmann Institute of Science, Rehovot, Israel

PERCY S. MANCHAND (337), Chemical Research Department, Hoffmann-LaRoche, Inc., Nutley, New Jersey

WALTER G. NIEHAUS, JR. (229), Department of Biochemistry and Nutrition, Virginia Polytechnic Institute and State University, Blacksburg, Virginia

YOSHIMOTO OHTA (1), Department of Chemistry, University of Washington, Seattle, Washington

RONALD J. PARRY (247), Department of Chemistry, Brandeis University, Waltham, Massachusetts

EUGENE G. SANDER (273), Department of Biochemistry, School of Medicine, West Virginia University, Morgantown, West Virginia

H. K. SCHNOES (299), Department of Biochemistry, College of Agricultural and Life Sciences, University of Wisconsin—Madison, Madison, Wisconsin

DANIEL D. SYRDAL (1), Department of Chemistry, University of Washington, Seattle, Washington

JAMES D. WHITE (337), Department of Chemistry, Oregon State University, Corvallis, Oregon

Foreword

What is bioorganic chemistry? It is the field of research in which organic chemists interested in natural product chemistry interact with biochemistry. For many decades the natural product chemist has been concerned with the way in which Nature makes organic molecules. In the absence of any information other than that provided by structure, conclusions had necessarily to be derived from structural analysis. Broad groups of natural products could be recognized, such as alkaloids, isoprenoids, and polyketides (acetogenins), which clearly had elements of structure indicating a common biosynthetic origin. Indeed, for alkaloids and terpenoids, structural work was greatly helped by such biogenetic hypothesis. Similarly, after A. J. Birch had made an extensive analysis of polyketides, the repeating structural element postulated also helped in the determination of structure.

The alternative, and complement, to the above analysis is to consider the chemical mechanisms whereby the units of structure are assembled into the final natural product. For example, alkaloid structure can often be analyzed in terms of anion–carbonium ion combination. Also, the later stages of biosynthesis of many alkaloids can be analyzed by the concept of phenolate radical coupling. In polyisoprenoids the critical mechanism for carbon–carbon bond formation is the carbonium ion–olefin interaction to give a carbon–carbon bond and regenerate a further carbonium ion.

The analysis of natural product structures in terms of either structural units or mechanisms of bond formation has been subjected to rigorous tests since radioactively labeled compounds became generally available. It is gratifying that, on the whole, the theories developed from structural and mechanistic analysis have been fully confirmed by *in vivo* experiments.

Organic chemists have always been fascinated by the possibility of imitating in the laboratory, but without the use of enzymes, the precise steps of a biosynthetic pathway. Such work may be called biogenetic-type, or biomimetic, synthesis. This type of synthesis is a proper activity for the bioorganic chemist and undoubtedly deserves much attention. Nearly all such efforts are, however, much less successful than Nature's synthetic activities using enzymes. It is well appreciated that Nature has solved the outstanding problem of synthetic chemistry, viz., how to obtain 100% yield and complete stereospecificity in a chemical synthesis. It, therefore, remains a major task for bioorganic chemists to understand the mechanism of enzyme action and the precise reason why an enzyme is so efficient. We are still far from the day when we can construct an organic molecule which will be as efficient a catalyst as an enzyme but which will not be based on the conventional polypeptide chain.

Much of contemporary bioorganic chemistry is presented in these volumes. It will be seen that much progress has been made, especially in the last two decades, but that there are still many fundamental problems left of great intellectual challenge and practical importance.

The world community of natural product chemists and biochemists will be grateful to the editor and to all the authors for the effort that they have expended to make this work an outstanding success.

DEREK BARTON
Chemistry Department
Imperial College of Science and Technology
London, England

Preface

Although natural scientists have always been concerned with the development and behavior of living systems, only in the twentieth century have investigators been in a position to study on a molecular level the intimate behavior of organic entities in biological environments. By mid-century, the form and function of various natural products were being defined, and complex biosynthetic reactions were even being simulated in the nonenzymatic laboratory. As the cinematographic focus on biomolecules sharpened, one heard increasingly the adjective *bioorganic* applied to the interdisciplinary area into which such activity falls.

In 1971, publication of a new journal, *Bioorganic Chemistry*, was begun. As a follow-up, what could be more timely and useful than a well-planned, multivolume collection of bioorganic review articles, solicited from carefully chosen professionals, surveying the entire field from all possible vantage points? This four-volume work contains a collection, but it did not originate in this manner.

As the journal *Bioorganic Chemistry* developed, the number and quality of regular, original research articles were maintained at an acceptable level. However, comprehensive review articles appeared only sporadically, despite their intrinsic value at a time when general interest in bioorganic chemistry was burgeoning. In order to enhance this function of the journal, as well as to mark the fifth anniversary of its birth, we originally planned to publish in 1976 a special issue comprised entirely of reviews by active practitioners. After contact with a handful of stalwart bioorganic chemists, about two hundred written invitations for reviews were mailed during late 1975 to appropriate, diverse scientists throughout

the world. The response was overwhelming! More than seventy preliminary acceptances were received within a few months, and it soon became
evident that the volume could not be handled adequately through publication by journal means. After consultation with representatives of Academic
Press, we agreed to publish the manuscripts in book form.

Although the stringency of journal deadlines disappeared, the weightier
matter of editorial treatment had to be reconsidered. Should contributions
be published in the same, piecemeal, random fashion as received? Such
practice would be acceptable for journal dissemination, but for book
purposes, broader, more orderly, and inclusive treatment might be desir-
able and also expected. Partly because of editorial indolence, but mostly
because of a predilection for maintaining the candor and spontaneity
which might be lost with increased editorial control, we decided not to
attempt coverage of all identifiable areas of bioorganic chemistry, not to
seek out preferentially the recognized leaders in particular areas, and
even not to utilize outside referees. Consequently, we present reviews
composed by scientists who were not coerced or pressured, but who
wrote freely on subjects they wanted to write about and treated them as
they wanted to, at the cost perhaps of a certain amount of objectivity and
restraint as well as proper coverage of some important bioorganic areas.

We turn now to the results of this publication project. Because of the
inevitable attrition for the usual reasons, fewer than the promised number
of reviews materialized: fifty-seven manuscripts were received in good
time and accepted by this office. Eight countries are represented by the
entire collection, which emanates almost entirely from academia, as
would be expected. A great variety of topics congregated—greater than
we had foreseen. Inclusion of all papers in one volume was impractical,
and thus the problem arose of logically dividing the heterogeneous material into several unified subsections, each suitable for one volume, a
problem compounded by the fact that an occasional author elected to
treat, in one manuscript, several unconnected topics happening to fall in
his purview. Therefore, perfect classification without discarding or dis-
secting bodies of material as received was simply not possible.

After some reflection and a few misconceptions, we evolved a plan for
division into four more or less scientifically integral sections; these, hap-
pily, also constitute approximately equal volumes of written material, an
aspect of some importance to the publisher. The enzyme–substrate interaction was expected to be a well-represented subject, and, in fact, too
many manuscripts on this subject for one proportionally sized volume
were received. Although the separation of enzyme action and substrate
behavior is contrived and not basically justifiable, it turned out that, for
the most part, a group of authors heavily emphasized the former, while
another concentrated on the latter. Accordingly, Volume I was entitled

"Enzyme Action," and Volume II "Substrate Behavior." Admittedly, in a few cases, articles could be considered appropriate for either volume.

A gratifyingly significant number of contributions dealt with the behavior of biologically important polymers and related matters, sent in by authors having quite different investigational approaches. In addition, several discourses were concerned with molecular aggregates, e.g., micelles. All of these were incorporated into Volume III, "Macro- and Multimolecular Systems."

Whatever papers did not belong in Volumes I–III were combined and constitute Volume IV. Fortunately, in these remaining papers some elements of unity could be discerned; in fact, their entire content falls into the following categories: "Electron Transfer and Energy Conversion (photosynthesis, porphyrins, NAD^+, cytochromes); Cofactors (coenzymes, NAD^+, metal ions); Probes (cytokinin behavior, steroid hormone action, peptidyl transferase reactivity)."

Finally, early in this enterprise, we asked Derek Barton to compose a Foreword. Sir Derek complied graciously, and in every volume his personalized view on the nature of bioorganic chemistry appears.

E. E. VAN TAMELEN

Contents of Other Volumes

VOLUME III Macro- and Multimolecular Systems

1

Studies in Sesquiterpene Biogenesis: Implications of Absolute Configuration, New Structural Types, and Efficient Chemical Simulation of Pathways

Niels H. Andersen, Yoshimoto Ohta, and Daniel D. Syrdal

INTRODUCTION

Nowhere does nature display such an uncanny synthetic expertise and variety as in the construction of sesquiterpenes from their acyclic precursor(s) of the farnesane skeleton. The sesquiterpenes are ubiquitous (or nearly so) [1] plant products and are known constituents of fungi [2,2a]; they have also been found in arthropods [3] and marine invertebrates [4]. Proposals concerning the natural construction of sesquiterpenes stem from the structure elucidation of essential oil constituents in the first half of this century and were first formulated as the biogenetic isoprene rule by Ruzicka [5]. The rule has no exceptions since it stands as the definition of a sesquiterpene. In 1959, Hendrickson [6] applied this principle to the biogenetic classification of sesquiterpenes, paying particular attention to the importance of steric and stereoelectronic factors in a series of 10- and 11-membered ring intermediates viewed as the cyclization products of *cis,trans*- or *trans,trans*-farnesyl pyrophosphate (FPP). These are shown in Scheme 1. The germacrene-related sesquiterpenes (such as eudesmanes, guaianes, and the cadalene-related ones)

1

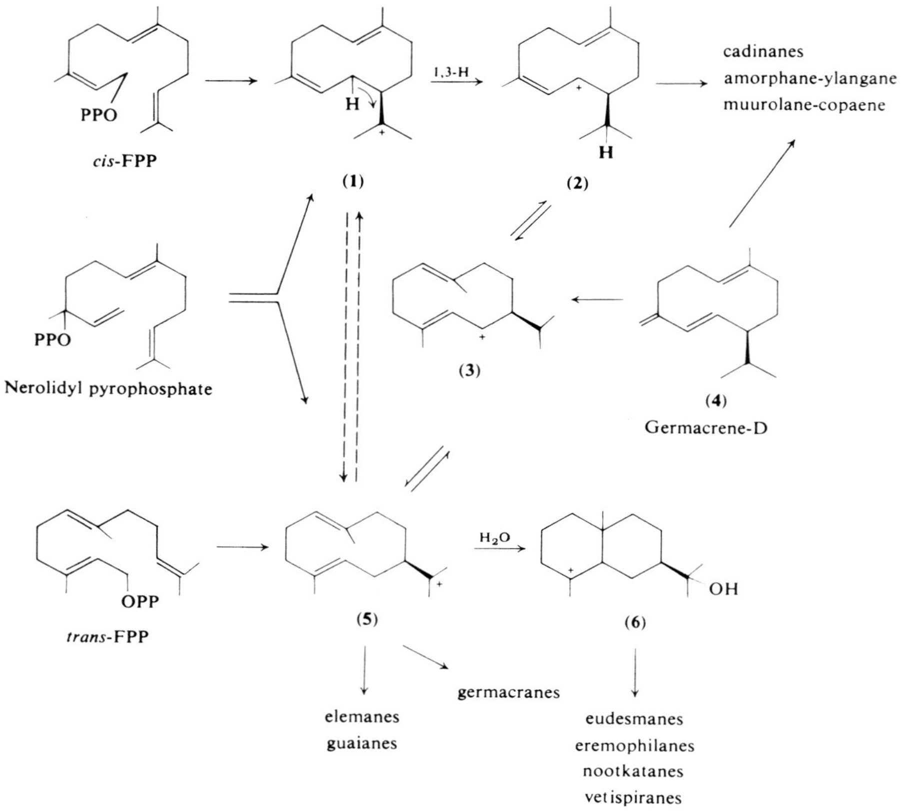

Scheme 1

have commonly been subdivided into *trans*-FPP- and *cis*-FPP-related classes [7].*

With the renaissance in the study of sesquiterpenes over the past 15 years, as modern separation methods and spectroscopy were applied to minor constituents of known oils and new natural sources in search of substances with pharmacological activity [9], the number of structural types to be fitted into this framework of biogenetic conjecture has increased dramatically.†
With this there has been an uncritical proliferation of biogenetic proposals. As an example, even though the stereochemical implications of Ruzicka's biogenetic postulates (as presented by Hendrickson [6] and others [8]) have

---

* Thus, the cadalenes are classed with bisabolene, cuparene, himachalene and its biogenetic products (longifolene, longipinene, etc.) as *cis*-FPP products.
† The number of sesquiterpene skeletons is rapidly approaching 200.

found wide acceptance, little attention was paid to possible implications of absolute stereochemical correlations until our entry into this field.

The pathways of sesquiterpene construction in plants—their significance to modern plants and to their ancient precursors—have been under investigation, both directly (by radiotracer techniques) and indirectly, in our laboratories for 8 years. Only the indirect evidence obtained from chemical studies is presented here. Our reasons for adopting, at least at the onset, a "chemical" approach to this topic have been stated previously [10]:

> To our knowledge, sesquiterpenes play no essential role in the metabolic pathways of plants and their frequently beneficial role in plants (insect control, as an example) can be considered as coincidence. The frequency of such coincidence in surviving species would, of course, increase due to the usual evolutionary forces. However, the amount of biochemical machinery mobilized in the synthesis of these substances should be minimal. For this reason we prefer biogenetic hypotheses employing the maximum amount of "chemical" reasoning rather than invoking enzymic control unnecessarily. The ideal (and almost certainly unrealistic) biogenesis would involve a single enzyme-controlled step generating a dissymmetric intermediate from which the wide variety of sesquiterpene types can be generated by chemical interconversions controlled by the usual steric and stereoelectronic effects associated with such rearrangements.

The distinct lines of indirect evidence supporting biogenetic proposals are congeneric occurrence patterns, absolute stereochemistry correlations, the isolation and structure elucidation of "missing links," and chemical simulation studies. Of these, the "missing links" are of particular importance since the relative merits of hypotheses are best judged on their predictive value. The chemical simulation studies, largely the study of acid-catalyzed transformations of sesquiterpenes, reveal the energetics of the reactions proposed as biosynthetic pathways and the points at which enzymatic control is essential for economy (yield maximization or selection between alternative rearrangements) and stereospecificity. These critical points are then those most subject to change during the evolution of plants. In addition, they provide a more detailed understanding of the mechanism of carbonium ion rearrangements and the requirements for hydride shifts in chemical systems.

THE NATURAL OCCURRENCE OF BIOGENETIC "MISSING LINKS"

In reviewing the literature on sesquiterpene constituents of tracheophytes through 1968, we were impressed by the nearly universal presence of a β-oriented isopropyl group in germacrene-related sesquiterpenes (see Scheme 1). The biogenetic scheme wherein these different types are derived separately from *trans*-FPP and *cis*-FPP (either directly or circuitously) offers no explanation for this correlation. These considerations led us to state an "absolute stereochemical homogeneity rule" for tracheophyte essential oils [11]. Rather than accept such an accumulation of coincidence, we applied "Occam's razor" and suggested that a single enzymatic cyclization produces one chiral germacrene, that differentiation into "*cis*-FPP" and "*trans*-FPP"-related classes occurs at a later stage,* and that nerolidyl pyrophosphate (NPP) (Scheme 1) might be the true substrate for the enzymatic cyclization. The NPP can afford either germacradienyl cation [(1) or (5)], and the chirality produced could represent asymmetric induction based on the chiral center already present in NPP rather than enzymatic control. We therefore chose, as a first project, to look for possible exceptions to the single chiral germacrene hypothesis.

Vetiver oils from Haiti or Java contain as the major ketonic constituents the vetivones (8) and (10) [12]. The chirality of the germacrene precursor in this case is masked by the sp^2 center at C-7. The structures of the vetivones can be rationalized as alternative alkyl migrations from ion (9).† The natural occurrence in other species of valencene, nootkatone, and valeranone could be taken as evidence for a 10-*epi*-eudesmane structure such as (7). However, the genesis of the vetivones from *ent*-germacrene-A [13] could not be excluded, particularly in light of the demonstrated occurrence of enantiomeric cadinenes (e.g., khusinol and khusilal) [14] in vetiver oils from different geographic areas. Our first studies were on the constituents of vetiver oil and those found in a local conifer, the Alaska yellow cedar (*Chamaecyparis nootkatensis*), the first known source of nootkatone [15]. Some of the new

---

* The stage at which such a *cis,trans* isomerization takes place cannot be stated with certainty. At the stage of germacrene-D this represents nothing more than a conformational change. The *cis,trans* isomerization indicated by (2) ⇌ (3) should also be relatively facile. The conversion of bicyclogermacrene to (+)-δ-cadinene on mild acid treatment [11a] suggests such a pathway:

$$\xrightarrow{\text{H}^+} \textbf{(3)} \longrightarrow \textbf{(2)} \longrightarrow$$

Bicyclogermacrene   (+)-δ-Cadinene

The normal occurrence of an isopropyl group in cadalanes and isopropenyl groups in eudesmanes suggests that the 6,11-hydride shift serves to direct germacrenes to the cadalane series.

† The presence of the 7,11- rather than 11,12-olefinic linkage could be viewed as alleviating the unfavorable steric interaction which would be associated with an axial (β) isopropenyl group.

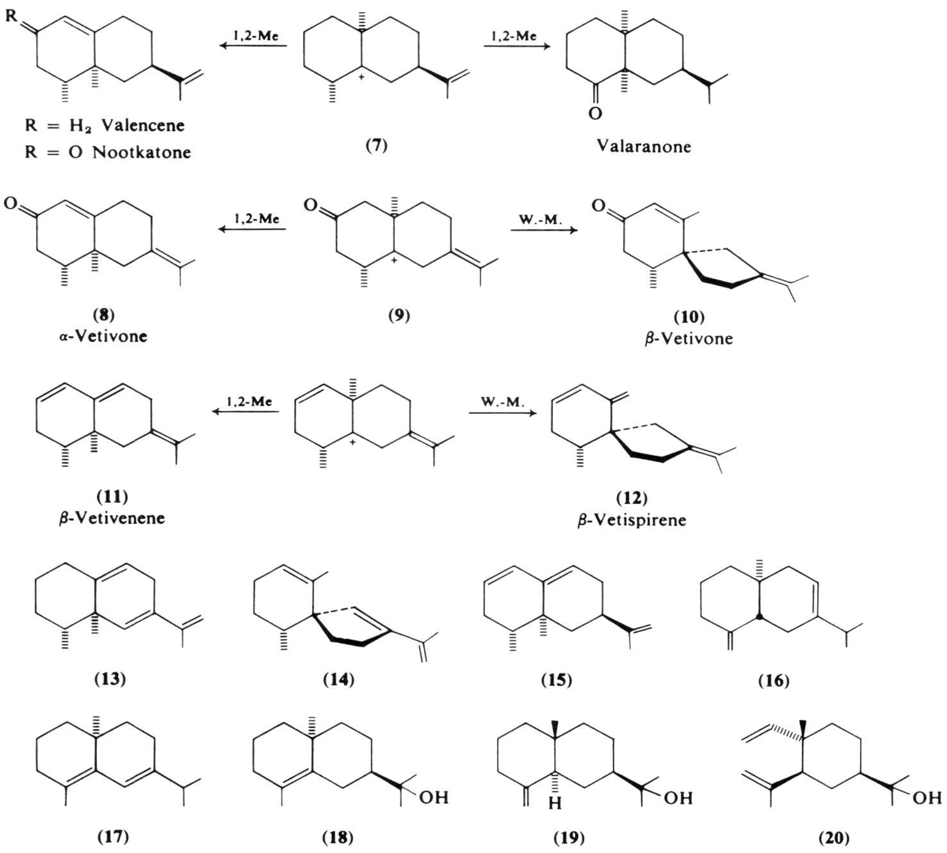

(7)

R = H₂ Valencene
R = O Nootkatone

Valaranone

(8)
α-Vetivone

(9)

(10)
β-Vetivone

(11)
β-Vetivenene

(12)
β-Vetispirene

(13)

(14)

(15)

(16)

(17)

(18)

(19)

(20)

Scheme 2

constituents [(11)–(20)] found in vetiver oil are shown in Scheme 2 (in which W.-M. is Wagner-Meerwein shift). The major bicyclic hydrocarbons (11)–(14) are biogenetically related pairs (as illustrated for β-vetivenene and β-veti-spirene) [16]. Other 10-*epi*-eudesmadienes [(16) and (17)] were also found. The occurrence of valencene and nootkatene (15) established the β orientation of the isopropyl group. The more usual bicyclization path was also represented by β-eudesmol (19) and elemol (20). Among the oxygenated components we also found 10-*epi*-γ-eudesmol (18) [17].*

* This has been confirmed by Kaiser and Naegeli [18]. The dehydration product of (18), 10-*epi*-selina-4,11-diene, was the first 10-*epi*-eudesmane to appear in the sesquiter-pene literature [19].

However, the occurrence of constituents bearing an α-isopropyl group in vetivone-rich vetiver oils was demonstrated when we established that zizanene (21) [20] is enantiomeric to the α-amorphene found in conifers. A biogenetic relationship between zizanene (21), ent-γ-amorphene (24) [17],* levojuneol (23), and cyclocopacamphene (25) [16,25,25a] is shown in Scheme 3. Alternate cyclobutyl–carbinyl bond scissions from cation (22) can provide the ent-amorphane and ent-eudesmane skeletons, respectively [20].

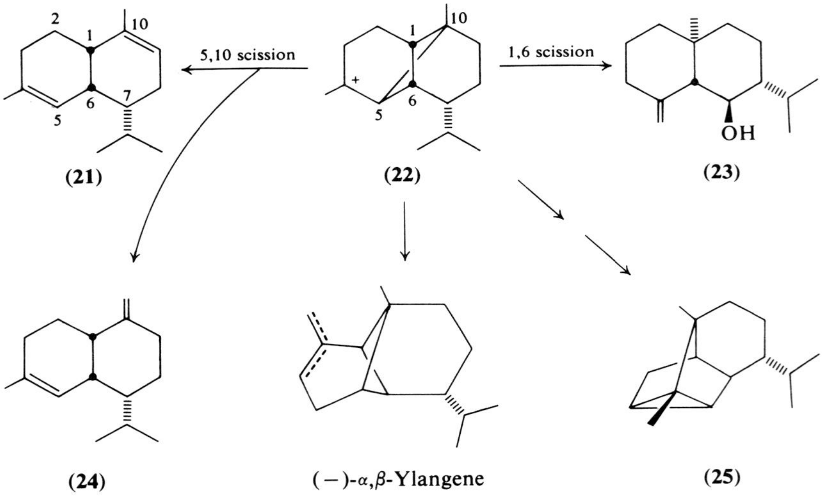

(21)   (22)   (23)

(24)   (−)-α,β-Ylangene   (25)

Scheme 3

In the course of these studies we also isolated additional members of the zizaane (tricyclovetivane) class of sesquiterpenes [5,25a], which, with the vetivones, make up the major part of the oil. They provided an unexpected connection with the concurrent work on the constituents of the Alaska yellow cedar. These studies had centered on the leaf oil rather than the wood oil, and in contrast to the literature accounts on the wood oil constituents (nootkatone and related eudalene-yielding compounds) the leaf oil contained less than 0.2% of components yielding eudalene on dehydrogenation [26].

* Nuclear magnetic resonance (nmr), gas chromatographic (gc), and circular dichroism (CD) comparisons of Herout's γ-amorphene and authentic γ-muurolene indicated that the original substance designated as γ-amorphene was an impure sample of the muurolene [21]. The same conclusion has been reached by others [22]. However, γ-amorphene has now been isolated, and δ-amorphene is produced on acid-catalyzed fragmentation of α-ylangene [23]. With the isolation of (24) from vetiver, both enantiomers are demonstrated natural products. An nmr and infrared (ir) comparison of a constituent of Pinus species with our substances indicates that γ-amorphene [most likely ent-(24)] occurs in conifers [24].

We found a number of substances common to conifer oils [10]: ($-$)-α-copaene, ($-$)-curcumenes (α, β, and γ), ($-$)-calamenene, and the dextrorotatory forms of β-bisabolene, nerolidol, longifolene, α-ylangene, and δ-cadinene. We also found a novel muuroladiene (26) (*l*-10α*H*-muurola-4,6-diene, or 10-*epi*-zonarene) [27,27a] and two diastereomeric substances with the spirane structure shown as (28), designated as α- and β-alaskene [28]. The alaskane (acorane) skeleton is that proposed in biogenetic proposals for both the

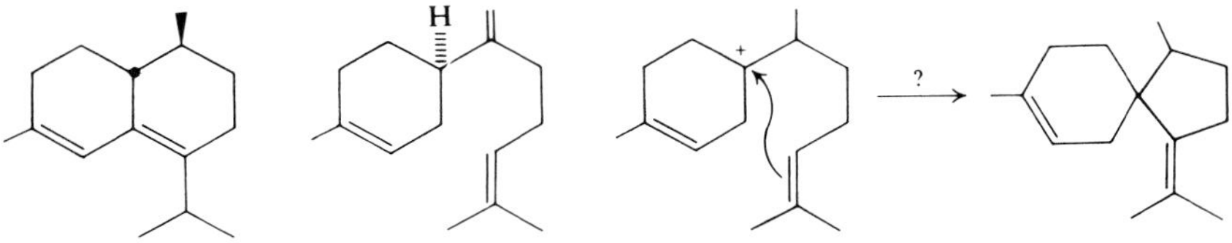

(26) (+)-β-Bisabolene (27) (28)

tricyclovetivanes and the cedranes. The only previous natural products of this skeleton were the acorones [29]. Two points should be noted. First, although the acorone skeleton had long been recognized as related to cedranes, it had, according to the literature [29], the incorrect absolute configuration at the methyl center for the production of the natural cedrenes [30,31]. The production of cedrenes would instead require what is now known as an

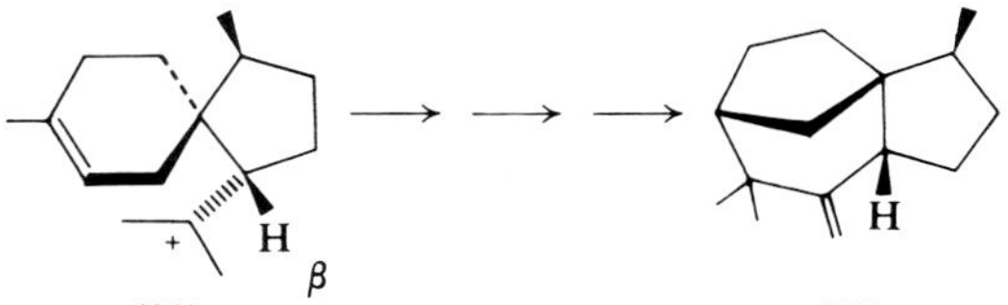

(30) β (32) Zizaene Tricyclovetivane

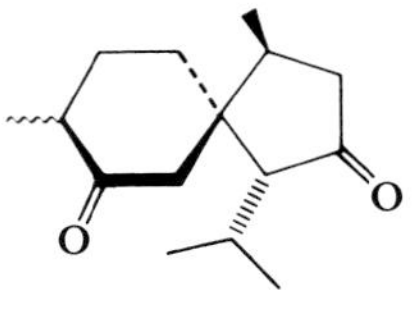

(29)

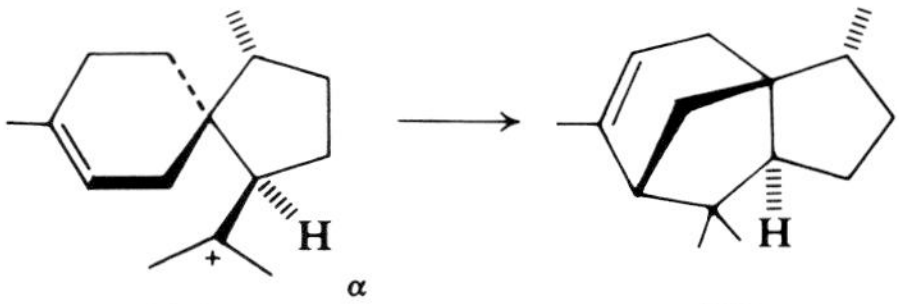

(31) α (33) α-Cedrene

Scheme 4 Known spiranes and related tricyclics—1968.

α-stereochemical arrangement.* Second, the acorone stereochemistry (29) is, in fact, related to zizaene or tricyclovetivene, as can be seen from the correspondence at the methyl center (Scheme 4), the center that would not be involved or inverted in any further transformation. We have since confirmed the absolute stereochemistry of the acorones [34]. That α-alaskene (34) represents the "missing link" in the α-cedrene biogenesis was quickly established when we cyclized it in formic acid, affording (−)-α-cedrene, identical in all respects to the natural product, in 85% yield. The reactions of β-alaskene in acid media proved to be more complicated and did not provide a way of deciding between possible structure assignments (39) and (40).†

The stereochemistry of β-alaskene (40) followed from correlation with the acorones. The alcohol (35) obtained by LiAlH₄ reduction of isoacorone [36], when dehydrated with phthallic anhydride, affords in addition to the major product (36) all other possible double isomers [including (37), in which all chiral centers were preserved], 2-*epi*-α-cedrene (38), and (+)-α-cedrene, the unnatural enantiomer. The latter confirmed the absolute stereochemistry of

---

* The α- and β-stereochemical designations for acoradienes and acoratrienes refer to relative, not absolute, stereochemistry. In α compounds the isopropenyl group is *cis* related to the cyclohexene double bond, allowing facile Markovnikov addition of the cation (31). The β series is the other C-1/C-7 diastereomer in which steric constraint should favor anti-Markovnikov cyclization. The nomenclature finds its origin in the names α- and β-acoradiene given to juniper oil constituents [32].

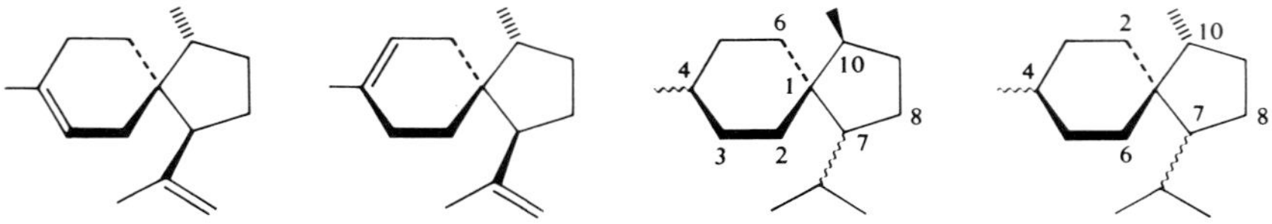

α-Acoradiene   β-Acoradiene   Acorane   Alaskane

In order to distinguish the enantiomeric series we have suggested that the name acorane be retained for one and that the series with the C-10 configuration of the cedranes be designated as alaskane. The skeletons and numbering are illustrated above. Tomita's "acoradienes" are related to natural cedrene, not to the acorones. The stereochemistry in these substances (at C-4, C-7, etc.) can be indicated by specifying the relationship (c denotes *cis*; t, *trans*) between the hydrogen at that center and the lowest-numbered carbon of the other ring. The C-1/C-10 stereochemical designation is inherent in the parent names, acorane and alaskane [33].

† Tomita, Isono, and Hirose [35] have, in an independent study of *Juniperus rigida*, isolated α-alaskene (called γ-acoradiene) and its diastereomer (39) (designated δ-acoradiene, enantiomeric to β-alaskene). The names are unfortunately confusing, and to retain some semblance of order it would be best to retain the name α-alaskene for (34) and name the rest semisystematically: α-acoradiene = 7cH-alaska-4,11-diene; β-acoradiene = 7cH-alaska-3,11-diene; β-alaskene = acora-3,7(11)-diene; δ-acoradiene = alaska-3,7(11)-diene [33].

the acorones. The hydrogenation of diene (37) afforded acorane (42), identical by all spectroscopic and chiroptical checks with the major hydrogenation product derived from β-alaskene.

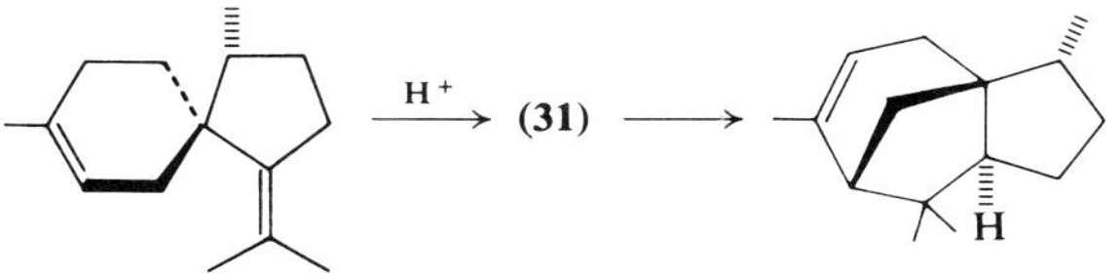

(34)
α-Alaskene

(33)
(−)-α-Cedrene

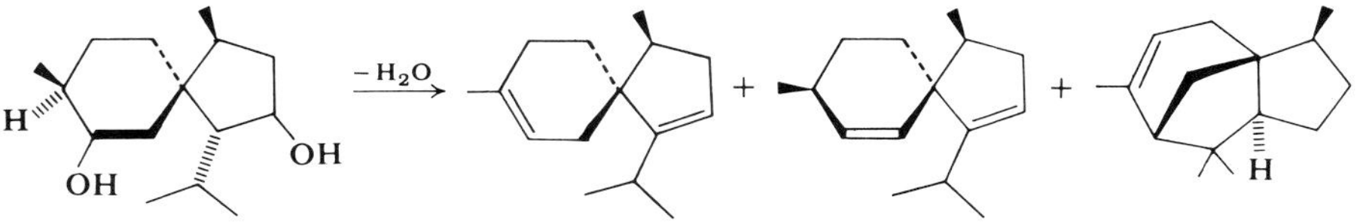

(35)   (36)   (37)   (38)

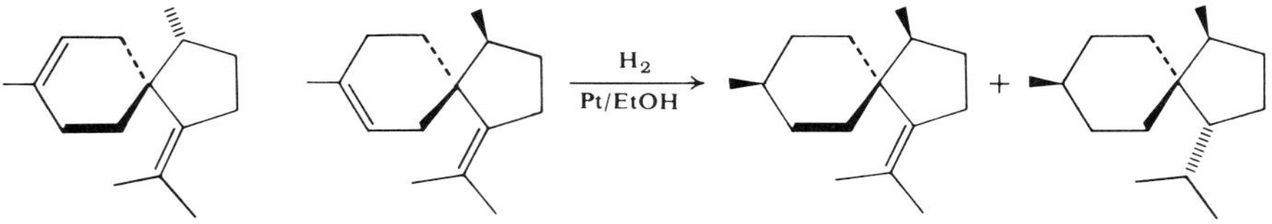

(39)   (40)   (41)   (42)

The Alaska cedar had provided not the 10-*epi*-eudesmanes that we sought, but the missing spirane intermediates for two important classes of tricyclic sesquiterpenes: the cedranes and the tricyclovetivanes. The alaskenes are of enantiomeric skeletal classes, with α-alaskene related to cedrene and β-alaskene having the stereochemical features requisite for the tricyclo-vetivanes [37]. This is, to our knowledge, the first record of a single plant producing two compounds that are formally double-bond isomers, each optically pure, but in enantiomeric forms. The biogenetic sequence to the tricyclovetivanes is shown in Scheme 5 [38].* Scheme 5 included, as inter-

* An alternative proposal [38a] was dismissed since vetiver oil produces sesquiterpene enantiomeric to hinesol (the vetispirenes):

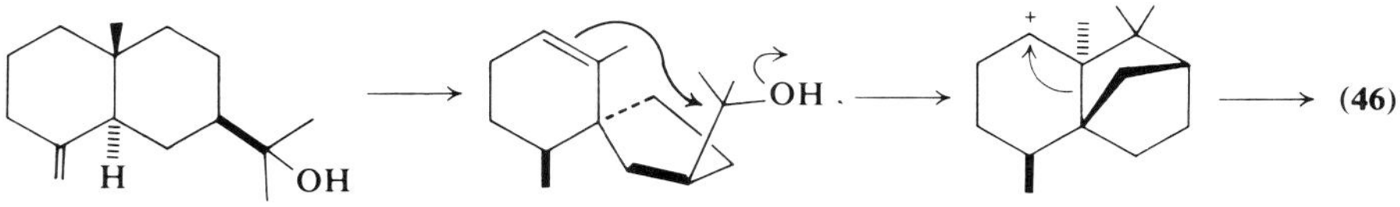

(46)

This proposal, although inconsistent with the known absolute configurations of vetiver constituents, served as the inspiration for a total synthesis of a tricyclovetivane sesquiterpene.

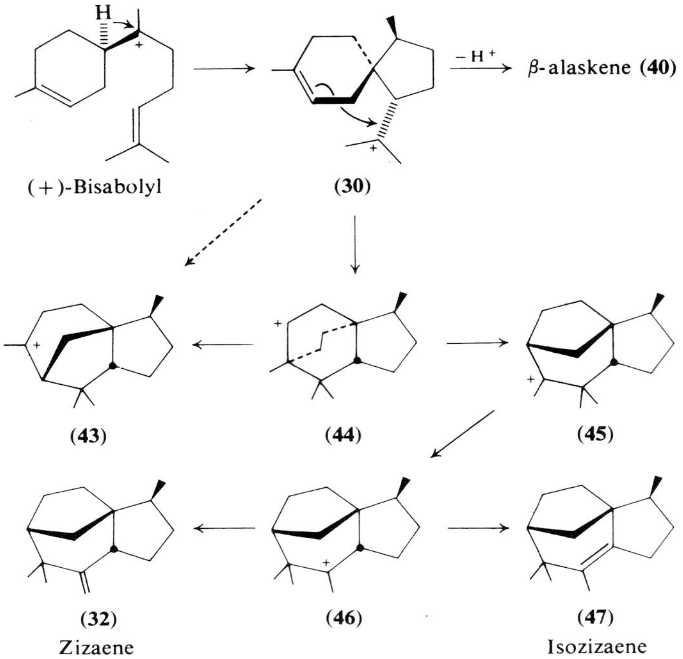

Scheme 5 A biogenetic proposal for zizaene.

mediates between the β-acorenyl cation and ion (**46**), two unknown skeletons [(**44**) and (**45**)]. If this were a proper biogenetic sequence, there was no reason to assume that the other skeletons were not represented in vetiver oil, and we set out to find these. A product derived from (**45**), the immediate precursor of the zizaane skeleton, was our first target. Deprotonation of (**45**) to the exocyclic double isomer seemed reasonable, and we therefore reexamined an earlier analysis of vetiver oil in which 25 still unidentified hydrocarbons had been isolated, with particular emphasis on exomethylene compounds. Our goal molecule was among these, in quite reasonable amounts, and we designated this substance prezizaene. The evidence for the structure assignment [39,40] need not be given here since the chemical simulation of this biogenetic pathway, discussed in the following section, provides an even more satisfying proof. One of the minor alcohols of vetiverol proved to have structure (**49**),* derived from ion (**44**) by water capture. Subsequent studies by others [18,42]

---

* On the basis of the work of Tomita and Hirose [41], this substance is designated as *ent*-allocedrol.

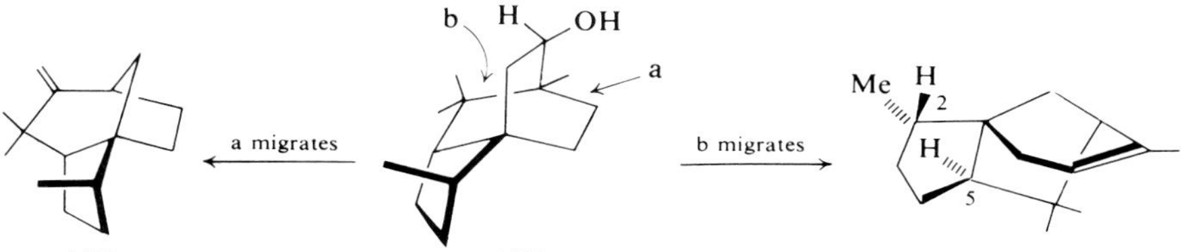

(48)
Prezizaene

(49)

(50)

have revealed that cedrene diastereomer (50) occurs in vetiver oil.* Additional studies of vetiver oil have confirmed the presence of other representatives of this biogenetic path: β-bisabolol (51) [18], acoradiene (36),† (+)-α-cedrene and oxygenated derivatives such as (53), and 7cH-acora-4,9-diene (52). Thus,

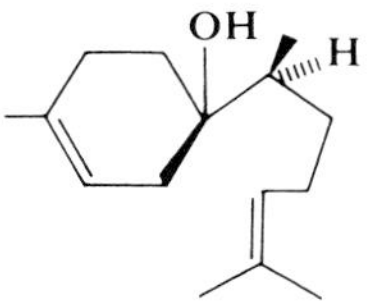

(51)‡

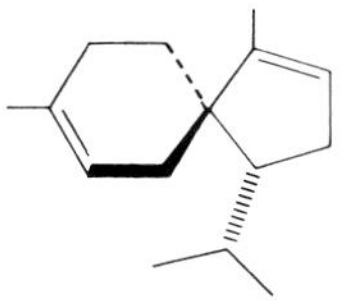

(52)†

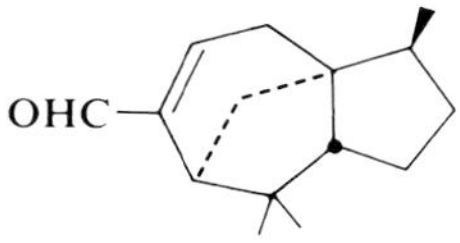

(53)

a detailed study of vetiver oil provided ample confirmation of the genesis of the major components in the structures of the molecular debris left behind in the course of the biosynthesis.

CHEMICAL SIMULATION OF THE BIOGENETIC SCHEME FOR CEDRENE AND THE VETIVER TRICYCLICS

The availability of this wide variety of novel allied sesquiterpenes led to an examination of their chemical interconversion in order to assess the energetics of the biosynthetic pathway. These studies constitute a chemical simulation of the biogenesis of both cedrene and zizaene proceeding from nerolidol.

Concerning the first step, the formation of the bisabolane skeleton, it has long been known that the dehydration of farnesol or nerolidol produces bis-abolenes [45]. On acid treatment cadalenes are eventually produced, but

* The exocyclic isomer of (50) is also known to be a natural product [43].

† In the original work [18] acoradienes (36) and (52) were assigned different structures. In the case of diene (36) we have compared the natural product with acorone-derived material. The now-established structures were first suggested on the basis of biogenetic conjecture [33]. That of diene (52) has been confirmed by partial synthesis from carotol [44].

‡ The stereochemistry has not been determined; that shown fits the biogenetic scheme.

studies of the time course *suggest* initial bisabolene formation and subsequent further cyclization rather than formation via germacrene [46]. Since we wished to promote spirocyclization of the bisabolenes we chose to examine milder conditions in which some degree of concert might be essential for reaction. As one criterion of concert would be asymmetric induction [(**54**) → (**55**)], we

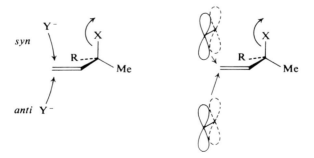

employed optically active nerolidol in all of our studies.*   An additional motive for this was our wish to elucidate the stereoselectivity of S_N2' reactions of biogenetic importance (Fig. 1). We found, in direct analogy to the solvolysis

Fig. 1   Stereochemistry of S_N2' reactions.

* A generous supply of (+)-nerolidol, donated by E. Klein (Dragoco), was used for these studies. In order to relate the work to the previous section dealing with the tricyclo-vetivanes the mirror-image results are quoted, as though (−)-nerolidol has been used.

of $(-)$-(R)-linaloyl phosphate, which yields $(+)$-(R)-α-terpineol [47], that derivatives of $(-)$-nerolidol afford $(+)$-β-bisabolene under a wide variety of conditions (Table 1). Participation of the 6,7 π bond was the rule rather than the exception. Even reactions proceeding through ion (56) yield significant quantities of bisabolene, and this pathway serves to dilute the products of the

TABLE 1

Elimination of HX from (R)-Nerolidyl Derivatives [48]

Run No.	X	Reagent, conditions	Product composition (%)[a]				Optical purity (%)[b]
			(59β)	(59α)	(60)	(62)	
1	HO	AlCl₃, ether, 25°C[c]	10	8	25	36	18
2	HO	Al₂O₃-pyr, 220°C	85	—	< 3	< 3	—
3	HO	Acidic Al₂O₃, 240°C	60	—	< 6	20	nd
4	HO	MeSO₂Cl, pyr	33	23	5	30	nd
5	Lio	MeSO₂Cl, pentane	14	20	25	35	16
6	MeSO₂O	Pentane, 25°C	25	18	25	20	12
7[d]	LiO	ClSO₂NCO, Et₂O, 25°C	25	14	29	19	37
8	Me₂CHCO₂	Al₂O₃, 230°C	90	—	—	—	—
9	i-Bu₂AlO	Basic Al₂O₃, 220°C	90	—	5	—	—
10	i-Bu₂AlO	220°C	70	15	10	—	nd
11	i-Bu₂AlO	i-Pr₂NLi, THF[e]	30	—	—	5	—
12	MeCHClCO₂	220°C	44	22	19	15	—
13	MeCHClCO₂	270°C (gc injector)	22	17	19	22	15
14	H₂O₃PO	270°C (gc injector)	60	25	5	5	—
15	H₂O₃PO[f]	1 N H₂SO₄/n-C₅H₁₂	3	—	50[g]	4	11
16	HO	DMSO, 140°C, 15 hr	—	—	35	45	0
17	HO	HCO₂H, neat, 3 min	1	1	26	26	16
18	HO	HCO₂H/n-C₅H₁₂, 2 min	5	5	33	26	32

[a] Refers to hydrocarbons and nerolidol. In some cases the yields were not above 40% since we wished to examine initial compositions.

[b] Asymmetric induction was always measured from the optical activity of β-bisabolene (nd, not determined).

[c] Other Lewis acids give comparable results. The reactivity sequence is AlCl₃ ≳ SnCl₄ > FeCl₃ > ZnCl₂. Optical purity was not determined in the other cases.

[d] Reaction repeated six times; the percentage of each component varied as follows: (59β), 14–27%; (59α), 11–15%; (60), 27–30%; (62), 11–14% of one isomer, 4–6% of the other. Optical purity of β-bisabolene 32–40%, three determinations.

[e] Major product is nerolidol.

[f] Nerolidol phosphate was prepared according to the procedure used for linalool and solvolyzed under identical conditions [47].

[g] Other major product is nerolidol; the minor alcoholic product can be assigned as bisabolol. Under these conditions some phosphate remains unreacted.

concerted reaction with racemic materials.* The only HX eliminations that give farnesenes as the major products are ester pyrolyses and dehydrations in the presence of alumina at elevated temperatures—processes that can reasonably be viewed as *cis* eliminations. Even in this category significant cyclization was observed in two cases: the pyrolysis of the nerolidyloxyallane and that of the α-chloroproprionate. In the latter case, the β-bisabolene produced (15–20% yield) was dextrorotatory, showing 15% of theoretically possible optical purity.†

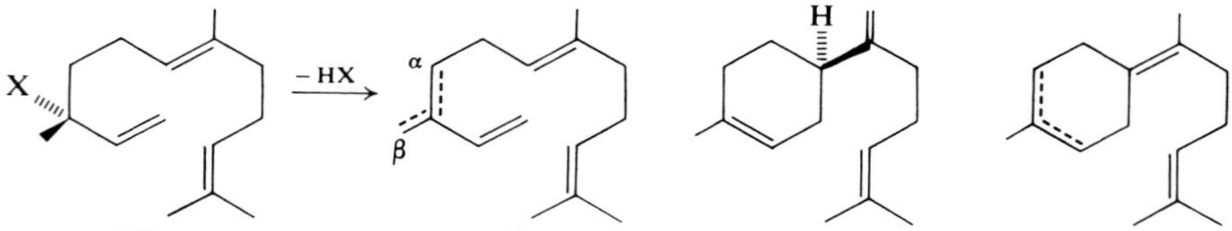

(58)	**(59)**	**(60)**	**(61)**
(−)-(R)-Nerolidol		(+)-β-Bisabolene	
X = OH			

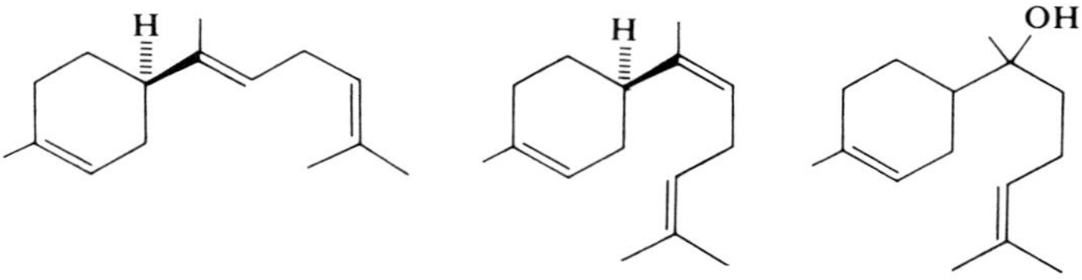

(62a)	**(62b)**	**(63)**
trans	*cis*	Bisabalol

(+)-α-Bisabolene

The sense of asymmetric induction observed throughout our studies is consistent with two of the four possible concerted pathways: *re* direction of attack on the 6,7 bond is excluded. Although Rittersdorf and Cramer [47] rationalized their result in the monoterpene case by the *si-syn* transi-

---

* Racemization after cyclization [via γ-bisabolene (61)] is not significant in these experiments. Acid equilibration of the bisabolene yields virtually pure γ-bisabolene [48]. The nerolidol cyclization experiments reported here yielded less than 3% γ-bisabolene in each case. The observed bisabolene isomer ratios represent kinetic deprotonation in the various media.

† The percent optical purity of β-bisabolene is given by

$$\% \text{ optical purity} = 100 \times \frac{14 \times \alpha_{300} \ (\beta\text{-bisabolene})}{\alpha_D \ (\text{nerolidol}) \times 420}$$

based on the assumed value of $\alpha_D = 14°$ for pure nerolidol and $\alpha_{300} = 420°$ for β-bisabolene. Since there is no evidence for the optical purity of natural nerolidol this is a conservative estimate of the minimum percent asymmetric induction.

tion state, there is no experimental basis for dismissing the *si-anti* S_N2' process.*†

Three of these cyclizations deserve comment. The solvolysis of nerolidol phosphate under the conditions described for the corresponding terpene

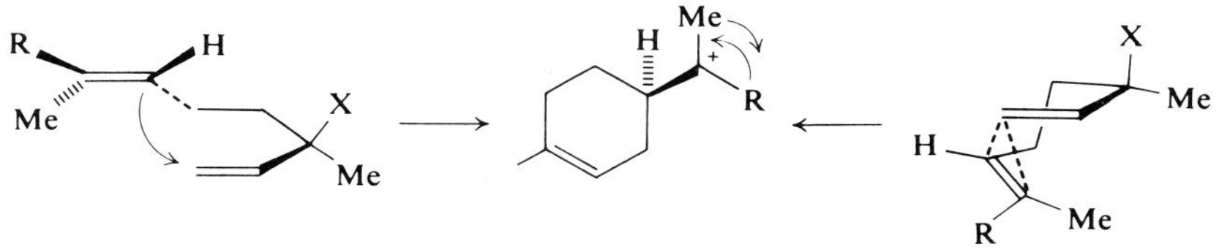

si-syn   *si-anti*

system gave, in our hands, very different results. In the literature study of linaloyl phosphate, the cyclic products were predominantly alcohols, and the asymmetric induction was measured from the optical activity of α-terpineol (40%). In the sesquiterpene case, we observed primarily olefinic cyclization products, and the optical purity of β-bisabolene was only 11%. Examining the two transition states for differences that might be expected for R = $CH_2CH_2C=CMe_2$ vs. R = Me is not particularly revealing. The decreased yield of alcohols by water capture can be explained by further interaction between π bonds in a coiled conformation with crossed double bonds such as (64). This also provides a rationale for the biogenesis of cuparene (66), which is always observed as a trace product in our nerolidol cyclizations.

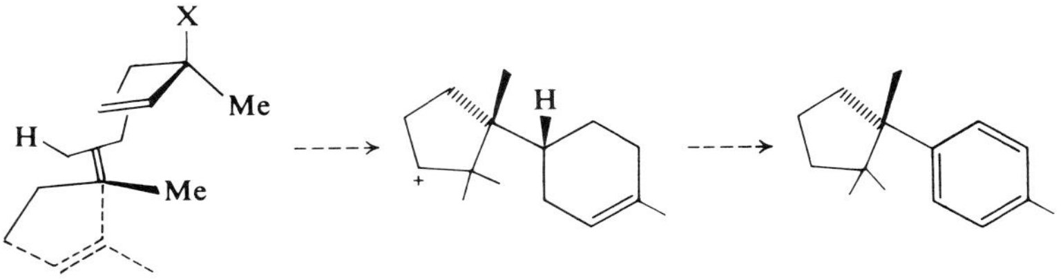

(64)   (65)   (66)

The reaction of the alkoxide of nerolidol with chlorosulfonyl isocyanate afforded the highest degree of asymmetric induction [48]. Quenching of the reaction revealed the intermediacy of anion (67), which decomposes slowly to olefins, chloride anion, CO_2, and sulfimide (isolated as the trimer). This mode

---

* This, no doubt, follows from the classic Stork experiment suggesting a *syn* relationship between the entering nucleophile and leaving group in a net S_N2' process [49]. The interpretation of this experiment has been questioned [50].

† A reexamination of the solvolysis of linalool *p*-nitrobenzoate using deuterium labeling has shown that cyclization occurs largely by the *anti* pathway [51].

of decomposition is like that reported for alkyl N-p-tosylcarbamates **(67)**–**(68)** [52].

As suggested from earlier studies [13a,45], nerolidol cyclizes rapidly on dissolution in formic acid. After 3 min of shaking at room temperature no

$$R{-}O{-}\underset{\underset{O}{\|}}{C}{-}\overset{\ominus}{\underset{\cdot\cdot}{N}}{-}SO_2{-}Cl \longrightarrow olefin + CO_2 + Cl^- + HN{=}SO_2$$

(67)   **(68)**

acyclic olefins could be detected. In trial studies we found that α- and β-farnesene were inert to short treatment under the conditions of cyclization runs 17 and 18 (see Table 1). We also determined that β-bisabolene was stable to short treatments but gave γ-bisabolene **(61)** and eventual bicyclic products on prolonged reaction. On this basis the absence of acyclic olefins in run 17 indicates that both ion **(56)** and the close ion pair cyclize prior to deprotonation. The lower degree of asymmetric induction most likely reflects racemization via dissociation of the nerolidol ion pair. In run 17 the material not accounted for in Table 1 was a mixture of bi- and tricyclic olefins and formates. The reaction can be moderated by using a two-phase system of 1:1 pentane–HCO_2H. In this case (run 18), the farnesenes were observed as minor products, some nerolidol remained, and all other products were minor ($<8\%$); β-bisabolene was found to have 32% of the theoretically possible optical rotation.

If chiral nerolidyl pyrophosphate is the true substrate for sesquiterpene-producing enzyme systems it appears that very little directing influence by the enzyme needs to be postulated to explain the chirality observed in natural sesquiterpenes. Chemical cyclizations of nerolidol can afford over 50% isolated yields of bisabolenes with substantial asymmetric induction. From here we proceeded to the unprecedented stages in biomimetic synthesis: spiroalkylation and subsequent tricyclic formation.

Rather than presenting our studies of acid-catalyzed cyclizations of mono- and bicyclic sesquiterpenes in detail we will present the results using three systems that have proved most informative: the two-phase systems n-decane/HCO_2H and n-decane/CF_3CO_2H and dilute CH_3SO_3H in trifluoroethanol (TFE). Since our aim was to synthesize known tricyclic sesquiterpenes we will first discuss their behavior in these acid media (Table 2).

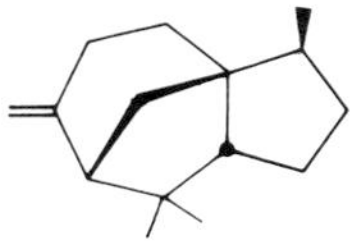

(69) 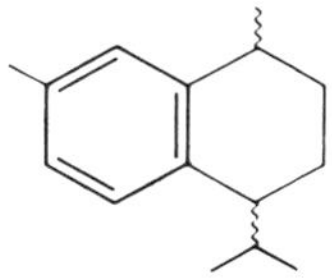 (70) 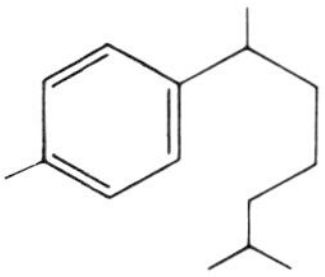 (71)

That zizaene, prezizaene, *ent*-allocedrol, and di-*epi*-β-cedrene (69) all afford with methanesulfonic acid the same equilibrium mixture of olefins (and tetracyclics) served to confirm their structural (and biogenetic) relationship.* It also eliminates from consideration a biomimetic synthesis of any of these substances employing strong acids.

TABLE 2

Reactions of Spiranes and Natural Tricyclics in Acid [a]

Compound	Structure	Acid medium		
		HCO_2H	CF_3CO_2H	CH_3SO_3H–TFE
α-Cedrene	(33)	nr	nr	nr
Zizaene	(32)	A	A → B	B
Prezizaene	(48)	A	A → B	B
ent-Allocedrol	(49)	nr [b]	—	B
Di-*epi*-β-cedrene	(69)	[c]	A → B	B
Acoradiene	(36)	[d]	D	—
α-Alaskene	(34)	C	C + D	—
β-Alaskene	(40)	[d]	D	—

[a] A indicates an equilibrium mixture (1:6:8) of three components designated as postzizaene-A, -B, and -C; B indicates an equilibrium mixture consisting of 20% (Z)-EQ-1, 29% (Z)-EQ-2, 4–5% each of postzizaene-B and -C, and 30% of (Z)-EQ-3; C indicates the formation of α-cedrene; D indicates the formation of isocalamenene (70); nr indicates no reaction [57].

[b] The corresponding formate is stable in HCO_2H.

[c] Forms equilibrium mixture A plus a formate.

[d] These reactions are discussed in the text.

With formic acid only zizaene and prezizaene behaved identically [39,53], producing with the same reaction profile a mixture of olefins designated as the postzizaenes. The time course for the reactions are depicted in Fig. 2. The structures of the postzizaenes are now established as those shown in Scheme 6. The methyl shift leading to ion (46) is rapid, and (46) partitions between the

* The components designated (Z)-EQ-1 and -2 appear to be tetracyclic. The (Z)-EQ-3 component is a tricyclic monoolefin of undefined structure, the product of some deep-seated rearrangement; its structure elucidation is in progress.

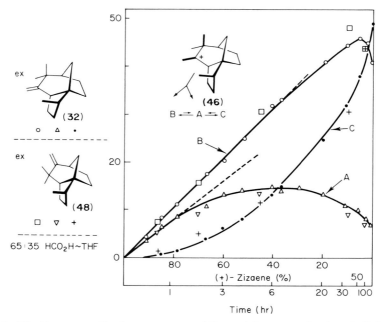

Fig. 2 The time course for the appearance of the postzizaenes on treatment of zizaene and prezizaene with 65:35 HCO_2H–THF at ambient temperature.

endocyclic isomer, postzizaene-B [isozizaene (**47**)], and postzizaene-A (**72**). Only a small portion of ion (**46**) [when produced from prezizaene (**48**)] deprotonates to zizaene (**32**). The reaction time course clearly shows that postzizaene-C (**73**), the eventual major product, is produced only by subsequent protonation of (**72**) from the β face. Using increasingly milder systems incorporating THF as a cosolvent, the sequence shown in the lower portion of Scheme 6 was demonstrated (the numbers over the arrows indicate the initial ratios of product appearance).

The structure assignments rest on an X-ray crystallographic study of diol (**74**) [53,54], the virtual enantiomeric relationship of the cyclopentanones (**77**) and (**78**) [53], and a chemical degradation of olefin (**72**) [55]. The epimeric relationship of the major components (**47**) and (**73**) was established by the synthesis of *epi*-zizaene (**75**) from the known ester (**76**) [56] and its reaction under the same conditions (Fig. 3). Now (**72**) and (**73**) are the initial products.*

---

* Despite a thorough search of trace components in these HCO_2H equilibrations, we failed to locate the methyl epimer of (**72**) (*iii*), even though the degradations of (**47**) and (**73**) reveal preferential α-face attack on the double bond. That ion (*i*) cannot rearrange

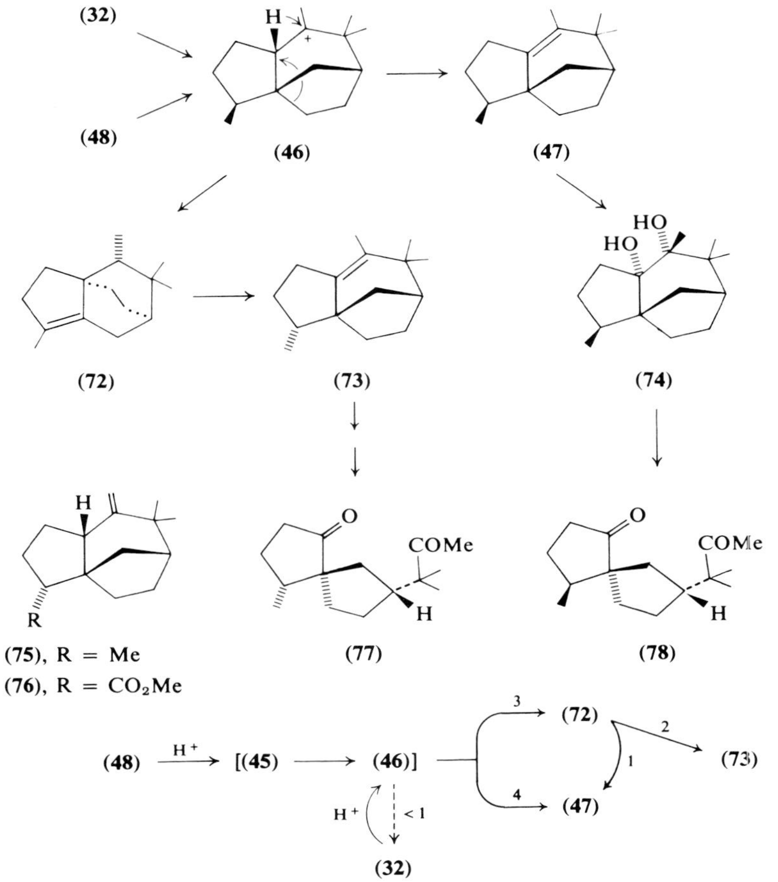

Scheme 6 The postzizaenes.

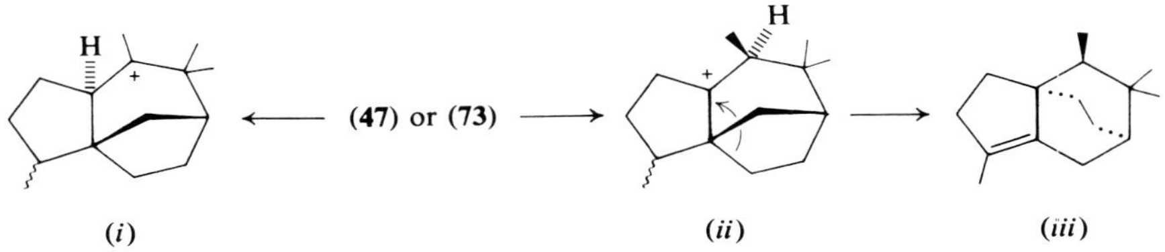

by a concerted ethano bridge shift is clear, but why should ion (*ii*) fail to rearrange? Our tentative conclusion is that the bridge shift only occurs from ion (**46**) in concert with the hydride shift.

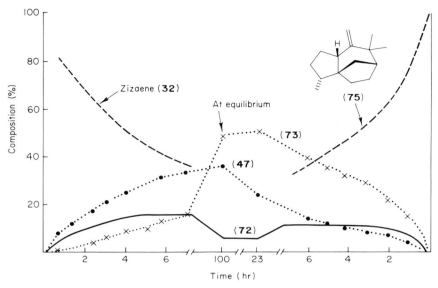

Fig. 3   The course of the rearrangements of zizaene (**32**) and *epi*-zizaene (**75**) in 1:1 *n*-decane–HCO₂H at room temperature.

The only natural product of the vetiver tricyclic series that could be expected as a product of direct cyclization of a monocyclic or spirane precursor was *ent*-allocedrol because of its stability as the formate in formic acid.

The prospects for a biomimetic synthesis of α-cedrene, however, were excellent, since it was stable under all of our cyclization conditions. Thus, we could expect γ-bisabolene (**61**) to cyclize via ion (**27**). Of the four spirane stereochemical series, we could not predict the fate of the β series (**79**): anti-Markovnikov cyclization or deprotonation (Scheme 7). The α series (**80**) should lead to the epimeric cedrenes (**33**) and (**38**), except possibly in CF₃CO₂H, where isocalamenene (**70**) could result. The latter proved to be a useful indication of the formation of spiradienes since all of them aromatize via 1,10 bond migration, affording isocalamenene (**70**) in CF₃CO₂H [57] or, under typical dehydrogenation conditions, isocadalene. The only remaining concern was the partitioning of ion (**27**) between spirocyclization and deprotonations yielding the curcumenes (**81**) and (**82**) or cadalenes.* The last

---

* The CF₃CO₂H reaction medium provided a means of detecting the intervention of the curcumenes since both (**81**) and (**82**) yield dihydro-*ar*-curcumene (**71**) rapidly and in

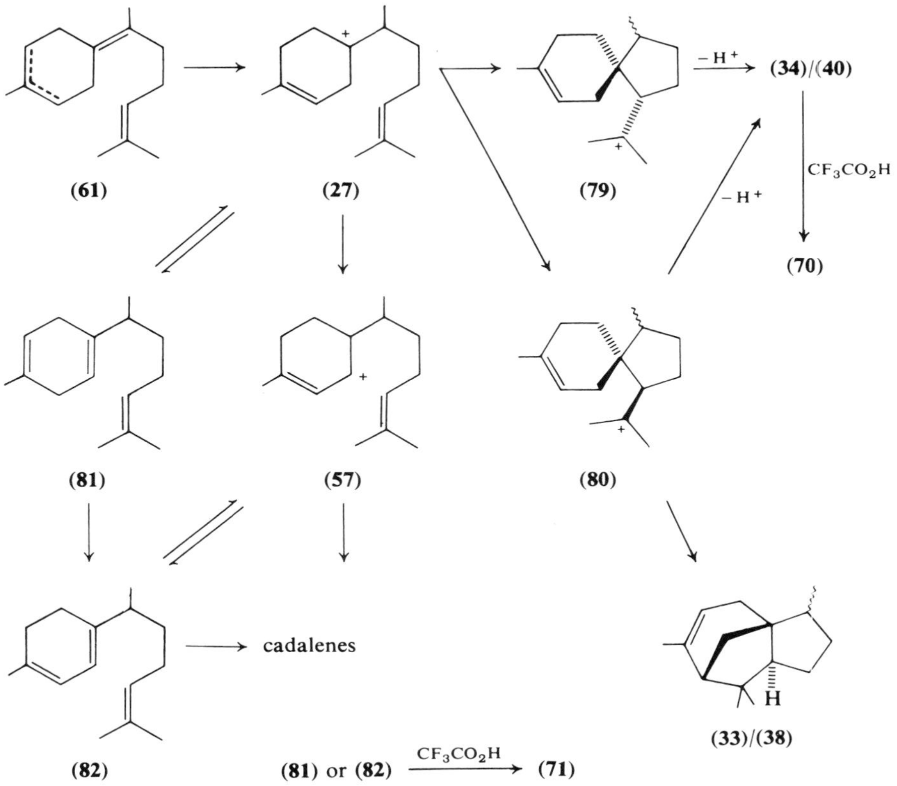

(61)   (27)   (79)   − H⁺ → (34)/(40)

CF₃CO₂H

(70)

(81)   (57)   (80)

(82)   (81) or (82) $\xrightarrow{\text{CF}_3\text{CO}_2\text{H}}$ (71)   cadalenes

(33)/(38)

Scheme 7 Pathways competing with spirocyclization.

question could be answered only by trial. When β-bisabolene in HCO_2H was heated to 80°C, a high-yield conversion to γ-bisabolene (61) was observed. Prolonged reaction gave an extremely complex mixture (>100 hydrocarbon components) which contained 17–19% of a mixture of the epimeric cedrenes (33) and (38) [48]. A nearly identical mixture could be obtained from

high yield ($>75\%$) in this medium. The reaction proceeds with partial retention of optical activity, indicating a hydride shift mechanism [57].

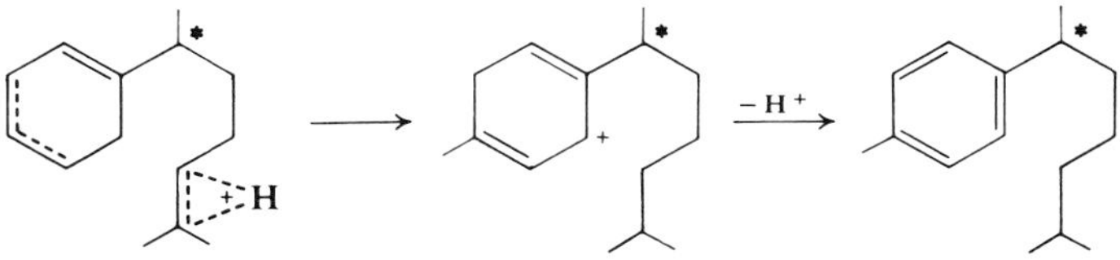

(71)

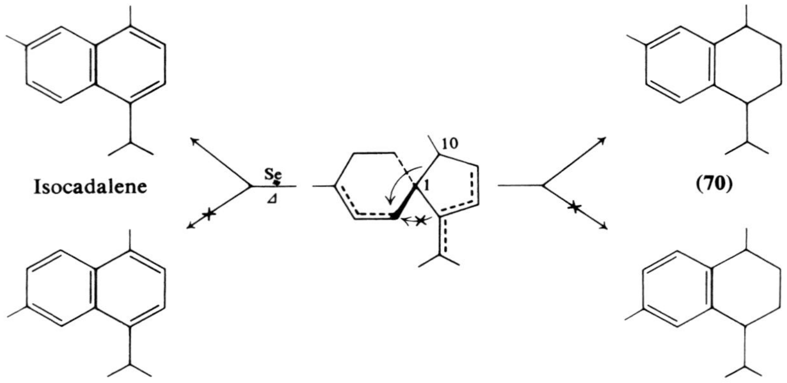

Isocadalene

(70)

(83)
Calamenene

β-curcumene (cedrene content, 15–16%), but γ-curcumene did not afford detectable amounts of the cedrenes (<2%) under these conditions. Apparently, in formic acid, ion (27) deprotonates toward the curcumenes rather than cyclizing to spiranes.

The results with CH_3SO_3H–TFE were only slightly better. The yield of cedrene mixture was 15–30% from γ-bisabolene, depending on the conditions employed. That left only the *a priori* least attractive possibility, CF_3CO_2H. In CF_3CO_2H the entire path to the cedrenes would have to occur without intermediate deprotonation: The curcumenes yield (71), and all of the acoradienes examined (even α-alaskene) afford (70) rapidly under these conditions. When (−)-β-bisabolene in decane was treated with CF_3CO_2H, 50% conversion to γ-bisabolene was observed in the 30-sec aliquot. What

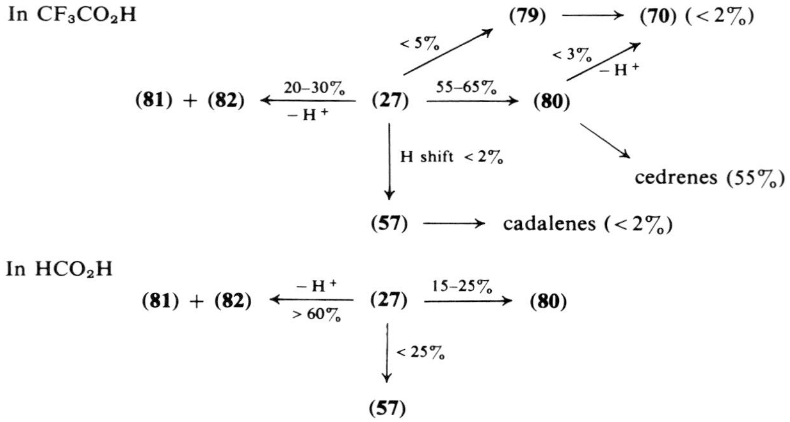

Scheme 8

followed was a pleasant surprise. After 10 min three products accounted for a 75% yield: 30% racemic α-cedrene (33), 25% racemic 2-*epi*-α-cedrene (38), and 20% dihydro-*ar*-curcumene (71) [48]. No other products were present in greater than 5% yield, and isocalamenene (70) and calamenene (23) were undetectable (<2%). From this we can define the kinetic fate of ion (27) in CF_3CO_2H and contrast it to that in HCO_2H (Scheme 8).

A truly biomimetic synthesis of the cedrene skeleton can thus be accomplished in 39% overall yield from nerolidyl phosphate by solvolysis followed by CF_3CO_2H treatment of β-bisabolene.

The most remarkable aspect was the stereoselectivity of the spirocyclization of ion (27). In excess of 92% of the spirane formation gives the α stereochemistry [(80), not (79)],* in which the cationic center produced can be stabilized by the *cis* π bond in the six-membered ring.

Is π-bond stabilization essential for spirocyclization? Is the process (27) → (80) → (33) + (38) fully concerted? Studies of the cyclization of some bisaboladienes in the same media (Scheme 9) provide a partial answer to the first question [57a].

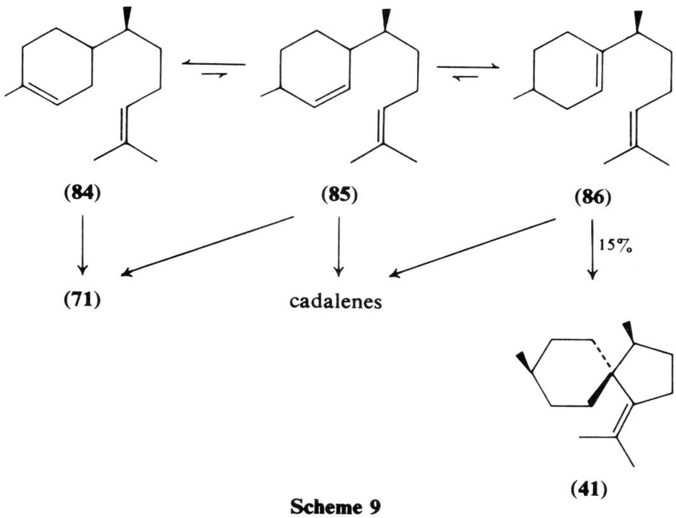

Scheme 9

Dienes (84), (85), and (86) were obtained by Birch reduction of the appropriate curcumenes, and each of these materials was subjected to CF_3CO_2H cyclization. Under these conditions one isomer (86) gave an identifiable spirane in 15% yield. The stereoisomer shown (41) was the only

---

* That is the "α-acoradiene"-related series. The α designation does not refer to the stereochemistry relative to the projection plane used in Scheme 7.

one that we could recognize positively since we had obtained it previously by semihydrogenation of alaskene. The other double-bond isomers go off to cadinenes or to curcumene-related materials, and apparently a significant portion of (**86**) is also so drained off by initial double-bond migration.

We then turned to the tricyclovetivane pathway. The observation that *ent*-allocedrol formate was stable to warm HCO_2H for prolonged periods suggested this medium for reproducing the anti-Markovnikov cyclization (**30**) → (**44**) in the biogenetic sequence. As shown in Scheme 10, there was no compelling reason to assume that the acoradienes available to us [(**36**) and β-alaskene (**40**)] would afford ion (**30**) in light of the alternative pathways emanating from (**40**) and presumably from (**36**). Even if ion (**30**) were generated during the course of the reaction, the barrier to cyclization and a short lifetime (rapid deprotonation) might prevent the formation of the

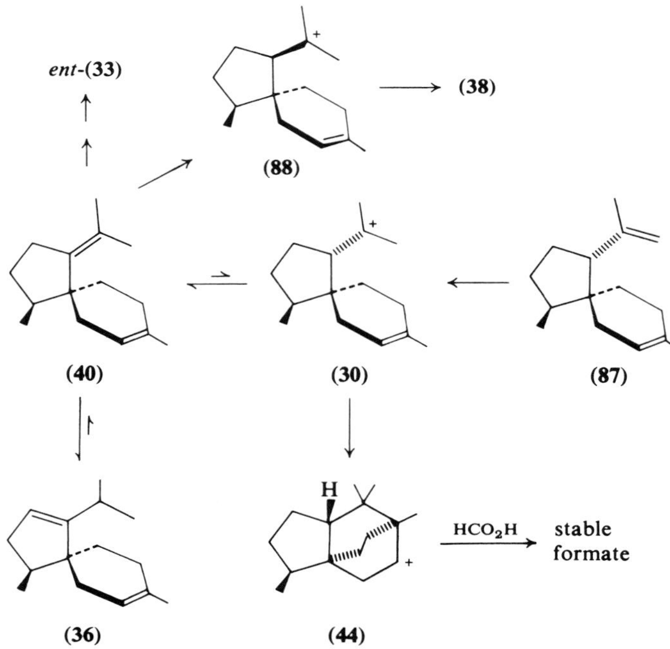

Scheme 10

allocedryl cation. Furthermore, ion (**44**) produced in the cyclization would be energy rich and might undergo further rearrangements before collisional deactivation would trap it as the stable formate.

We were thus not surprised to find that acoradiene (**36**) afforded, at best, a very poor yield of the desired formate ($\sim 3\%$) under a variety of conditions.* With β-alaskene (40), gc analysis revealed up to 6% of the desired formate under some conditions, the diastereomeric cedrenes (38) and ent-(33) in comparable yields, and unknown olefins and formates. Due to the minute quantities of β-alaskene available, we were unable to confirm the gc-assigned structure by spectroscopic comparisons. All in all, these studies did not supply the information concerning the energetics of the cyclization (30) → (44) required for an assessment of the role of enzymes in the biosynthesis of the vetiver tricyclics.

Fortunately, Tomita et al. were guided by similar considerations in their study of the acid-catalyzed transformations of "β-acoradiene" [the enantiomer of (87)], and the recent publication [41] of the full report provides a gc analysis of the product obtained in HCO_2H at 90°C. The gc conditions and standardization procedures employed were such that assignments could be based on retention data gathered during the course of our studies. The results of these workers are clarified below:

$$(87) \xrightarrow[90°C]{HCO_2H} \text{allocedryl formate} + (47) + (73) + ent\text{-}(33)$$
$$20\% \qquad 15\% \quad 4\% \qquad 10\%$$

Even in the case of olefin (87) alternative cyclizations (e.g., to α-cedrene) need not imply rapid deprotonation of ion (30); rather, they could result from a prior shift of the other double bond. At least 40% of the product results from ion (44). The contrast with the results obtained with β-alaskene reveals that anti-Markovnikov cyclization competes favorably with deprotonation of ion (30). Since the facile cyclization of ion (88) in this medium has been demonstrated we now know that β-alaskene does not protonate at the 7 position readily. In contrast, α-alaskene protonates preferentially, and stereoselectively, at C-7, producing α-cedrene in high yield.

The HCO_2H cyclization of (87) yields products [(47) and (73)] of further rearrangements (but fails to give prezizaene or zizaene), even though allocedryl formate is stable to the reaction conditions. Ion (44) is, as suspected, produced in a vibrationally excited form that is prone to further reaction [(44) → (45) → (46) → (47)].

The only biogenetic step that had not been realized at this stage, a synthesis of prezizaene (48), had to be accomplished under nonacidic conditions. Although the established stereochemistry of ent-allocedrol from vetiver oil is that required to produce the cedrenes with a trans-pentalene unit by assisted solvolysis, we have yet to observe that process: the reverse [(43) → (44)] has

---

* Due to the low yield, the structure assignment is based only on gc retention data. The major product is an unidentified tricyclic formate (up to 80% yield) of a tertiary methylcarbinol displaying one additional methyl singlet and two methyl doublets in its nmr spectrum.

been demonstrated in TFE containing methanesulfonic acid. When *ent*-allocedrol is converted to the mesylate (89) and subjected to pyrolysis (100°–120°C) in the presence of basic alumina, prezizaene (48)—identical by nmr, gc, and CD with the natural product—constituted 64% of the olefinic product (obtained in 90 + % yield). Interestingly, under these nonacidic conditions, (47) and (Z)-EQ-2 [but not (73), (Z)-EQ-1, or (Z)-EQ-3] were significant minor products.

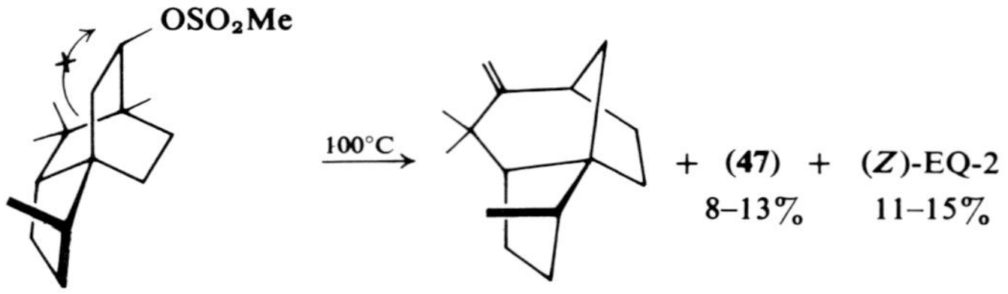

$$OSO_2Me$$

$$\xrightarrow{100°C}$$

+ (47) + (Z)-EQ-2
8–13%   11–15%

(89)

64%
(48)

With this transformation each stage of the biogenesis of the vetiver tricyclics has been demonstrated in purely chemical systems. Nature's catalysts are not as magical as they appear to be at first glance. What then are the distinctions of the enzymatic systems?

In the cedrene series the absence of 2-*epi*-cedrenes in nature is the major difference. If the natural cedrene biosynthesis proceeds via a germacrene rather than bisabolene (Scheme 11) the natural pattern can be rationalized [48]. A shift of C-7 in the muurolene-copaene intermediate (90), perhaps concerted with the hydride shift, proceeding via bridged intermediate (91) would retain the C-7 chirality and produce cation (31), the immediate precursor of cedrene (in both chemical synthesis and biosynthesis), after a 1,4 hydride shift [illustrated in (92)].

For the tricyclovetivane series a comparable biogenetic proposal would require an *ent*-germacrene and fails for lack of a sterically accessible 1,4 hydride shift to place the charge at the isopropyl group. Cation (27) (or the corresponding nonchiral ion pairs) is known to give *ent*-(31) and (88), not the required β-configured (relative) ion (30). If the sequence is initiated from a chiral or enzyme-bound β-bisabolyl equivalent (93), a hydride shift (95) would yield (94), a specific chiral conformer of (27), which would yield the tricyclovetivane precursor [cation (30)] either directly or via a chamigrane intermediate (96).

As it stands the only point at which an enzymatic-directing influence must be postulated is the formation of ion (30) leading to the tricyclovetivanes. That directing influence is nothing more than a specific orientation of ion (27), or the selection, as substrate, of a specific chiral β-bisabolyl intermediate.

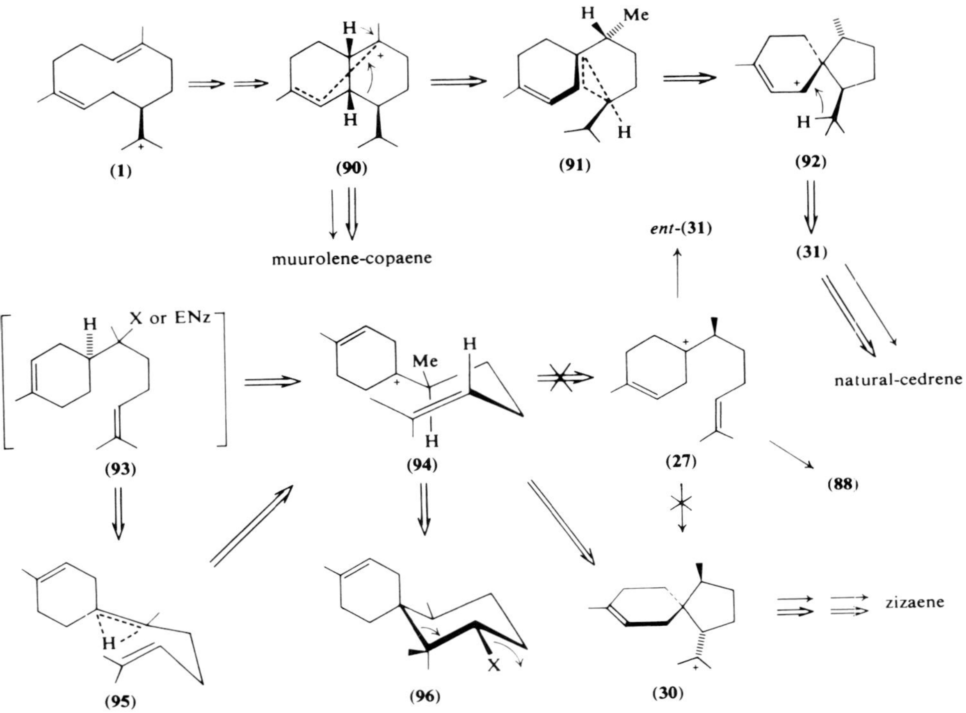

Scheme 11   Cedrene and kizaene—current biogenetic proposals. Double arrows indicate biosynthetic steps. Single arrows indicate observed steps *in vitro*.

Every other stage in the biosynthesis of both series of tricyclics represents the natural chemical preference of the systems involved as demonstrated by *in vitro* transformations.

SESQUITERPENE CONSTITUENTS OF LIVERWORTS

We began our discussion of biogenetic "missing links" by noting the consistent correlation in absolute configuration in sesquiterpenes derived from vascular plants. In those cases in which fungi and marine organisms have been shown to elaborate the same sesquiterpenes as vascular plants the substances have been found to be enantiomeric [2,2a,4]. This observation has been strengthened by recent work on *ent*-cadinenes from fungi [2a]. With the reported isolation of (−)-longiborneol (**107**) and (−)-longifolene (**108**) from a liverwort, *Scapania undulata* [58], the production of strictly enantiomeric sesquiterpenes from taxa of green plants in phylogenetic line to higher plants was established. Our interest in "enantiomeric" sesquiterpenes and a letter from Dr. S. Huneck inquiring about gc analyses of sesquiterpene oils prompted a joint study of liverwort essential oils.

The first oil examined was that of *Barbilophozia barbata*. The hydrocarbon portion had three major components, one of which was immediately recognized as (−)-α-alaskene (**34**). The other two proved to be novel sesquiterpenes now designated as α- and β-barbatene [(**98**) and (**99**)] [59,60]. The same

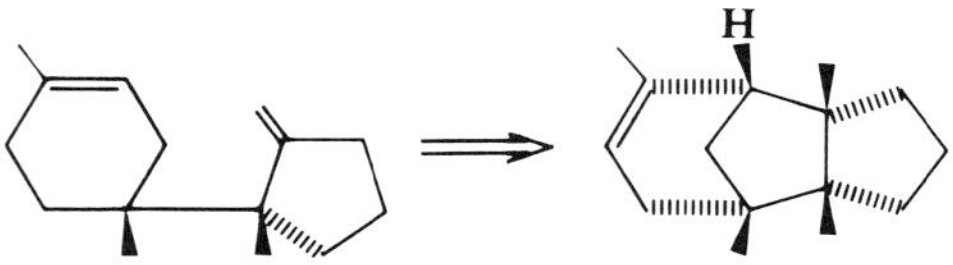

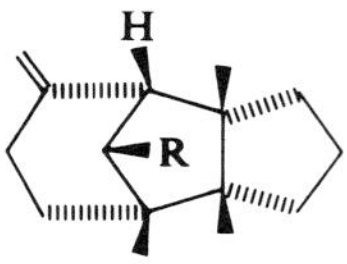

(**97**)   (**98**)   (**99**), R = H
                                   (**100**), R = OH

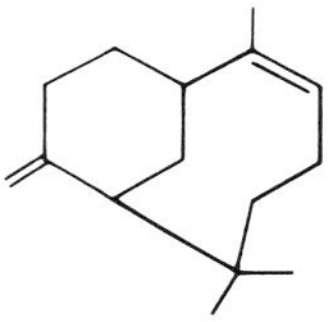

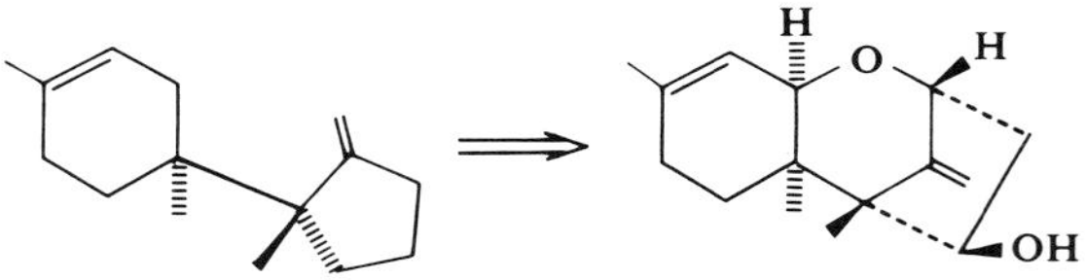

(**101**)   (**102**)

substances were isolated by Matsuo and assigned different structures [61]. An alcohol, gymnomitrol (**100**), was isolated from *Gymnomitrion obtusum*, and its structure has been established by independent studies [62]. Our subsequent studies [63] have revealed that one or another of these three substances occurs in 22 of 23 species of liverworts (order Jungermanniales). The barbatenes frequently co-occur with a substance called bazzanene [59,64], tentatively assigned structure (**101**) by Matsuo [64]. The nmr, ir, and mass spectral data for bazzanene are essentially the same as those reported for trichodiene (**102**) [65], the precursor of the trichothecanes. The circular dichroism displayed by bazzanene is inconsistent with either a *cis*- or *trans*-cyclooctene as implied in (**101**). On reaction with CF_3CO_2H, bazzanene yields cuparene (**66**). These results, together with the attractiveness of (**97**) → barbatene as a biogenetic proposal, led us to suggest structure (**97**) for bazzanene [66]. Studies to confirm this are in progress.

In the course of our studies of liverwort essential oils [63] we have found only sesquiterpenes derived from *ent*-germacrenes, the enantiomers of α-selinene, β-selinene, γ-cadinene, δ-cadinene, and β-elemene among them. We have also found (−)-α- and (−)-β-bisabolenes and representatives of the further stages of the γ-humulene family of sesquiterpenes; (−)-longiborneol and (−)-longifolene are common constituents of many *Scapania* species. This biogenetic sequence is illustrated in Scheme 12 together with the structures of longipinanol **(109)**, β-longipinene **(110)**, (+)-γ-himachalene **(111)**, and γ₂-himachalene **(112)**, which are new natural products in this family. In all occurrences in liverworts, these substances have the absolute stereochemistry shown, enantiomeric to the products in conifers. (−)-Longifolene has also been found in fungi [2a].

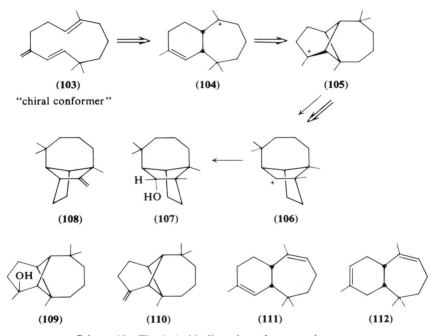

(103)
"chiral conformer"

(104)

(105)

(108)

(107)

(106)

(109)

(110)

(111)

(112)

Scheme 12 The (−) chirality γ-humulene sesquiterpenes.

Recently, we have identified sesquiterpene α-methylene-γ-butyrolactones in a number of liverwort extracts.* These, too, proved to be *ent*-germacrene derived and thus provided an opportunity to establish whether the cyto-

---

* A Δ⁴-eudesman-6,12-olide has been reported to occur in liverworts of the genus *Frullania* [67] and appears to be associated with a dermatitis of woodcutters. *Frullania* liverworts grow on the bark of living trees. The lactone from *F. dilatata* was shown to have structure (*iv*), an *ent*-santonin, but that from *F. tamarisci* is the optically pure enantiomer of (*iv*). The latter would be the first established case of a germacrene-derived sesquiterpene from a liverwort with a β-isopropyl group. In light of the habitat of these

toxicity of this class of compounds against carcinoma cells shows a chiral specificity.

A series of lactones [(113)–(115)] obtained from *Diplophyllum albicans* was based on the *ent*-eudesm-4-ene skeleton and displayed the first case of 9-hydroxylation in a natural eudesmane. Even the basic structure, diplophyllin (115), was unknown, in either enantiomeric form, in nature.* Benešová, Samek, and Vašíčková report that European samples of this liverwort species elaborate an isomeric lactone, diplophyllolide-A (117) [69]. The enantiomeric forms of both *Diplophyllum* lactones can be produced from isoalantolactone (119) on treatment with CH_3SO_3H in TFE. Only a preliminary study of the relative cytotoxicity of these substances against human epidermoid carcinoma in KB cell culture has been completed. Diplophyllin displays an ED_{50} of 1.4–12 µg/ml, which is in the same range as the vernolepin compounds. The synthetic enantiomer displays only one-third to one-tenth the activity of the natural compound. It is unlikely that this is the result of unknown impurities in the natural material since the synthetic and natural compounds display identical melting behavior, have mirror-image CD spectra (within experimental error), and were both subjected to identical acid treatment prior to final purification. An unexpected finding was that the previously noted activity of alantolactone was due to contamination with isoalantolactone (119) (ED_{50} = 2.4 µg/ml). Further exploration of the cytotoxicity of liverwort products is in progress.

The study of liverwort sesquiterpenes has also uncovered at least eight novel sesquiterpene skeletons, but here discovery precedes structure elucidation by an uncanny margin and we can only guess that further biogenetic "missing links" are already under investigation.

Coming around full circle, we now return to the intermediacy of the 4,5-*cis*-germacrenyl cation (1) in sesquiterpene biosynthesis, but in the mirror-image form in liverworts and fungi. Recent work at the Eidgenössische Technische Hochschule (Zurich) clearly established the stereospecific hydride

---

* The proof of structure and other studies are the subject of a paper by Ohta, Andersen, and Liu [68].

---

liverworts it is possible that in the latter case the liverwort utilizes a precursor produced by its vascular plant host.

(*iv*)

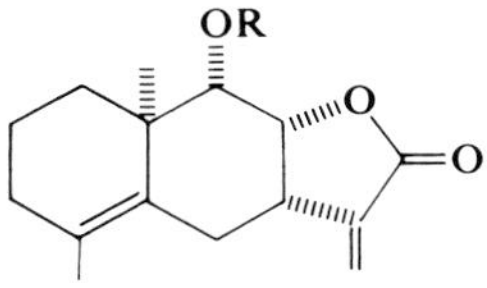

(113), R = OH
(114), R = OAc

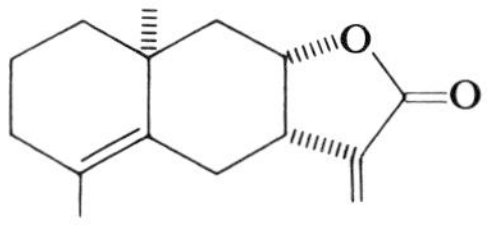

(115)

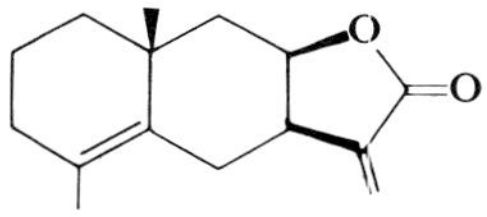

(116)

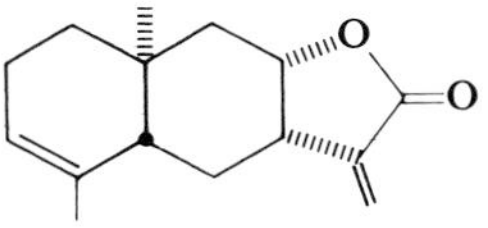

(117)

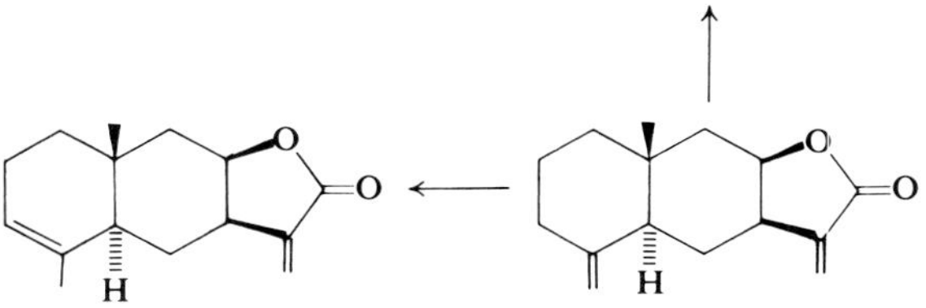

(118)

(119)

shift (C-6 → C-11) between *ent*-(1) and (−)-sativene (121) [2a], but ion *ent*-(1) was still not represented by a natural product in which the cationic charge remained at C-11. A representative of this class, helmiscapene (122), has been found in *Helminthosporium* species [70] and in *Scapania undulata* [71]. The structure of helmiscapene rests on conversion to (+)-δ-selinene and the nonidentity of helmiscapene by gc and nmr to the other three α-selinene diastereomers.* *cis*-α-Selinene was prepared from santonin by a hydrogenation reaction known to yield a *cis*-fused decalin [73].

* The structure assignment has also been confirmed [70] by the synthesis of the enantiomeric substances. Dehydration of the known dihydrooccidentalol (*v*) affords *ent*-helmiscapene (*vi*) and the isopropylidene isomer.

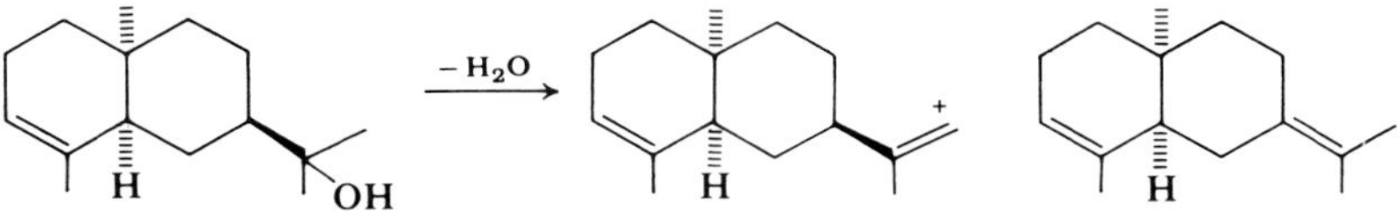

$$-H_2O \longrightarrow$$

(v)

(vi)

Another study that serves to clarify the relationship between *ent*-(1) and helmiscapene has just been completed. Another of the minor hydrocarbons of *H. sativum* has been identified as *cis*-germacene-A (*vii*) by comparison with the racemic material produced by total synthesis [72].

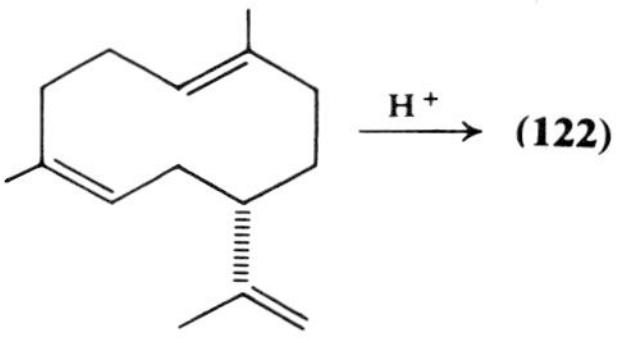

$$\xrightarrow{\text{H}^+} \quad (122)$$

(vii)

Acid treatment of racemic (*vii*) provides a good entry to *rac*-helmiscapene [72].

(+)-δ-Selinene   α-Selinene   10-*epi*-α-Selinene [19]   *cis*-α-Selinene

A comparison of compositions of the sesquiterpenic oils of these two unrelated species shows other similarities (Scheme 13), as if in each case the biosynthetic process begins with the production of a bridged cyclization intermediate (120) having the *cis*-4,5 bond and able to collapse into either the γ-humulene or the germacrene series. The germacrene series, represented by sativene (121) and in greater quantities by *seco*-sativenes such as victoxinine, is the "intended" path for *Helminthosporium*, with longifolene and helmiscapene representing errors in which an intermediate escapes the molding influence of the enzyme. In *Scapania undulata* a similar enzyme system aims toward the γ-humulene group with a comparable ratio of errors. An additional attraction of this unified medium-ring intermediate hypothesis is the consistent correlation of absolute configuration among germacrene- and humulene-related compounds in both vascular and nonvascular plants. Only detailed radiotracer biosynthetic experiments will provide tests of this hypothesis.

The configurational distinction between "primitive" taxa and the vascular plants deserves more than passing mention. At what stage in plant phylogeny did this dramatic reversal in optical specificity occur,* and why? Clearly the evolution of an effectively enantiomeric cyclizing enzyme cannot be explained readily by one or two mutations.

We can, however, present a rationale for the evolutionary success of the early mirror-image plants. Assuming that, in primitive marine organisms, sesquiterpene pathways were developed for some vital internal regulatory function, it is possible that the sesquiterpenes were replaced in these roles by more elaborate and specific molecules (di- and triterpenes?). The retention of the pathways then suggests an ecological value to the plants and the fact that the pathways are not energetically demanding, as indicated by our chemical simulation studies.

The availability of sesquiterpenoids in the marine environment had led to their use by arthropods, and in this development we can see a rationale for the dramatic reversal in optical specificity that apparently occurred somewhere

---

* We are currently examining other bryophyte essential oils and will extend this search to ferns, lycopods, etc., in search of the point or points at which the present-day optical specificity of vascular plant sesquiterpenes arose in nature.

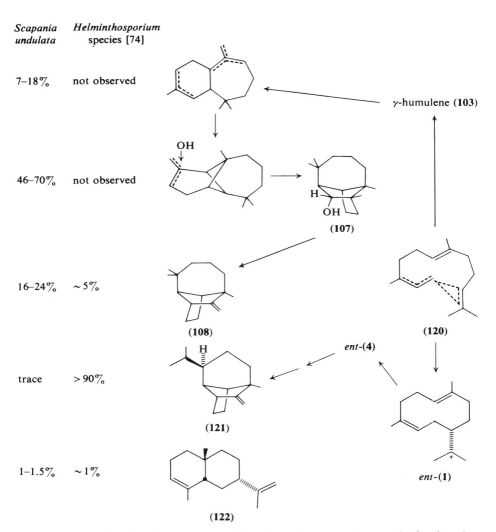

Scapania undulata	*Helminthosporium* species [74]
7–18%	not observed
46–70%	not observed
16–24%	~5%
trace	>90%
1–1.5%	~1%

Scheme 13 The unified medium-ring intermediate hypothesis as seen in fungi and liverwort oil compositions.

between the most primitive land plants and the more recently evolved ones. Arthropod predation became more significant in the land environment (insects) and served to select plants producing sesquiterpenes enantiomeric to those of marine organisms.

CONCLUDING REMARKS

We have attempted to put an 8-year study of sesquiterpene chemistry into a biogenetic perspective, without major deviation from chronological sequencing. Throughout, we have been pleasantly surprised by the trail of evidence that nature has left us in the constitution of the essential oils when the examination is carried to the necessary depth, the imperfections in individual threads giving us the best comprehension of the fabric.

It is clear that virtually all sesquiterpene biosynthetic steps can be simulated by *in vitro* chemical transformations. The carbocationic rearrangements and cyclizations employed in biomimetic transformations are most readily understood as "concerted" processes with specific stereoelectronic effects associated with conformational preferences, even if the kinetic data for the reactions do not meet the stricter definition of "concertedness" used by the physical organic chemist. The carbocationic conceptual basis for rationalizing terpenoid chemistry will continue to provide valuable insights as long as absolute and relative configurational correlations are not violated.

The test of biogenetic proposals in the next 8 years will come from tracer studies. Here, studies in nonvascular plants will undoubtedly lead the way.

ACKNOWLEDGMENTS

This research has drawn support from grant funds from the National Institutes of Health (GM), the Research Corporation, the Petroleum Research Fund, and from funds associated with a Sloan Fellowship and Dreyfus Foundation Teacher-Scholar Award to N.H.A. This past support and the continuing NIH support (Grant GM 18143) were essential for rapid progress in the area. We also wish to acknowledge the contributions of other collaborators whose work in essential oil composition and structure elucidation was an integral part of the chemical simulation studies presented: C.-B. Liu, Dr. S. E. Smith, B. Bottino, Dr. A. Moore, C.-L. W. Tseng, P. Bissonnette, B. Shunk, C. M. Kramer, C. Graham, C. R. Costin, W. E. Dasher, and K. Allison. Special acknowledgment goes to Dr. Siegfried Huneck (Institut für Biochemie der Pflanzen, Halle/Saale DDR-401) for leading us to a study of liverwort constituents.

REFERENCES

1. V. Herout, *in* "Aspects of Terpenoid Chemistry and Biochemistry" (T. W. Goodwin, ed.), pp. 53–94. Academic Press, New York, 1971. There appears to be no demonstrated occurrence of sesquiterpene in the two most primitive kingdoms, Monera and Protista, but this may represent only the lack of a directed search.
2. D. H. R. Barton and N. H. Werstiuk, *J. Chem. Soc. C* p. 148 (1968); N. N. Gerber, *Phytochemistry* **10**, 185 (1971); P. de Mayo and R. E. Williams, *J. Am. Chem. Soc.* **87**, 3275 (1965); S. Nozoe and Y. Machida, *Tetrahedron Lett.* p. 2671 (1970).

2a. D. Arigoni, *Pure Appl. Chem.* 141, 219 (1975); F. Dorn, P. Bernasconi, and D. Arigoni, *Chimia* 29, 25 (1975), and references therein.

3. V. Herout, in "Aspects of Terpenoid Chemistry and Biochemistry" (T. W. Goodwin, ed.), pp. 89–90. Academic Press, New York, 1971; note also the juvenile hormone compounds.

4. A. J. Weinheimer *et al.*, *Chem. Commun.* p. 1070 (1968); *Tetrahedron Lett.* p. 3315 (1969); A. J. Weinheimer, W. W. Youngblood, P. H. Washecheck, T. K. B. Karns, and L. S. Ciereszko, *ibid.* p. 497 (1970); F. Semmler and K. Spornitz, *Chem. Ber.* 46, 4025 (1913).

5. L. Ruzicka, A. Eschenmoser, and H. Heusser, *Experientia* 9, 357 (1953); L. Ruzicka, A. Eschenmoser, O. Jeger, and D. Arigoni, *Helv. Chim. Acta* 38, 1890 (1955); L. Ruzicka, *Proc. Chem. Soc., London* p. 341 (1959); *Pure Appl. Chem.* 6, 493 (1963).

6. J. B. Hendrickson, *Tetrahedron* 7, 82 (1959).

7. Such a system was used by Parker, Roberts, and Ramage in their timely review [8]. See also L. A. Smedman, E. Zavarin, and R. Teranishi, *Phytochemistry* 8, 1457 (1969).

8. W. Parker, J. S. Roberts, and R. Ramage, *Q. Rev., Chem. Soc.* 21, 331 (1967).

9. M. Martin-Smith and W. E. Sneader, *Prog. Drug Res.* 13, 11–100 (1969). For recent cytotoxic sesquiterpene lactones, see, for example, the work of S. M. Kupthan, R. Hemingway, D. Werner, and A. Harim, *J. Org. Chem.* 34, 3903 (1969). See also S. M. Kupthan, R. Doskotch, S. Keely, Jr., and C. Huffard, *Chem. Commun.* p. 1137 (1972); S. M. Kupthan, R. Doskotch, S. Keely, Jr., C. Huffard, and F. El-Feraly, *Phytochemistry* 14, 769 (1975).

10. N. H. Andersen and D. D. Syrdal, *Phytochemistry* 9, 1325 (1970).

11. N. H. Andersen, *Phytochemistry* 9, 145 (1970).

11a. K. Nishimura, N. Shinoda, and Y. Hirose, *Tetrahedron Lett.* p. 2263 (1969).

12. J. A. Marshall and N. H. Andersen, *Tetrahedron Lett.* p. 1611 (1967); J. A. Marshall, N. H. Andersen, and P. C. Johnson, *J. Am. Chem. Soc.* 89, 2748 (1967); J. A. Marshall and P. C. Johnson, *ibid.* p. 2750.

13. A known natural product found in marine invertebrates also elaborating *ent*-selinenes; see Semmler and Spornitz [4].

14. See Andersen [11], and citations therein.

15. W. D. MacLeod, Jr., *Tetrahedron Lett.* p. 4779 (1965).

16. N. H. Andersen, M. S. Falcone, and D. D. Syrdal, *Tetrahedron Lett.* p. 1759 (1970).

17. Structure elucidation and isolation from vetiver by Y. Ohta (unpublished).

18. R. Kaiser and P. Naegeli, *Tetrahedron Lett.* p. 2009 (1972).

19. E. Klein and W. Rojan, *Tetrahedron Lett.* p. 279 (1970).

20. N. H. Andersen, *Tetrahedron Lett.* p. 4651 (1970).

21. Unpublished work with D. P. Svedberg.

22. L. H. Briggs and G. W. White, *Aust. J. Chem.* 26, 2229 (1973).

23. Our sample of γ-amorphene corresponds by ir to that of bicyclic sesquiterpene No. 3 of Ulahov *et al.* [*Collect. Czech. Chem. Commun.* 32, 809 (1967)] and by nmr and ir to a sample derived from *Amorpha fruticosa* by Ohta and Hirose. For δ-amorphene, see Y. Ohta, K. Ohara, and Y. Kirose, *Tetrahedron Lett.* p. 4181 (1968).

24. K. Snajberk and E. Zavarin, *Phytochemistry* 14, 2025 (1975).

25. N. H. Andersen, *Tetrahedron Lett.* p. 1755 (1970).

25a. F. Kido, R. Sakuma, H. Uda, and A. Yoshikoshi, *Tetrahedron Lett.* p. 3169 (1969); A. Homma, M. Kato, M.-D. Wu, and A. Yoshikoshi, *ibid.* p. 231 (1970).

26. N. H. Andersen, M. S. Falcone, and D. D. Syrdal, *Phytochemistry* 9, 1341 (1970).

27. N. H. Andersen, D. D. Syrdal, B. M. Lawrence, S. J. Terhune, and J. W. Hogg,

Phytochemistry **12**, 827 (1973); W. Fenical, J. J. Sims, R. M. Wing, and P. C. Radlick, *ibid.* **11**, 1161 (1972).

27a. Y. Ohta and Y. Hirose, *Chem. Lett. Jpn.* p. 263 (1972).

28. N. H. Andersen and D. D. Syrdal, *Tetrahedron Lett.* p. 2277 (1970).

29. C. E. McEachan, A. T. Phail, and G. A. Sim, *J. Chem. Soc. C* p. 579 (1966); J. Vrkoč, J. Jonáš, V. Herout, and F. Šorm, *Collect. Czech. Chem. Commun.* **29**, 539 (1964), and references therein.

30. G. Buchi, R. E. Erickson, and N. Wakabayashi, *J. Am. Chem. Soc.* **83**, 927 (1961).

31. G. Stork and F. H. Clarke, Jr., *J. Am. Chem. Soc.* **83**, 3114 (1961); I. G. Guest, C. R. Hughes, R. Ramage, and A. Sattar, *J. Chem. Soc., Chem. Commun.* p. 526 (1973).

32. B. Tomita and Y. Hirose, *Tetrahedron Lett.* p. 143 (1970).

33. J. A. Marshall, S. F. Brady, and N. H. Andersen, *Fortschr. Chem. Org. Naturst.* **31**, 283 (1974).

34. N. H. Anderson and D. D. Syrdal, *Tetrahedron Lett.* p. 899 (1972).

35. B. Tomita, T. Isono, and Y. Hirose, *Tetrahedron Lett.* p. 1371 (1970).

36. Samples of authentic acorone and isoacorone were the gifts of Professor V. Herout.

37. The relative and absolute stereochemistry of the tricyclovetivanes has been confirmed in our laboratories, reported in part in Andersen [25]. See also R. M. Coates, R. F. Farney, S. M. Johnson, and I. C. Paul, *Chem. Commun.* p. 999 (1969); N. Hanayama, F. Kido, R. Tanaka (née Sakuma), H. Uda, and A. Yoshikoshi, *Tetrahedron* **29**, 945 (1973).

38. F. Kido, H. Uda, and A. Yoshikoshi, *Tetrahedron Lett.* p. 2815 (1967); I. C. Nigam, H. Komae, G. A. Neville, C. Redecka, and S. K. Pahnikar, *ibid.* p. 2497 (1968).

38a. D. F. MacSweeney, R. Ramage, and A. Sattar, *Tetrahedron Lett.* p. 557 (1970).

39. N. H. Andersen and M. S. Falcone, *Chem. Ind. (London)* p. 62 (1971).

40. Unpublished studies with S. E. Smith.

41. B. Tomita and Y. Hirose, *Phytochemistry* **12**, 1409 (1973).

42. Note that 2,5-di-*epi*-cedrene (50) was originally viewed as a Δ^2-cedrene by Kaiser and Naegeli [18]. S. K. Paknikar, S. V. Bhatwadekar, and K. K. Chakravarti [*Tetrahedron Lett.* p. 2973 (1975)] suggested structure (50) and the name $(-)$-α-funebrene based on its presumed enantiomeric relationship to the substance of undefined rotation obtained from *Cupressus funebris* [J. W. Kirtany and S. K. Paknikar, *IUPAC Symp. Chem. Nat. Prod., 8th, 1972*, Abstract C-44 (1972); *Ind. J. Chem.* **11**, 508 (1973)].

43. T. Norin, S. Sundin, P. Karlsson, P. Kierkegaard, A. M. Pilloti, and A. G. Wiehager, *Tetrahedron Lett.* p. 17 (1973).

44. L. H. Zalkow and M. G. Clower, Jr., *Tetrahedron Lett.* p. 75 (1975).

45. F. Semmler and K. Spornitz, *Chem. Ber.* **46**, 4025 (1913); L. Ruzicka and E. Capato, *Helv. Chim. Acta* **8**, 259 (1925).

46. See C. D. Gutsche, J. R. Maycock, and C. T. Chang, *Tetrahedron* **24**, 859 (1968), and references therein. See also Ohta and Hirose [27a].

47. W. Rittersdorf and F. Cramer, *Tetrahedron* **24**, 43 (1968).

48. These cyclization studies have been reported in part. N. H. Andersen and D. D. Syrdal, *Tetrahedron Lett.* p. 2455 (1972).

49. G. Stork and W. N. White, *J. Am. Chem. Soc.* **78**, 4609 (1956).

50. F. G. Bordwell, *Acc. Chem. Res.* **3**, 281 (1970).

51. S. Godtfredsen and D. Arigoni, personal communication.

52. L. Roach and W. Daly, *Chem. Commun.* p. 606 (1970).

53. N. H. Andersen, S. E. Smith, and Y. Ohta, *J. Chem. Soc., Chem. Commun.* p. 447 (1973).
54. Reported in the Ph.D. thesis of S. E. Smith, University of Washington, Seattle (1972).
55. Unpublished studies by Y. Ohta.
56. F. Kido, H. Uda, and A. Yoshikoshi, *Tetrahedron Lett.* p. 1247 (1968); N. Hanayama, F. Kido, R. Sakuma, H. Uda, and A. Yoshikoshi, *ibid.* p. 6099.
57. N. H. Andersen, D. D. Syrdal, and in part C. Graham, *Tetrahedron Lett.* p. 903 (1972).
57a. Reported in the Ph.D. thesis of D. D. Syrdal, University of Washington, Seattle (1971).
58. S. Huneck and E. Klein, *Phytochemistry* **6**, 383 (1967).
59. N. H. Andersen and S. Huneck, *Phytochemistry* **12**, 1818 (1973).
60. N. H. Andersen, C. R. Costin, C. M. Kramer, Y. Ohta, and S. Huneck, *Phytochemistry* **12**, 2709 (1973).
61. A. Matsuo, T. Maeda, M. Nakayama, and S. Hayashi, *Tetrahedron Lett.* p. 4131 (1973); A. Matsuo, H. Nozaki, M. Nakayawa, Y. Kushi, S. Hayashi, and N. Kamijo, *ibid.* p. 241 (1975).
62. J. D. Connolly, A. E. Harding, and I. M. S. Thornton, *Chem. Commun.* p. 1320 (1972).
63. Only a small portion of these studies have been reported. N. H. Andersen, B. Shunk, and C. R. Costin, *Experientia* **29**, 645 (1973); N. H. Andersen, B. J. Bottino, A. Moore, and J. R. Shaw, *J. Am. Chem. Soc.* **96**, 603 (1974); see also Andersen *et al.* [59,60].
64. S. Hayashi and A. Matsuo, *Experientia* **25**, 1139 (1969); A. Matsuo, *Tetrahedron* **27**, 2757 (1971); see also Andersen and Huneck [59].
65. S. Nozoe and Y. Machida, *Tetrahedron Lett.* p. 2671 (1970); *ibid.* p. 1969 (1972).
66. Dr. Y. Asakawa (Dept. of Chemistry, Faculty of Science, University of Hiroshima) has also concluded that bazzanene is not represented by structure (101) but has a trichodiene skeleton.
67. The report of *Frullania tamarisci* constituents appears in H. Knocke, G. Ourisson, G. W. Perold, J. Foussereau, and J. Maleville, *Science* **166**, 239 (1969). The report of (*iv*) is a note in proof (with J. C. Muller) in the same article.
68. Y. Ohta, N. H. Andersen, and in part C.-B. Liu, *Tetrahedron* **33**, 617 (1977).
69. V. Benešova, Z. Samek, and S. Vašíčková, *Collect. Czech. Chem. Commun.* **40**, 1966 (1975).
70. Unpublished studies of F. Dorn and D. Arigoni.
71. Unpublished studies of C.-L. W. Tseng and N. H. Andersen.
72. Unpublished studies of F. Dorn, R. E. K. Winter, and D. Arigoni.
73. T. Nozoe, T. Asao, M. Ando, and K. Takase, *Tetrahedron Lett.* p. 2821 (1967).
74. Data taken from the Ph.D. dissertation of F. Dorn, Eidgenössische Technische Hochschule, Zurich (1975).

2

Mechanisms for Proton Transfer in Carbonyl and Acyl Group Reactions

Ronald E. Barnett

INTRODUCTION

Reversible addition of nucleophiles to the carbonyl group is one of the most studied classes of reactions. Understanding of the mechanisms of these reactions is of considerable interest to both chemists and biochemists, particularly with respect to unraveling the chemistry occurring at the active site of an enzyme.

Addition reactions of nucleophiles to carbonyl groups and, more generally, to acyl groups are subject to catalysis by general acids and bases. Acyl group reactions of the type shown in Eq. (1) almost always proceed through a tetrahedral addition intermediate [1], and so carbonyl addition reactions can serve as models for the more complex acyl substitution reactions.

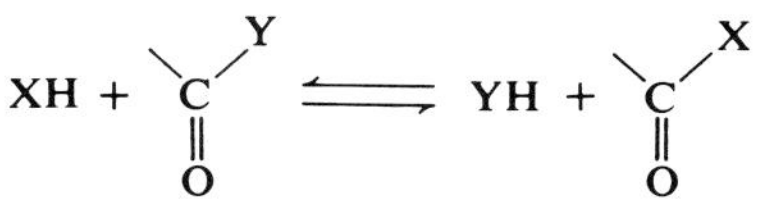

$$XH + \underset{\underset{O}{\|}}{\overset{Y}{\underset{}{\diagdown}C\diagup}} \; \rightleftharpoons \; YH + \underset{\underset{O}{\|}}{\overset{X}{\underset{}{\diagdown}C\diagup}} \tag{1}$$

CARBONYL ADDITION REACTIONS

Generalized schemes for the additon of neutral and anionic nucleophiles to carbonyl groups are shown in Eqs. (2) and (3). The mechanism for a given

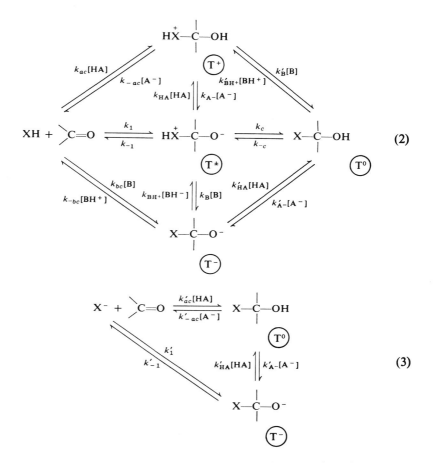

$$(2)$$

$$(3)$$

nucleophile depends on the lifetimes and the relative energies of the tetrahedral intermediates. As a general rule the stronger the nucleophile and the
more reactive the carbonyl group, the greater will be the stability of the
tetrahedral intermediates (T). It follows that electron-donating groups in the
nucleophile and electron-withdrawing groups attached to the carbonyl will
increase the stability of the intermediates. Steric effects are also important,
with bulky groups attached to the carbonyl destabilizing the tetrahedral
adduct.

Typically, strong nucleophiles such as alkoxide ions, basic aliphatic amines,
and basic thiol anions add to the carbonyl group without catalysis [2].
Equilibration of the various ionic forms of the tetrahedral adduct is fast

compared to elimination of the nucleophile to give reactants. For weak nucleophiles the intermediate that would be formed from uncatalyzed attack would be too unstable to exist as a discrete intermediate. For these reactions proton transfer must be concerted with bond formation to carbon, and so general acid and general base catalysis is observed. The Brönsted plots are usually linear over a wide pK_a range of catalysts, with a slope in the range of 0.2–0.8.

For example, consider the addition of thiol anions of varying basicity to acetaldehyde to form a hemithioacetal [3,4]. The addition of basic thiol anions such as the ethanethiol anion is uncatalyzed [k_1' is rate determining in Eq. (3)]. As the basicity of the attacking anion is decreased the lifetime of T^- also decreases due to an increase in k_{-1}', the rate constant for expulsion of the nucleophile from T^-. Extrapolation of the values of k_{-1}' for the basic anions to weakly basic anions leads to the prediction that the lifetime of T^- will be less than 10^{-13} sec if the adduct is formed from a thiol with a pK_a of less than 6, and so T^- cannot exist. For the weakly basic thiol anions T^0 is formed directly in a general-acid-catalyzed step. Hemithioacetal formation between thioacetate ion ($pK_a = 3.2$ for thioacetic acid) and acetaldehyde is general acid catalyzed with $\alpha = 0.20$ and is linear over a range of at least eight pK_a units. According to the scheme of Eq. (3), k_{ac}' is rate determining and T^- is too unstable to exist.

A similar situation exists for nitrogen nucleophiles. The attack of basic aliphatic amines, such as *tert*-butylamine, on aldehydes is uncatalyzed, with k_1 of Eq. (2) being rate determining [5]. Equilibration of the ionic forms of the tetrahedral intermediate is fast compared to k_{-1} [6]. From extrapolation of the estimated values of k_{-1} for a series of adducts formed between various weakly basic amines and *p*-chlorobenzaldehyde one can predict that, if the conjugate acid of the amine has a pK_a less than approximately 0, the lifetime of $T^{\pm}$ will be less than 10^{-13} sec and so it cannot exist [7–9]. Addition reactions of such amines would have to occur with a concerted proton transfer. An example of this situation is seen with the addition of urea to acetaldehyde [10]. The pK_a of the conjugate acid of urea is 0.2. However, this is for protonation on oxygen. The nitrogen of urea should be even less basic, and so the adduct $T^{\pm}$ should not exist. Attack of urea must be concerted with proton transfer. As expected, the reaction is subject to general acid catalysis so that T^+ is formed directly and k_{ac} of Eq. (2) is rate determining. The Brönsted coefficient α is 0.46.

For nucleophiles of intermediate strength the mechanism for the addition reaction depends on the lifetimes of the different ionic forms of the tetrahedral adduct and their relative energies compared to reactants and products. Depending on the reaction any of the steps shown in Eqs. (2) and (3) could be rate limiting. In addition, various types of "preassociation" mechanisms

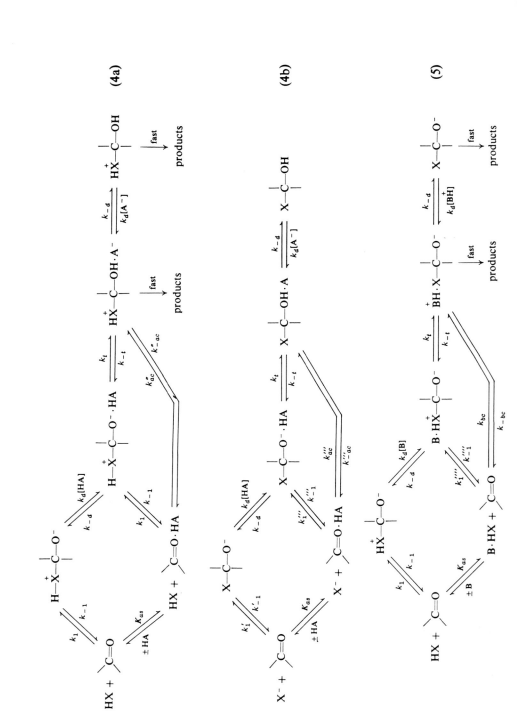

must be considered [Eqs. (4) and (5)]. The equilibrium constant K_{as} is for the formation of the encounter complex between the catalyst and one of the reactants. It will be assumed to have a value of approximately 0.02 M^{-1}. The constants k_d and k_{-d} are rate constants for diffusion, with $k_d = 2 \times 10^9$ M^{-1} sec^{-1} and $k_{-d} = 1 \times 10^{11}$ sec^{-1} for nonsolvent catalysts [10]. If the catalyst is hydronium ion then $k_d = 5 \times 10^{10}$ M^{-1} sec^{-1}, and if the catalyst is hydroxide ion then $k_d = 2 \times 10^{10}$ M^{-1} sec^{-1}. The constants k_t and k_{-t} are for simple proton transfer within the encounter complex. If the proton transfer is thermodynamically favorable then k_t or k_{-t} has no activation energy and a value of approximately 10^{13} sec^{-1}. The relative energies of the ionic forms of the tetrahedral adduct can be determined if all of the pK_a values are known. In some cases these can be determined directly, but in most cases they must be estimated using a $\sigma_I - \rho_I$ correlation for substituted alcohols and ammonium ions [11,12]. For example, the measured pK_a of (1) is 12.7; the calculated value is 12.6. The measured pK_a of (2) is 9.3; the calculated value is 8.8.

$$H_3C-\overset{\underset{\displaystyle CH_3}{|}}{\overset{\displaystyle CH_3}{\overset{|}{\overset{+}{N}}}}-CH_2-OH$$

(1)   (2)

In this chapter a concerted mechanism is considered to be one in which bond formation to carbon and proton transfer occur in the same step. A stepwise mechanism is one in which bond formation to carbon and proton transfer occur in different steps. A concerted mechanism always shows general acid or general base catalysis. The Brönsted plot for such a reaction is linear, or shows only gradual curvature, over a wide range of catalyst pK_a values. In a stepwise mechanism a simple proton transfer may be rate limiting. To understand the consequences of this for the behavior of such a reaction it is necessary to examine the characteristics of simple proton transfers [Eq. (6)].

$$B + HA \underset{k_{-d}}{\overset{k_d}{\rightleftharpoons}} B \cdot HA \underset{k_{-t}}{\overset{k_t}{\rightleftharpoons}} \overset{+}{B}H \cdot A^- \underset{k_d}{\overset{k_{-d}}{\rightleftharpoons}} BH^+ + A^- \qquad (6)$$

If the proton transfer is thermodynamically favorable, that is, if HA is a stronger acid than BH$^+$, the rate-determining step will be the diffusion together of B and HA. If a series of different acids HA$_i$, all of which are stronger than BH$^+$, is used, the rate constants for proton transfer to B will all be the same because they will be determined by the rate of diffusion. The Brönsted coefficient α for general acid catalysis in this case would be 0.00. If the HA$_i$ are weaker acids than BH$^+$ then the rate constant for proton transfer k_{HA} will be equal to $K_{as}K_{HA_i}k_{-d}/K_{BH^+}$, where K_{HA_i} and K_{BH^+} are the acid ionization constants of HA_i and BH^+, respectively. For this situation $\alpha = 1.00$. If the acidity of HA_i is similar to that of BH^+ the observed rate of transfer is slower than would be predicted by a simple diffusion mechanism. A satisfactory fit to the data is usually obtained if it is assumed that in the region of $\Delta pK_a = 0$, $\log k_t = 10 + 0.5(pK_{BH^+} - pK_{HA_i})$ [8]. The rate law for simple proton transfer is given in Eq. (7),

$$k_{HA} = \frac{K_{as}k_{-d}k_t}{k_{-d} + k_t[1 + (K_{BH^+}/K_{HA})]} = \frac{k_d k_t}{k_{-d} + k_t[1 + (K_{BH^+}/K_{HA})]} \quad (7)$$

where $k_t/k_{-t} = K_{HA}/K_{BH^+}$. The equations for general-base-catalyzed simple proton transfer are analogous. The Brönsted plot for simple proton transfer is biphasic with limiting slopes of 0 and 1 and a relatively sharp transition region (Fig. 1). Carbonyl addition reactions that have a simple proton transfer as the rate-determining step give this type of Brönsted plot. A biphasic Brönsted plot can also be observed for preassociation mechanisms and concerted mechanisms. Consider a reaction that goes by one of the pathways in the scheme of Eq. (4a). The complete rate law is shown in Eq. (8).

$$v = [HX]\left[\begin{array}{c} \diagdown \\ \diagup \end{array}C{=}O\right][HA]k_{-d}$$

$$\times \left\{ \frac{k_t k_d k_1 + k_{ac}'' K_{as} k_{-1} k_{-d}}{k_{-1} + k_d[HA]} + K_{as}(k_1'' k_t + k_{ac}'' k_{-1}'' + k_{ac}'' k_t) \right\} \quad (8)$$

$$\times \left[\left\{ k_{-1}'' + \frac{k_{-1}k_{-d}}{k_{-1} + k_d[HA]} + \frac{k_t(k_{-ac}'' + k_{-d})}{k_{-t} + k_{ac}'' + k_{-d}} \right\} \left\{ k_{-t} + k_{-ac}'' + k_{-d} \right\} \right]^{-1}$$

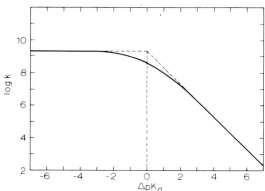

Fig. 1 Brönsted plot for simple diffusion-limited proton transfer between a base and a series of acids. The curve is calculated from Eq. (7), with $k_d = 2 \times 10^9$ M^{-1} sec^{-1}, $k_{-d} = 10^{11}$ sec^{-1}, and $\log k_t = 10 + 0.5(pK_{BH^+} - pK_{HA_i})$.

For a stepwise mechanism with rate-determining simple proton transfer the rate law reduces to Eq. (9a). For a preassociation mechanism the rate law is shown in Eq. (9b). A concerted mechanism that will also give a biphasic Brönsted plot has the rate law of Eq. (9c).

$$k_{HA} = \frac{K_1 k_d k_t}{k_{-d} + k_t + k_{-t}} \tag{9a}$$

$$= \frac{K_{as} k_{-d} k_t k_1''}{k_{-1}''(k_{-t} + k_{-d}) + k_t k_{-d}} \tag{9b}$$

$$= \frac{K_{as} k_{ac}'' k_{-d}}{k_{-ac}'' + k_{-d}} \tag{9c}$$

The expected Brönsted plots for these mechanisms are shown in Fig. 2. Note that for the stepwise mechanism the break in the Brönsted plot corresponds to the pK_a of T^+, while for the preassociation mechanism and the concerted mechanism the breaks in the Brönsted plot are shifted. Also note that the transition in slope in the Brönsted plots is broader for the preassociation mechanism and narrower for the concerted mechanism than for the stepwise mechanism. When the catalyst is a weak acid the rate-determining step is the

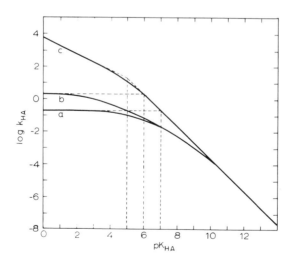

Fig. 2 Brönsted plots for general acid catalysis. (a) Stepwise mechanism [Eq. (9a)], with $K_1 = 10^{-10}$ M^{-1}, $k_d = 2 \times 10^9$ M^{-1} sec^{-1}, $k_{-d} = 10^{11}$ sec^{-1}, and $\log k_t = 13.5 - 0.5\,pK_a$; (b) preassociation mechanism [Eq. (9b)], with $K_{as} = 0.02$ M^{-1}, $k_1'' = 10^2$ sec^{-1}, $k_{-1}'' = 10^{12}$ sec^{-1}, $k_{-d} = 10^{11}$ sec^{-1}, and $\log k_t = 13.5 - 0.5\,pK_a$; (c) concerted mechanism [Eq. (9c)], with $K_{as} = 0.02$ M^{-1}, $\log k_{-ac}'' = 8.5 + 0.5\,pK_a$, $k_{ac}'' = 10^{-3}K_{HA}k_{-ac}''$, and $k_{-d} = 10^{11}$ sec^{-1}.

diffusion away of A^- and T^+ for all three mechanisms. Analogous behavior is seen for general base catalysis.

The factors affecting which reaction pathway predominates can be conveniently examined by analysis of structural effects on the acid-catalyzed addition of weakly basic amines to substituted benzaldehydes to form aminols [Eq. (10)]. The predominant paths for these reactions are the stepwise

$$
RNH_2 + \quad \overset{H}{\underset{\displaystyle \bigcirc_{X}}{C}}{=}O \quad \rightleftharpoons \quad R-NH-\overset{H}{\underset{\displaystyle \bigcirc_{X}}{C}}-OH \tag{10}
$$

and concerted mechanisms. The amines studied varied in the pK_a of their conjugate acids from 0.55 to 6.15 [7–9]. The reaction of methoxyamine $(pK_{BH^+} = 4.73)$ with p-chlorobenzaldehyde occurs by both the stepwise and concerted paths. The lifetime of $T^{\pm}$ is long enough so that a preassociation mechanism is not observed $(k_{-1} = 3 \times 10^8 \ \text{sec}^{-1})$. A preassociation mechanism is required only if $k_{-1} > k_{-d}$. At very low pH the rate-determining step is k_{ac} of Eq. (2) or k''_{ac} of Eq. (4a), where H_3O^+ is the catalyzing acid. As the pH is increased the acid-catalyzed term diminishes and k_1, the uncatalyzed attack of the amine, dominates. As the pH is further increased, $k_d[H_3O^+]$ becomes small compared to k_{-1} and so a change in rate-determining step to diffusion-limited proton transfer occurs. General acid catalysis is observed with a biphasic Brönsted plot expected for the stepwise mechanism. Catalysis by water as a general acid is considerably faster than would be predicted for conversion of $T^{\pm}$ to T^+. Since water can act bifunctionally it can catalyze the direct conversion of $T^{\pm}$ to T^0, probably through a transition state such as (3).

$$
\begin{array}{c}
R \\
\overset{\delta^+}{\underset{N}{H}} \cdots H \cdots \overset{H}{\underset{O}{}} \\
\overset{|}{\underset{C}{}} \overset{}{\underset{O}{}} \cdots H \\
\overset{}{\underset{\delta^-}{}}
\end{array}
$$

(3)

Other bifunctional acids, such as bicarbonate, can also act via k_c of Eq. (2). For water as a catalyst in the methoxyamine reaction, $k_c = 6 \times 10^6 \ \text{sec}^{-1}$. It is important to note that even though $T^{\pm}$ is moderately stable, the concerted pathway is still important. Similar behavior is seen for phenylhydrazine p-sulfonate, the conjugate acid of which has a pK_a of 4.90.

2-Methyl-3-thiosemicarbazide is a much weaker base, with $pK_{BH^+} = 1.20$. The intermediate $T^\pm$ is considerably less stable, with $k_{-1} = 5 \times 10^{11}$ sec^{-1}, and so the stepwise mechanism is not observed. The rate-determining step is k_{ac} or k''_{ac}, with a Brönsted coefficient α of 0.2. Even though $T^\pm$ has a finite lifetime, a preassociation mechanism is not observed for general acid catalysis. The reaction is also general base catalyzed. One might expect that since general acid catalysis occurs by a concerted mechanism, so would the general base catalysis. However, the concerted mechanism is not observed, but instead the reaction proceeds by a preassociation pathway with k'''_1 or k_{-d} being rate determining depending on the strength of the catalyst [Eq. (5)]. The observation of a preassociation mechanism clearly demonstrates that $T^\pm$ has sufficient stability to exist, so why is general acid catalysis concerted? Jencks [13] has proposed the following rule, which he has named the libido rule, to help answer questions of whether a general-acid- or general-base-catalyzed reaction should proceed by a concerted or stepwise pathway:

> Concerted general acid–base catalysis of complex reactions in aqueous solution can occur only (a) at sites that undergo a large change in pK in the course of the reaction and (b) when this change in pK converts an unfavorable to a favorable proton transfer with respect to the catalyst, i.e., the pK of the catalyst is intermediate between the initial and final pK values of the substrate site.

For general acid catalysis the proton transfer site is converted from a pK_a of -8 for the protonated carbonyl to 8.3 for hydroxyl ionization of T^+. The rule predicts a concerted mechanism for catalysts with pK_a values between -8 and 8.3, which is also the experimental observation. As the pK_a of the catalysts becomes close to that for T^+, the mechanism should switch to the preassociation pathway. Since only strongly acidic catalysts were studied this prediction could not be tested. For general base catalysis the proton transfer site is converted from a pK_a of 17 to a pK_a of 3.1. Most of the catalysts used had pK_a values of their conjugate acids near the pK_a of T^+ and so the rule would predict a stepwise mechanism, as observed.

With semicarbazide as the base ($pK_{BH^+} = 3.86$) the stepwise and concerted pathways are of nearly equal energy. The pathway that predominates depends on the reactivity of the aldehyde. The more reactive the aldehyde, the greater the stability of $T^\pm$, and the greater the importance of the stepwise pathway. For reactions with p-nitro-, m-nitro-, and p-chlorobenzaldehyde, the stepwise mechanism predominates. In reactions with p-methyl- and p-methoxy-benzaldehyde the concerted mechanism predominates. It had been previously noted that the curvature in the Brönsted plot for the reaction of semicarbazide with p-chlorobenzaldehyde suggested that a stepwise pathway is important [14].

The reaction of hydroxylamine ($pK_{BH^+} = 6.15$) with p-chlorobenzaldehyde represents a special case. Although stepwise general acid catalysis is expected from the stability of $T^\pm$, it is not observed. Stepwise catalysis is not observed most likely because of the exceptionally large value of k_c for water. It has been estimated to be approximately 10^9 sec^{-1}, while it would normally be expected to be in the range of 10^7 sec^{-1} [15–18]. The pH-independent attack of hydroxylamine may not be a water-catalyzed reaction at all, but may be due to the reaction of Eq. (11).

$$
\begin{array}{ccc}
\underset{\substack{| \\ -\text{C}-\text{O}^-}}{\overset{\displaystyle \text{H}_2\overset{+}{\text{N}}}{\diagdown}} \overset{\text{O}}{\diagup} \text{H}
& \rightleftharpoons &
\underset{\substack{| \\ -\text{C}-\text{OH} \\ |}}{\overset{\substack{\text{H} \\ \overset{+}{|}\diagup\text{O}^- \\ \text{HN} \\ |}}{}}
& \rightleftharpoons &
\underset{\substack{| \\ -\text{C}-\text{OH} \\ |}}{\overset{\substack{\diagup\text{OH} \\ \text{HN} \\ |}}{}}
\end{array}
\tag{11}
$$

An example of a reaction that is concerted for strong acid catalysts and diffusion limited for weak acid catalysts [Eq. (9c)] is the reaction of weakly basic thiol anions with acetaldehyde [4]. As discussed above, the intermediate T$^-$ formed from the uncatalyzed attack of the thiol anion is too unstable to have a finite existence, and so attack must be concerted with proton transfer. However, if the catalyst is a weak acid, the reverse reaction will be catalyzed by a base and so will be quite fast. For very weak acids such as water the reverse reaction is so fast that k_{-d}, the diffusion-limited separation of the conjugate base of the catalyst and T^0, is the rate-determining step. The Brönsted plot has the appearance of the curve for the concerted mechanism in Fig. 2.

ACYL SUBSTITUTION REACTIONS

Compared to simple carbonyl reactions, the tetrahedral intermediate formed in acyl substitution reactions is generally less stable. This is because the intermediate has two groups with unshared pairs of electrons to expel the leaving group [compare (4) with (5)]. In reactions with very good leaving groups a tetrahedral intermediate may not even be formed.

$$
\begin{array}{cc}
-\overset{\displaystyle |}{\underset{\displaystyle |}{\text{C}}}-\text{X} \qquad & \qquad -\text{N}-\overset{\displaystyle |}{\underset{\displaystyle |}{\text{C}}}-\text{X} \\
\text{OH} & \ \text{H} \quad \text{OH} \\
(4) & (5)
\end{array}
$$

In most acyl group substitution reactions at least one proton transfer must occur during the course of the reaction. Since the tetrahedral intermediate is usually fairly unstable the mechanism is usually stepwise with rate-limiting proton transfer, requires preassociation of the catalyst, or is concerted.

An example of a change in mechanism from a concerted to a stepwise mechanism with a change in the stability of the tetrahedral intermediate is seen in the hydrolysis of thiol esters [19–21]. The hydrolysis of ethyl trifluoro-thiol acetate (6) is general base catalyzed, and there is kinetic evidence for an intermediate in the reaction, with the formation of the intermediate being general base catalyzed. The Brönsted coefficient β is 0.19 for carboxylate ions. The most likely mechanism is proton removal from the attacking water molecule, which is concerted carbon–oxygen bond formation [Eq. (12)]. The

$$H_2O + CF_3{-}\underset{\underset{O}{\|}}{C}{-}S{-}C_2H_5 \quad \underset{k_{-1}[BH^+]}{\overset{k_1[B]}{\rightleftharpoons}}$$

(6)

$$CF_3{-}\underset{\underset{O^{\ominus}}{|}}{\overset{\overset{OH}{|}}{C}}{-}S{-}C_2H_5 \quad \longrightarrow \quad CF_3\overset{\overset{O}{\|}}{C}OH + C_2H_5S^{\ominus} \quad (12)$$

stability of the tetrahedral intermediate (7) should be increased as more electron-withdrawing groups are attached to the carbonyl. The hydrolysis of methyl S-trifluoroacetylmercaptoacetate, which has a more electron-with-drawing thiol leaving group (8), is also subject to general base catalysis. In this case the Brönsted plot is sharply curved for the most weakly basic catalysts. For the most basic carboxylate catalysts $\beta = 0.02$, which is experi-mentally indistinguishable from zero. The best estimate of the pK_a for protonation of the oxygen is -3.4. The break in the Brönsted plot occurs for catalysts with pK_a values near -1, which is much higher than the estimated pK_a. This is consistent with a preassociation mechanism with $k_{-1} = 5 \times 10^{11}$ sec^{-1} [Eq. (13)]. This is analogous to the mechanism proposed for the reaction of 2-methylthiosemicarbazide with p-chlorobenzaldehyde discussed earlier.

$$F_3C{-}\underset{\underset{O^-}{|}}{\overset{\overset{\overset{H\underset{+}{\diagdown}\underset{O}{|}\diagup H}{}}{}}{C}}{-}SR \qquad CF_3{-}\underset{\underset{O}{\|}}{C}{-}S{-}CH_2{-}\underset{\underset{O}{\|}}{C}{-}OCH_3$$

(7)   **(8)**

$$CF_3\underset{\underset{O}{\|}}{C}SCH_2\underset{\underset{O}{\|}}{C}OCH_3 + HOH\cdot B \quad \underset{k_{-1}}{\overset{k_1}{\rightleftharpoons}} \quad CF_3\underset{\underset{O^-}{|}}{\overset{\overset{\overset{H\underset{+}{\diagdown}\underset{O}{|}\diagup H\cdot B}{}}{}}{C}}{-}SR \quad \underset{k_{-t}}{\overset{k_t}{\rightleftharpoons}} \quad F_3C{-}\underset{\underset{O^-}{|}}{\overset{\overset{\overset{H\diagdown\underset{O\cdot H\overset{+}{B}}{|}}{}}{}}{C}}{-}SR \quad (13)$$

$$\downarrow$$

products

The first example of an acyl group reaction that must have a simple diffusion-limited proton transfer as the rate-determining step is the aminolysis of a thiol ester [22–28]. The mechanism for this reaction is shown in Eq. (14). The evidence for this mechanism is the following. The hydrolysis of 2-methyl-Δ^2-thiazoline (9) proceeds by way of a neutral intermediate, which partitions to give S-acetylmercaptoethylamine (10) and N-acetylmercaptoethylamine (11). Below pH 2 approximately equal amounts of thiol ester and amide are

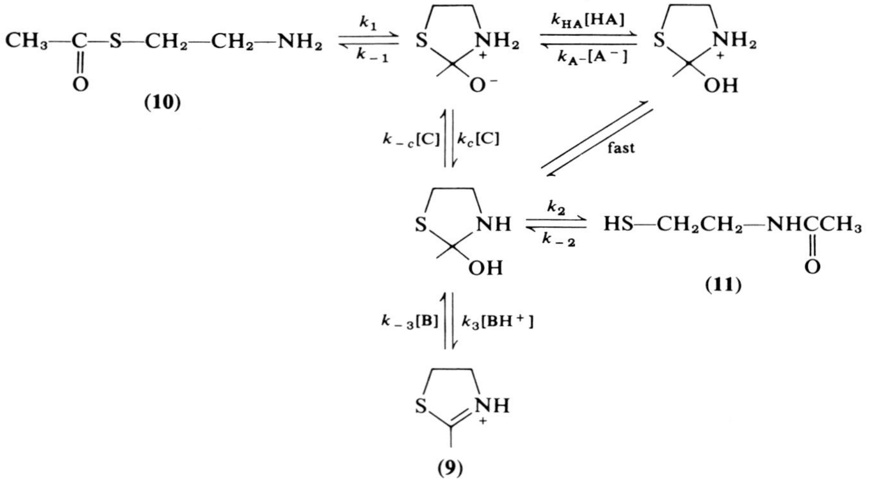

$$CH_3—\underset{\underset{O}{\|}}{C}—S—CH_2—CH_2—NH_2 \underset{k_{-1}}{\overset{k_1}{\rightleftharpoons}} \quad (10)$$

(14)

(9)

produced, with the ratio of products being independent of pH. As the pH is increased above 2 the fraction of thiol ester formed drops to near zero. The decrease in the amount of thiol ester formed is not a result of the conversion of initially formed thiol ester to amide and so it must be due to a change in rate-determining step in the partitioning of the intermediate formed in thiazoline hydrolysis. In thiol ester aminolysis there is also a neutral intermediate formed, and a change in rate-determining step occurs at the same pH as the drop off in the yield of thiol ester produced in thiazoline hydrolysis. These data produce a dilemma with respect to a simple two-step mechanism for thiol ester aminolysis. The data for thiazoline hydrolysis at low pH require that the transition states for the conversion of the neutral intermediate to thiol ester and to amide have the same charge, and so a change in rate-determining step could never occur with a change in pH for a simple two-step mechanism. Since a change in rate-determining step does occur with a change in pH there must be a third sequential step and a second kinetically significant intermediate in thiol ester aminolysis. Since only two of the steps can be bond making or bond breaking to carbon the third step must be a proton transfer step. In support of this conclusion the Brönsted plot for general acid catalysis of the reaction is biphasic. For strong acid catalysts α is near zero, while for weak acid catalysts α is near 1.0. The break in the Brönsted plot occurs with catalysts with a pK_a of 7.4. The estimated pK_a of the intermediate is 7.2. Similar behavior is seen for the aminolysis of the chloroacetyl thiol ester (12). Because of the inductive effect of the chloro group the break in the Brönsted plot should occur 1.4–1.6 pK_a units lower. The observed break occurs for catalysts with a pK_a of 6.9, or 1.5 pK_a units lower.

$$\text{ClCH}_2\text{—}\underset{\underset{\text{O}}{\|}}{\text{C}}\text{—S—CH}_2\text{—CH}_2\text{—NH}_2$$

(12)

Further evidence supporting the mechanism comes from the viscosity dependence of the reaction. If a reaction is diffusion limited, as would be the case for a simple proton transfer, $k_d = C/\eta$, where C is a constant and η is the viscosity of the reaction medium. Reactions that are not diffusion limited, such as a concerted or preassociation mechanism, are independent of the viscosity. The rate constant for catalysis by H_3O^+ was found to be inversely proportional to the viscosity of the medium over a 10-fold viscosity range.

Direct conversion of $T^{\pm}$ to T^0 occurs with bifunctional catalysts, giving an enhanced rate constant for weak acid catalysts which would normally have a rate constant smaller than the diffusion-limited rate constant. For example, the rate constants for catalysis by bicarbonate and by water show positive deviations of 50 and 2×10^4, respectively, from the Brönsted plot.

The estimated value of k_{-1} is 7×10^8 sec^{-1} for $T^{\pm}$ obtained from (10). The intermediate obtained from (13) should be less stable since the leaving group is better, and nitrogen provides more driving force for expulsion of the leaving group than sulfur [Eq. (15)].

(13)

The conversion of (13) to (14) is subject to both general acid and general base catalysis [29–31]. The Brönsted plot for general base catalysis is shown in Fig. 3. The plot is biphasic, with $\beta = 0.05$ for the four most basic amines. The estimated pK_a of $T^{\pm}$ is 5.2, and the solid line in Fig. 3 is calculated for a diffusion-limited stepwise mechanism. The good agreement between the data and the calculated curve eliminates a preassociation mechanism, and so $k_{-1} < 10^{11}$ sec^{-1}. Since k_{-1} does not become rate determining at the highest

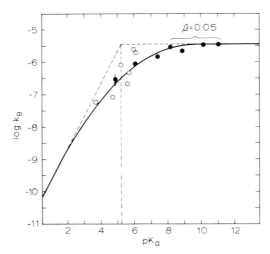

Fig. 3 General-base-catalyzed aminolysis of (**13**). The solid circles are for a series of aliphatic tertiary amines from Blanco and Barnett [31], and the open circles are for miscellaneous catalysts from Robinson and Jencks [29]. The curve is calculated for a stepwise proton transfer from an acid with a pK_a of 5.2.

$$(13) \qquad\qquad (14)$$

hydroxide concentration used for the conversion of (**14**) to (**13**), $k_{-1} > 10^{10}$ sec^{-1}. The intermediate formed in amide aminolysis has less stability than the intermediate formed in thiol ester aminolysis but still enough stability so that a stepwise mechanism can be observed.

Even though $T^{\pm}$ is stable enough so that a stepwise mechanism is observed for general base catalysis, general acid catalysis is concerted. The Brönsted coefficient α is 0.57 for tertiary ammonium ions, and the rate constants are in excess of what could be expected for a diffusion-limited stepwise mechanism (Fig. 4). The overall mechanism is summarized in Eq. (16). This reaction illustrates that even if the intermediates for the stepwise mechanism exist, the concerted mechanism may still be observed. The factors that determine whether a concerted or a stepwise mechanism will be the lowest-energy pathway when both routes are available are not yet fully understood.

Stepwise diffusion-limited proton transfer and preassociation mechanisms

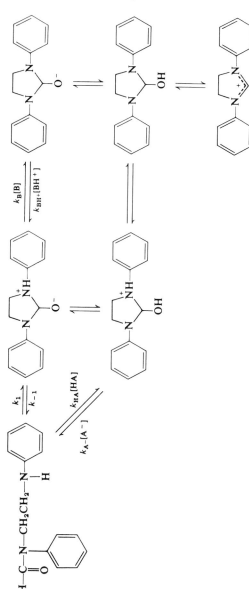

(16)

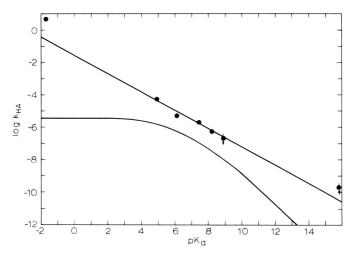

Fig. 4 General-acid-catalyzed aminolysis of **(13)** by tertiary ammonium ions [31]. The least-squares line was calculated with the solvent points excluded. The curve is calculated for a stepwise proton transfer to a base whose conjugate acid has a pK_a of 6.7.

are commonly observed for a variety of acyl group reactions [12,20,21,26–28, 31–34], such as ester aminolysis, amide hydrolysis, imidate hydrolysis, amide aminolysis, and thiol ester hydrolysis. Mechanisms in which proton transfer is concerted with making or breaking of bonds to carbon appear to be relatively rare, being limited to acyl compounds with extremely good leaving groups. The prevalence of reactions with rate-determining stepwise proton transfers must force a revision of the ideas on the nature of general acid–general base catalysis at an enzyme surface.

ACKNOWLEDGMENT

This research was supported in part by the National Science Foundation.

REFERENCES

1. W. P. Jencks, "Catalysis in Chemistry and Enzymology." McGraw-Hill, New York, 1969.
2. L. do Amaral, W. A. Sandstrom, and E. H. Cordes, *J. Am. Chem. Soc.* **88**, 2225 (1966).
3. G. E. Lienhard and W. P. Jencks, *J. Am. Chem. Soc.* **88**, 3982 (1966).
4. R. E. Barnett and W. P. Jencks, *J. Am. Chem. Soc.* **91**, 6758 (1969).
5. E. H. Cordes and W. P. Jencks, *J. Am. Chem. Soc.* **85**, 2843 (1963).

6. J. Hine and F. C. Kokesh, *J. Am. Chem. Soc.* **92**, 4383 (1970).
7. J. M. Sayer, B. Pinsky, A. Schonbrunn, and W. Washstien, *J. Am. Chem. Soc.* **95**, 7998 (1974).
8. S. Rosenberg, S. M. Silver, J. M. Sayer, and W. P. Jencks, *J. Am. Chem. Soc.* **96**, 7986 (1974).
9. J. M. Sayer and W. P. Jencks, *J. Am. Chem. Soc.* **95**, 5637 (1973).
10. M. Eigen, *Angew. Chem., Int. Ed. Engl.* **3**, 1 (1964).
11. R. E. Barnett, Ph.D. thesis, Brandeis University, Waltham, Massachusetts (1969).
12. J. Fox and W. P. Jencks, *J. Am. Chem. Soc.* **96**, 1436 (1974).
13. W. P. Jencks, *J. Am. Chem. Soc.* **94**, 4731 (1972).
14. R. E. Barnett, *Acc. Chem. Res.* **6**, 41 (1973).
15. J. E. Riemann and W. P. Jencks, *J. Am. Chem. Soc.* **88**, 3973 (1966).
16. E. Grunwald and S. Meiboom, *J. Am. Chem. Soc.* **85**, 522 (1963).
17. E. Grunwald and S. Meiboom, *J. Am. Chem. Soc.* **85**, 2047 (1963).
18. Z. Luz and S. Meiboom, *J. Am. Chem. Soc.* **85**, 3923 (1963).
19. L. R. Fedor and T. C. Bruice, *J. Am. Chem. Soc.* **87**, 4138 (1965).
20. R. E. Barnett and W. P. Jencks, *J. Org. Chem.* **34**, 2777 (1969).
21. R. J. Zygmunt and R. E. Barnett, *J. Am. Chem. Soc.* **94**, 1996 (1972).
22. R. B. Martin, S. Lowey, E. L. Elson, and J. T. Edsall, *J. Am. Chem. Soc.* **81**, 5089 (1959).
23. R. B. Martin and A. Parcell, *J. Am. Chem. Soc.* **83**, 4830 (1961).
24. R. B. Martin and R. I. Hedrick, *J. Am. Chem. Soc.* **84**, 106 (1962).
25. R. B. Martin, R. I. Hedrick, and A. Parcell, *J. Org. Chem.* **29**, 3197 (1964).
26. R. E. Barnett and W. P. Jencks, *J. Am. Chem. Soc.* **90**, 4199 (1968).
27. R. E. Barnett and W. P. Jencks, *J. Am. Chem. Soc.* **91**, 2358 (1969).
28. C. Cerjan and R. E. Barnett, *J. Phys. Chem.* **76**, 1192 (1972).
29. D. R. Robinson and W. P. Jencks, *J. Am. Chem. Soc.* **89**, 7088 (1967).
30. D. R. Robinson, *J. Am. Chem. Soc.* **92**, 3138 (1970).
31. R. Blanco and R. E. Barnett, unpublished observations.
32. M. I. Page and W. P. Jencks, *J. Am. Chem. Soc.* **94**, 8828 (1972).
33. A. C. Satterthwait and W. P. Jencks, *J. Am. Chem. Soc.* **96**, 7018 (1974).
34. A. C. Satterthwait and W. P. Jencks, *J. Am. Chem. Soc.* **96**, 7031 (1974).

The Withanolides—A Group of Natural Steroids

Erwin Glotter, Isaac Kirson, David Lavie, and Arieh Abraham

INTRODUCTION

In the early 1960's an investigation aimed at the structure elucidation of withaferin A was initiated in our laboratory at the Weizmann Institute. This compound had been previously isolated [1] from the leaves of *Withania somnifera*, a perennial plant of the Solanaceae family that is widely distributed along the shores of the Mediterranean Sea, in India, South Africa, and several other countries throughout the world.

Various therapeutic properties have been attributed to this plant since ancient times [2]. Native South African tribes use the roots, as well as the leaves, for a wide variety of medicinal purposes [3]: "A decoction of the roots is administered by the Transvaal Sotho to tone up the uterus in a woman who habitually miscarries, and in order to remove retained conception products." In Basutoland, the plant is used as a remedy against intestinal parasites introduced by witchcraft. Interestingly, the Xhosa apply the fresh juice of the leaves to anthrax pustules and also use the plant for the preservation of meat. Such therapeutic uses eventually attracted the attention of the European settlers, who administered a leaf paste to erysipelas. Indian popular medicine proposed the use of this plant as a hypnotic and sedative [4].

Most of the studies performed at the beginning of this century [5] did not confirm the therapeutic properties attributed to this plant; although several compounds were isolated, at different times, their structures remained unknown, and even their correct elemental compositions were not determined. More recently it has been shown [1,6] that a crystalline compound, identified as withaferin A (**1**),* possesses antibacterial activity, mainly against acid-fast bacilli and gram-positive microorganisms.

THE WITHANOLIDES

After the elucidation of the structure for withaferin A [7–7b], a series of related compounds was isolated from *W. somnifera* and other species of the Solanaceae. They were shown to constitute a new group of steroidal lactones derived from an ergostane-type skeleton in which C-22 and C-26 form a six-membered ring lactone. For convenience, the basic structure shown below was designated as the "withanolide" skeleton [8].

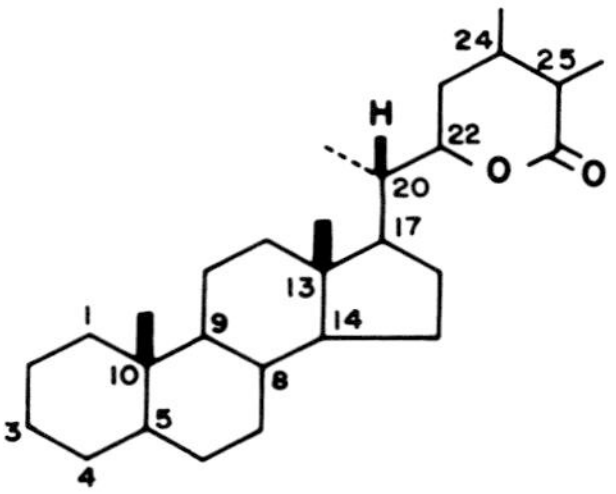

Most of the withanolides investigated so far are 2-en-1-ones and have the side-chain lactone unsaturated (Δ^{24}). One compound (**30**), however, was found to have ring A saturated with C-1—OH instead of the corresponding ketone. The lactone side chain of certain compounds was saturated [(**8**) and (**36**)] or at a lower oxidation level [22 → 26-lactol, (**28**) and (**29**)]. Two other groups of compounds, the physalins [9–10] and the Nic derivatives [11–12a], which have recently been characterized, are biogenetically related to the withanolides.

In order to facilitate discussion of the biogenetic aspects of the problem, the withanolides are grouped according to certain structural features: the substitution pattern of the withanolide skeleton, the stereochemistry of the side chain, and the naturally occurring modifications of the basic framework.

The "regular" withanolides presented in Tables 1–3 are compounds with an unmodified carbocyclic skeleton; they can be subdivided into three groups: (a) compounds unsubstituted at C-20 (Table 1) [7,7b,8,11,13–25]; (b) compounds possessing a hydroxy group at C-20 (Table 2) [16,19,23,26–30];

* Structures (**1**)–(**30**) can be found in Table 1; (**31**)–(**49**) in Table 2; (**50**)–(**56**) in Table 3; (**57**)–(**62**) in Table 4; (**63**)–(**65**) in Table 5; and (**66**)–(**70**) in Table 6.

TABLE 1
20-Deoxywithanolides

Formula	Name	Melting point (°C)	$[\alpha]_D$ (deg)	Reference
	Withaferin A	243–245	+144	7
	27-Deoxywithaferin A	268–269	+101.5	8
	14α-Hydroxy-27-deoxywithaferin A	265–267	+67	13
	17α-Hydroxy-27-deoxywithaferin A	195–196 (4-acetate)	175	14
	Withacnistin	130–135	+123	7b
		239–240	+122	14
		256–259	+98	15
		252	+32.5	8,16

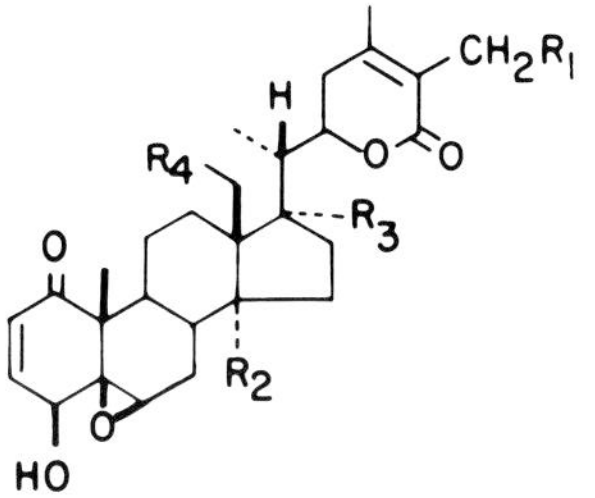

(1) $R_1 = OH$; $R_2 = R_3 = R_4 = H$
(2) $R_1 = R_2 = R_3 = R_4 = H$
(3) $R_2 = OH$; $R_1 = R_3 = R_4 = H$
(4) $R_3 = OH$; $R_1 = R_2 = R_4 = H$

(5) $R_4 = OAc$; $R_1 = R_2 = R_3 = H$
(6) Δ^{14}; $R_1 = R_3 = R_4 = H$
(7) Δ^{14}; $R_1 = OH$; $R_3 = R_4 = H$
(8) 24,25-dihydro (24S,25R); $R_1 = R_2 = R_3 = R_4 = H$

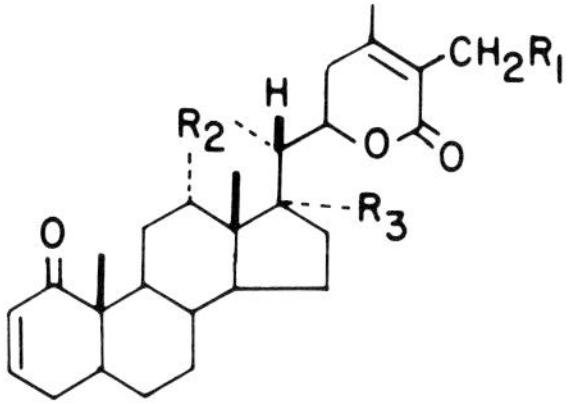

Table 1 (*continued*)

Formula	Name	Melting point (°C)	$[\alpha]_D$(deg)	Reference
(9) 5β,6β-epoxy; R_1 = OH; R_2 = R_3 = H	Jaborosalactone A	215–220	+95	17,18
(10) 5β,6β-epoxy; R_1 = R_3 = OH; R_2 = H	17α-Hydroxyjaborosalactone A	214–215	+49	19
(11) Δ^4; 6β-OH; R_1 = OH; R_2 = R_3 = H	Jaborosalactone B	227–229		17,18
(12) 5β-OH; 6α-Cl; R_1 = OH; R_2 = R_3 = H	Jaborosalactone C	218–220		20
(13) 5α-OH, 6β-OH; R_1 = OH; R_2 = R_3 = H	Jaborosalactone D	277–279		20
(14) 5α-Cl; 6β-OH; R_1 = OH; R_2 = R_3 = H	Jaborosalactone E	255–258		20
(15) 5α-OH; 6β-OH; R_1 = R_2 = OH; R_3 = H	Jaborosalactone F	268–270		21

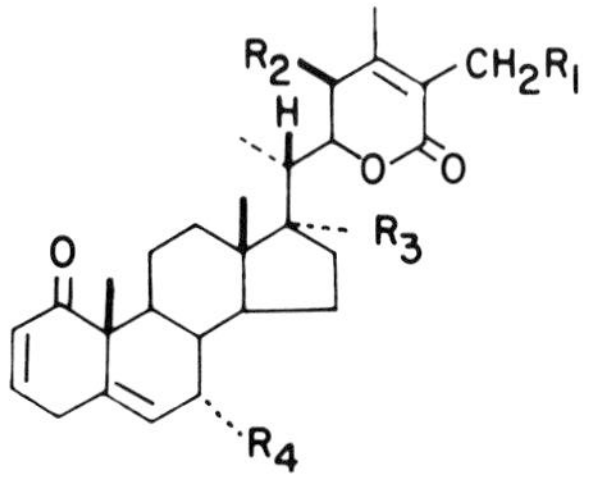 Formula	Name	Melting point (°C)	$[\alpha]_D$(deg)	Reference
(16) R_1 = R_3 = OH; R_2 = R_4 = H		205 (27-acetate)	+74	14
(17) R_1 = R_4 = OH; R_2 = R_3 = H		208–209 (7,27-diacetate)	−119	14
(18) R_1 = R_2 = R_3 = OH; R_4 = H	Withanolide Q	200–202	−6.6	22
(19) $\Delta^{8(14)}$; R_1 = R_3 = OH; R_2 = R_4 = H		224–225	+48	19
(19a) $\Delta^{8(14)}$; R_3 = OH; R_1 = R_2 = R_4 = H		210–211	+56.5	19
(20) $\Delta^{8(14)}$; 4β-OH; R_3 = OH; R_1 = R_2 = R_4 = H	Withanolide O	201–202	+112.5	23

(21) Δ^{14}; $R_1 = R_3 = OH$; $R_2 = R_4 = H$ · Withanolide N · Amorphous (27-acetate) · +31 · 23

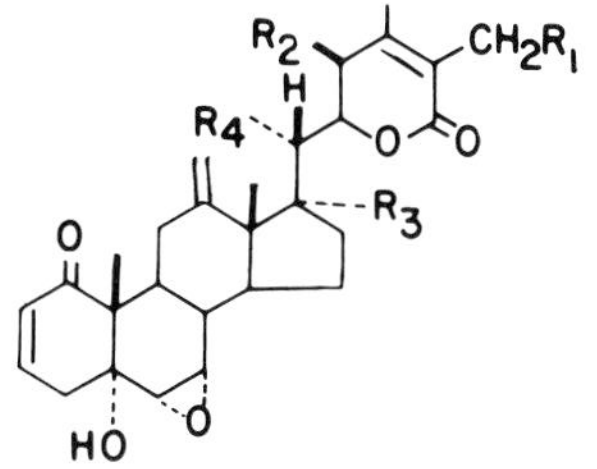

(22) $R_3 = OH$; $R_1 = R_2 = H$; $R_4 = H_2$ · Withanone · 275–276 · +81 · 14
(23) $R_1 = OH$; $R_2 = R_3 = H$; $R_4 = H_2$ · · 230–232 (27-acetate) · +73 · 14
(24) $R_1 = R_2 = R_3 = H$; $R_4 = H_2$ · Lyciumsubstanz B · 260–265 · +107 · 24
(25) $R_4 = O$; $R_1 = R_2 = R_3 = H$ · Withanicandrin · 267–269 · +105 · 25
(26) $R_2 = OH$; $R_1 = R_3 = H$; $R_4 = H_2$ · Withanolide R · 182–184 (23-acetate) · −69 · 22

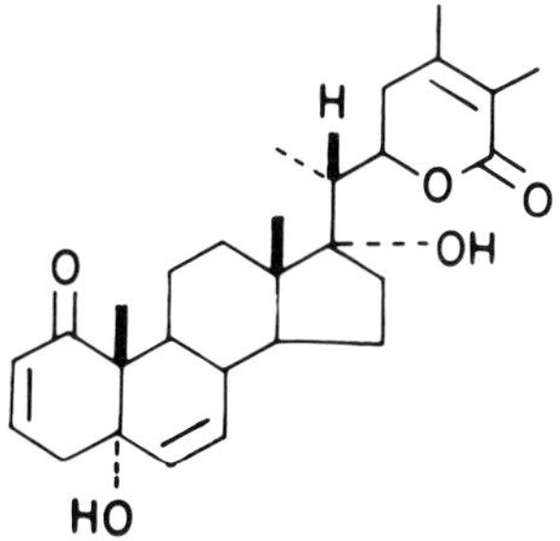

Table 1 (*continued*)

Formula	Name	Melting point (°C)	$[\alpha]_D$(deg)	Reference
(27) 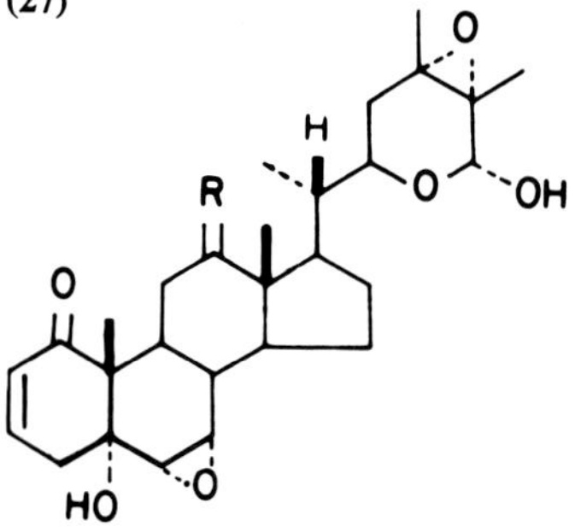		238–239 (decomposes)	+75.5	14
(28) R = H$_2$	Nic-3	262 (26-acetate)		11
(29) R = O	Nic-7			11
(30) 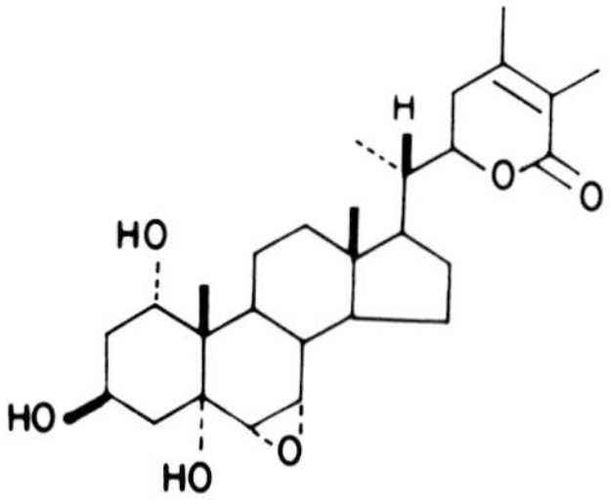		256–258 (3-acetate)	+71.5	14

TABLE 2
20-Hydroxywithanolides

Formula 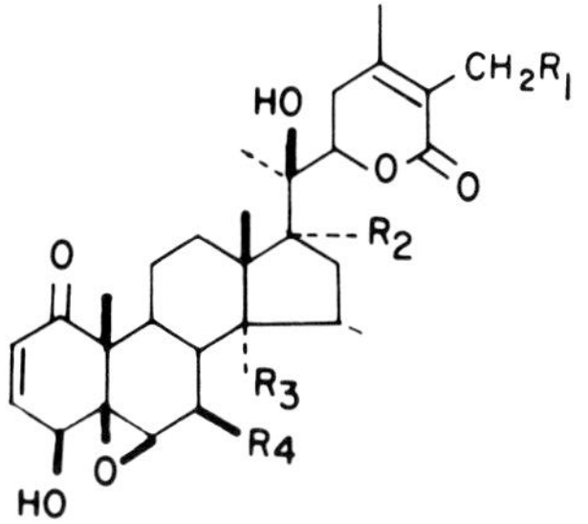	Name	Melting point (°C)	$[\alpha]_D$(deg)	Reference
	Withanolide D	253–255	+80	26
	27-Hydroxywithanolide D	202 (4,27-diacetate)	+17	23
	14α-Hydroxywithanolide D	211–213 (4-acetate)	−18	23
	17α-Hydroxywithanolide D	196–197 (4-acetate)	+2	23
	7β-Hydroxywithanolide D	260	118.5	27
		173–175		28
		275	+14	16

(31) $R_1 = R_2 = R_3 = R_4 = H$

(32) $R_1 = OH$; $R_2 = R_3 = R_4 = H$

(33) $R_3 = OH$; $R_1 = R_2 = R_4 = H$

(34) $R_2 = OH$; $R_1 = R_3 = R_4 = H$

(35) $R_4 = OH$; $R_1 = R_2 = R_3 = H$

(35a) $R_4 = OAc$; $R_1 = R_2 = R_3 = H$

(36) 24,25-dihydro (24S,25R); $R_1 = R_2 = R_3 = R_4 = H$

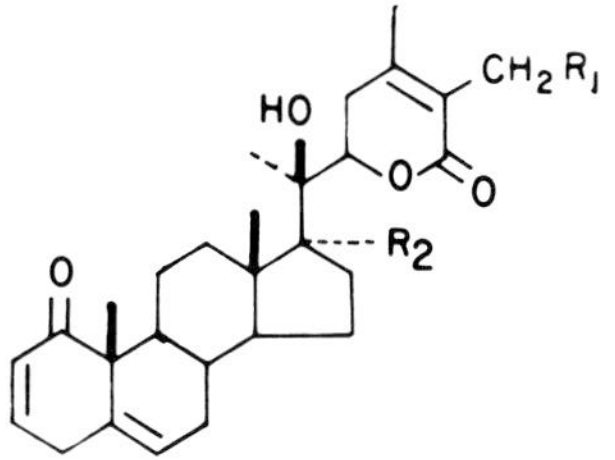

TABLE 2 (*continued*)

Formula	Name	Melting point (°C)	$[\alpha]_D$(deg)	Reference
(37) $\Delta^{8(14)}$; $R_1 = R_2 = H$	Withanolide G	194–195	+52.5	29
(38) $\Delta^{8(14)}$; $R_2 = OH$; $R_1 = H$	Withanolide J	215–216	+32.5	29
(39) $\Delta^{8(14)}$; $R_1 = OH$; $R_2 = H$	Withanolide H	141–142	+35.5	29
(40) $\Delta^{8(14)}$; 4β-OH; $R_1 = R_2 = H$		230 (27-acetate)	+110.6	19
(41) Δ^{14}; $R_1 = H$; $R_2 = OH$	Withanolide L	213 (4-acetate)	+9.5	29
(42) Δ^{14}; $R_1 = R_2 = H$		184–185	+7.5	19
(43) 4β-OH; 7β-OH; $R_1 = R_2 = H$		173–175	+118	27,28
(44) $14\alpha,15\alpha$-epoxy; $R_2 = OH$; $R_1 = H$	Withanolide M	240	+44.6	29

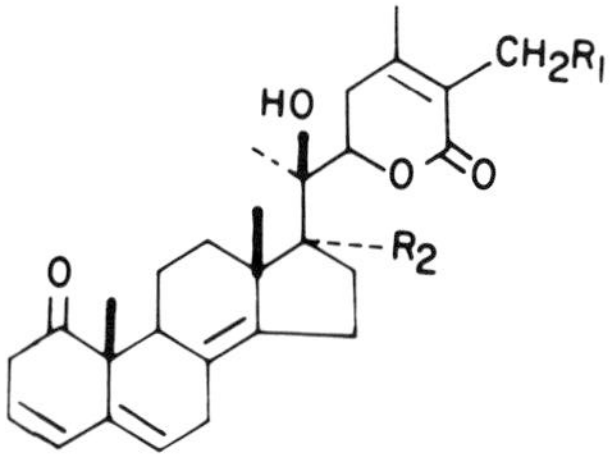

(45) $R_1 = R_2 = H$ | Withanolide I | 184 | +118 | 29

64

(46) $R_2 = OH$; $R_1 = H$

Withanolide K   218–219   +92   29

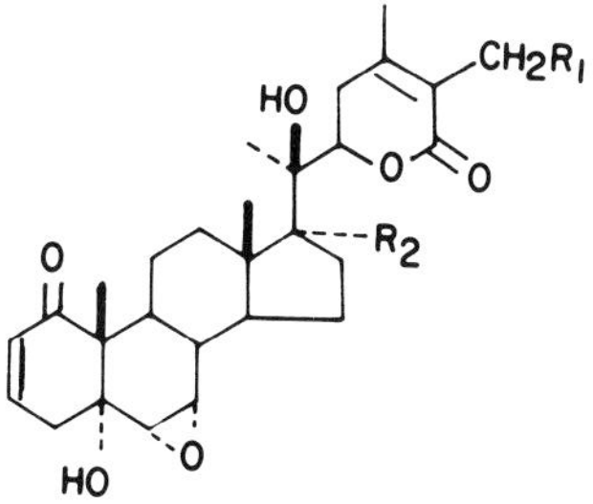

(47) $R_1 = R_2 = H$ 282–284   +102   30

(48) $R_2 = OH$; $R_1 = H$ 257   +60.5   19

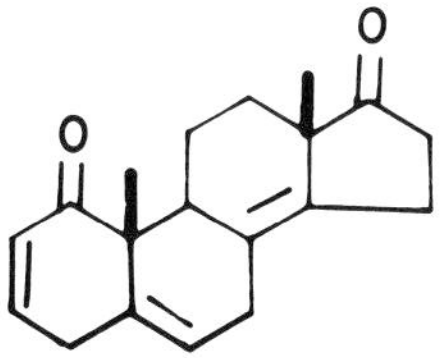

(49)   19

TABLE 3

Withanolides with a 17α Side Chain

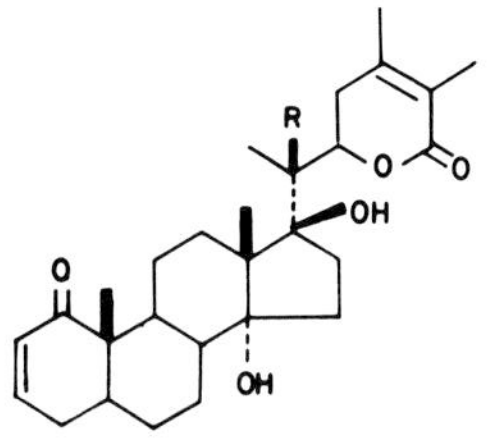

Formula	Name	Melting point (°C)	$[\alpha]_D$ (deg)	Reference
	Withanolide E	167–168	+102	31,32
	Withanolide F			31,32
	4β-Hydroxywithanolide E	197–199	+96 (CH₃OH)	33
	Withanolide S	272 (decomposes)	+97.5 (CH₃OH)	32
	Withanolide P	216–217	+51	32

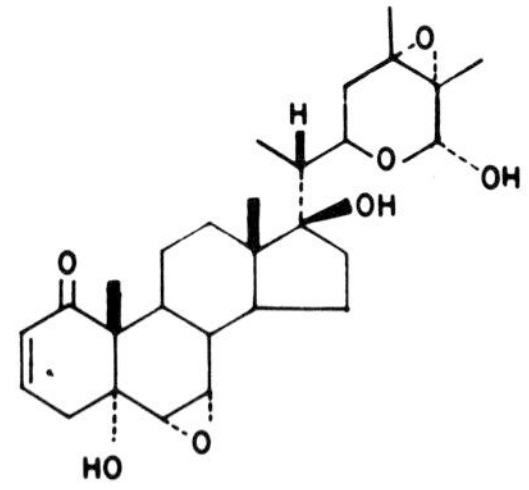

(50) 5β,6β-epoxy; R = OH

(51) Δ⁵; R = OH

(52) 5β,6β-epoxy; 4β-OH; R = OH

(53) 5α-OH; 6β-OH; R = OH

(54) Δ⁵; R = H

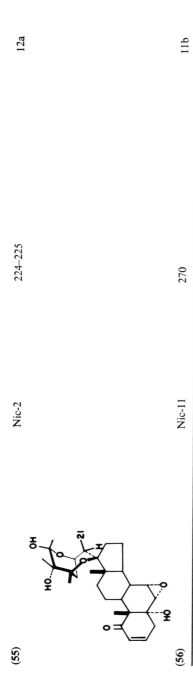

(55)　　　　　　Nic-2　　　　　224–225　　　　12a

(56)　　　　　　Nic-11　　　　　270　　　　　11b

TABLE 4
Physalins

Formula	Name	Melting point (°C)	$[\alpha]_D$(deg)	Reference
(57) R = OH	Physalin A	265–266	−126 (EtOH)	9,9a
(58) R = H	Physalin C	274–277	−160 (acetone)	9b
(59)	Physalin B	269–272	−150 (EtOH)	9,9a
(60) 5β,6β-epoxy	5β,6β-Epoxyphysalin B	262–264	−60.7	34
(61) 5α-OH; 6β-OH	Physalin D			10,34a
(62) 5α-OH; 7α-OH	Physalin E	305–307	−83	9e

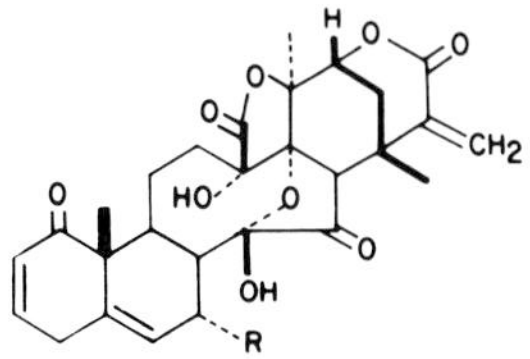

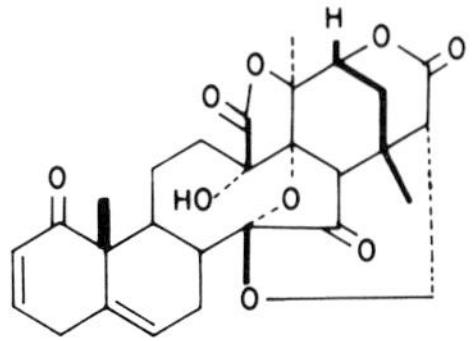

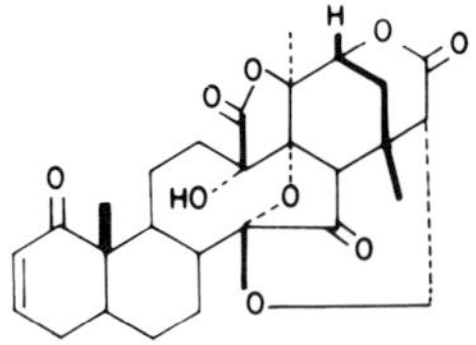

TABLE 5

Withaphysalins

Formula	Name	Melting point (°C)	$[\alpha]_D$(deg)	Reference
(63) 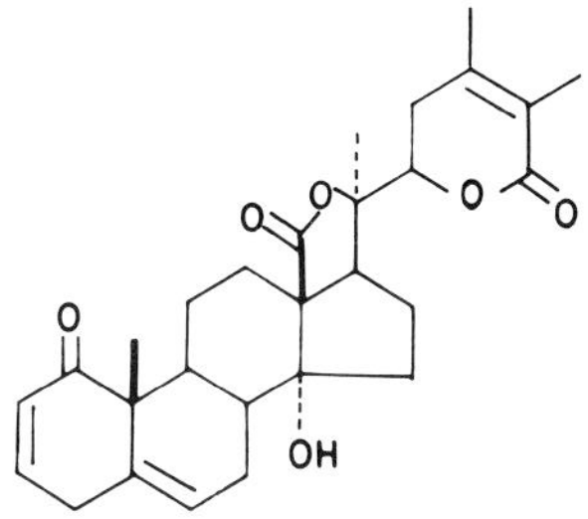	Withaphysalin A	222–223	+43.6	34
(64) 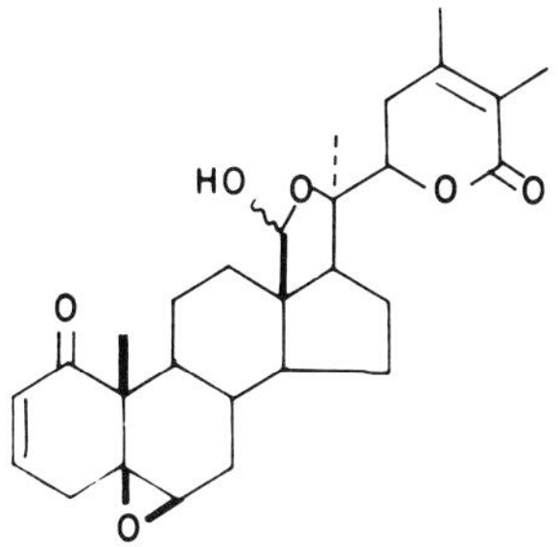	Withaphysalin B	161–162		34

Table 5 (*continued*)

Formula	Name	Melting point (°C)	$[\alpha]_D$(deg)	Reference
(65) 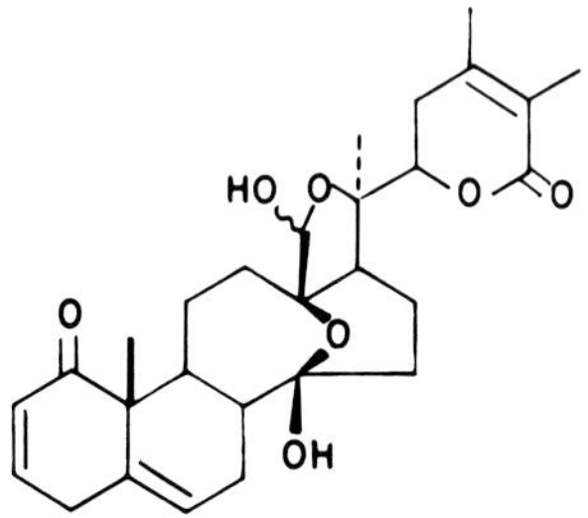	Withaphysalin C	202–203	+33.4	34,35

TABLE 6
Compounds with Aromatic Ring D

Formula	Name	Melting point (°C)	$[\alpha]_D$(deg)	Reference
(66) R 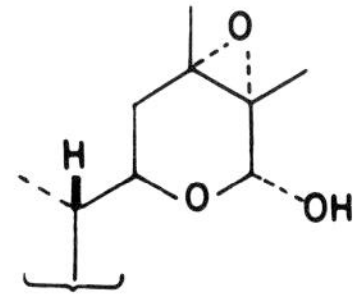 =	Nic-1 (nicandrenone)	138		11a
(67) R 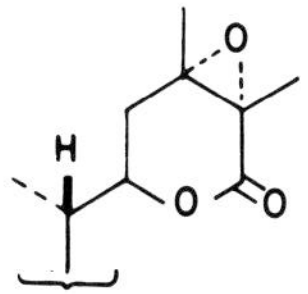 =	Nic-1-lactone	230–232	+34.6	36

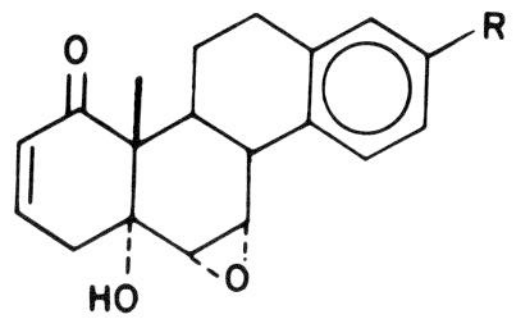

Table 6 (*continued*)

Formula	Name	Melting point (°C)	$[\alpha]_D$(deg)	Reference
(68) R 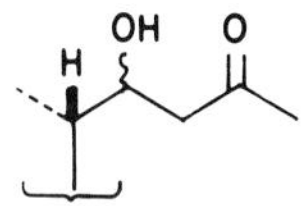 =	Nic-17	86		11a
(69) R 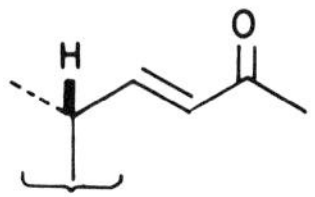 =	Nic-12	175		11a
(70) R  =	Nic-10	207		11a

(c) compounds in which the C-17 side chain is α oriented (Table 3) [11b,12a, 31–33]. Compounds which have a δ-lactol [(28) and (29)] instead of a δ-lactone, or even lack the side chain (49), have been included in the above groups.

The physalins, isolated from several *Physalis* species, are presented in Table 4 [9–9b,9e,10,34,34a]. The common structural features of these compounds are the oxidative cleavage of the C-13—C-14 bond and the closure of a new six-membered ring by C-16—C-24 bond formation.

Several compounds designated as withaphysalins, which have recently been isolated [34,35] from *Physalis minima*, are grouped together in Table 5 since they represent intermediate stages in the biogenetic pathway from the withanolides to the physalins.

In Table 6 [11a,36] are presented the first known steroids in which ring D is aromatic [11–12a], being formed in the plant through a *D*-homo rearrangement involving the incorporation of C-18 into the ring and its aromatization.

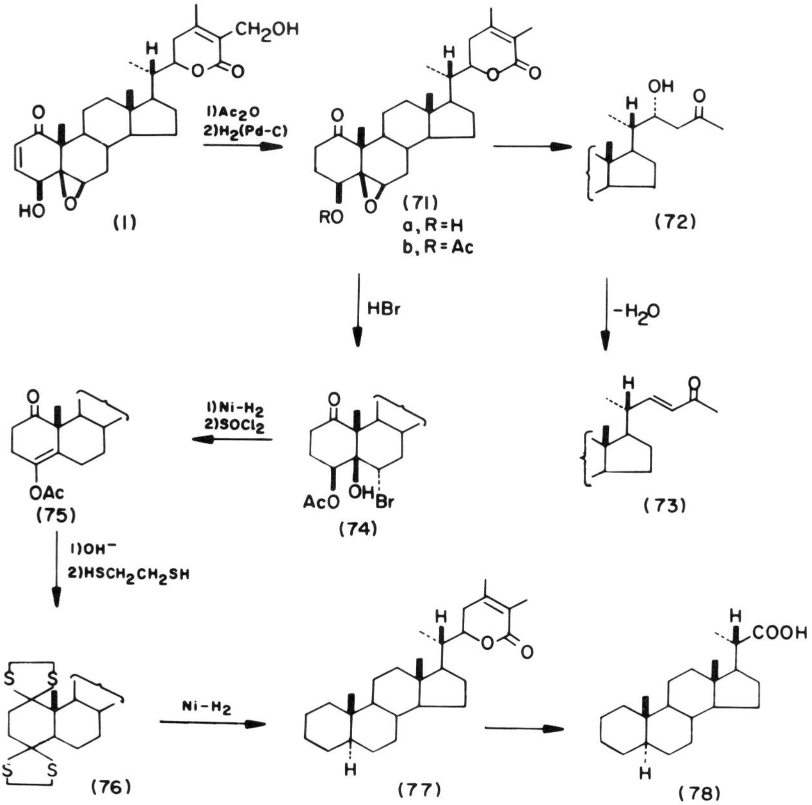

Fig. 1 Structure determination of withaferin A.

The Structure of Withaferin A and Related Compounds

The structure assigned to the carbocyclic skeleton of withaferin A (1) is based on its degradation to bisnor-5α-cholanic acid (78). The position and orientation of the various functional groups were inferred by interpretation of the spectral data as well as by chemical transformations [7–7b,37], which are succinctly illustrated in Fig. 1.

The stereochemistry at C-22 was established by comparison of its circular dichroism (CD) (positive band at ~250 nm) with that of sorbic acid [16,18]. All the withanolides isolated so far are $22\beta_F$ compounds. Interpretation of the aromatic-solvent-induced shifts in withaferin A and several derivatives [38], as well as other nuclear magnetic resonance (nmr) criteria, led to the conclusion that ring A in 1-keto-4β-hydroxy-5β,6β-epoxysteroids and in the

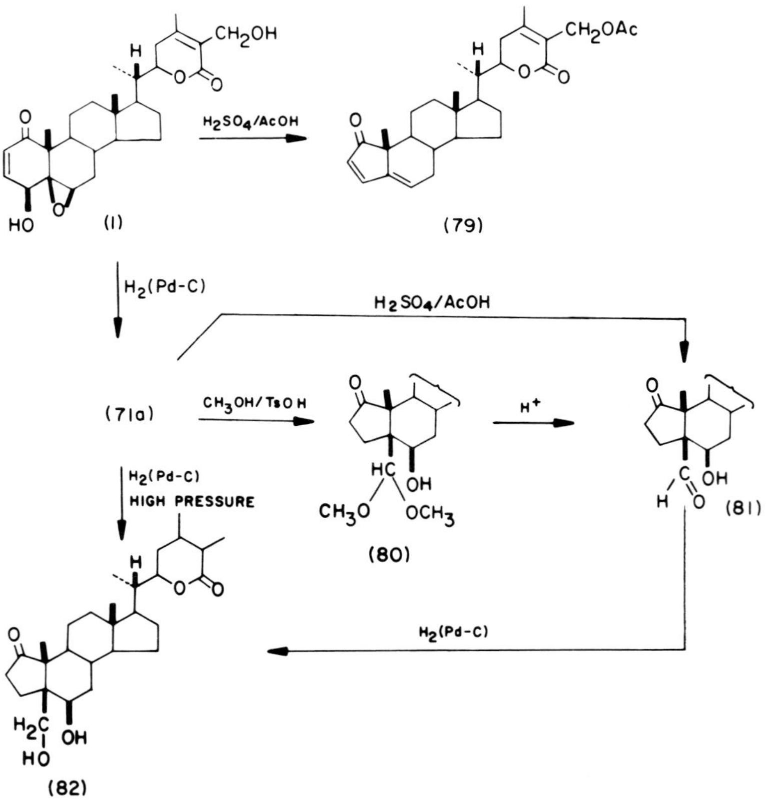

Fig. 2   Rearrangements in rings A and B in withaferin A.

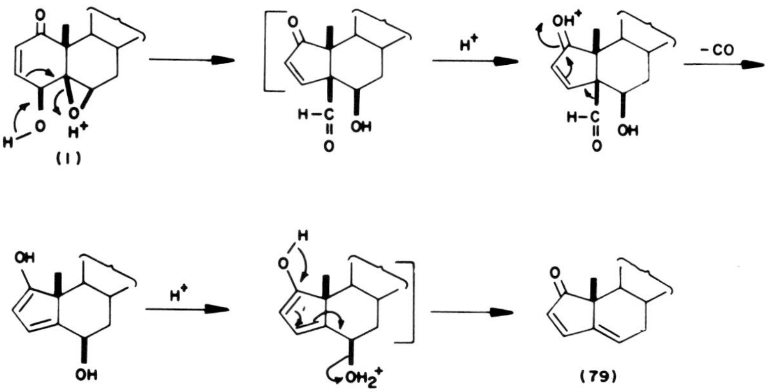

Fig. 3 Possible mechanism for the pinacol-type rearrangement.

corresponding 2-ene derivatives has a boat conformation in which both C-1 and C-4 are oriented upward. The structure of withaferin A was confirmed by an X-ray analysis of its 4-acetate-27-*p*-bromobenzoate derivative [39].

During the chemical work on the structure of (1) and its derivatives, difficulties were encountered [40] due to a rearrangement taking place in the presence of acidic reagents, leading to the formation of A-nor derivatives (Fig. 2).

It is worth noting that such a rearrangement does not occur in the absence of the C-1 ketone. For instance, $5\beta,6\beta$-epoxycholestan-4β-ol yields under similar conditions cholest-5-en-4-one [41]. The pinacol-type rearrangement has been interpreted [40] as shown in Fig. 3.

In the dihydroderivative (71a), the reaction stops at the first step, yielding the ring A-contracted aldehyde (81) (Fig. 2).

Jaborosalactone A (9), isolated from *Jaborosa integrifolia* [17,18], a South American Solanaceae, is 4-deoxywithaferin A; the other jaborosalactones [(11)–(15)] [20,21,42] are elaborated in nature through secondary processes in which the sensitive $5\beta,6\beta$-epoxide is involved.

The main withanolide (22) occurring in a variety of *Withania somnifera* growing in India was found to have the same physical constants as an uncharacterized compound called withanone [4], earlier isolated from this plant species. The structure assignment for (22) is based primarily on the reactions described in Fig. 4. The substitution pattern of rings A and B in this compound can be rationalized as shown in the hypothetical biogenetic scheme (see Fig. 17).

The degradation of the side chain of withaferin A (1), which allowed the structure assignment of the δ-lactone, has not been repeated in other compounds. The presence of such a structural feature was inferred by the analysis of the spectral data, which are as follows: in the ultraviolet, maximum absorp-

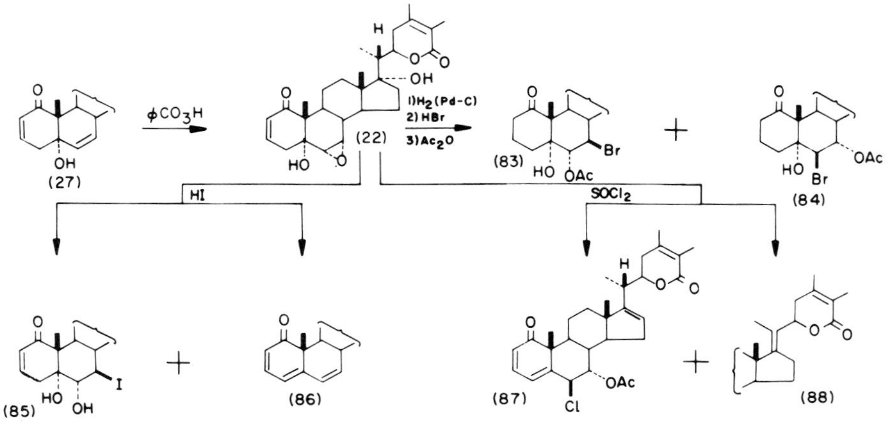

Fig. 4   Structure determination of withanone (**22**).

tion at ∼210 nm for the 27-hydroxy derivatives and at ∼226 nm for the 27-deoxy derivatives; in the infrared, a lactone carbonyl band at ∼1690 cm^{-1}; in the CD, a positive band at ∼250 nm ($\Delta\varepsilon \sim 2.5$). The information is corroborated by the nmr data: In the 27-hydroxy derivatives, there are two signals for a vinylic methyl group ($\delta \sim 2.1$) and for a CH_2OH methylene group ($\delta \sim 4.3$), whereas in the 27-deoxy derivatives, the two vinylic methyl groups appear as a broadened signal at $\delta \sim 1.95$. The signal of the 22-H appears as a double triplet in the compounds that are unsubstituted at C-20 and as a double doublet in those possessing a 20-OH structure. The chemical shift of this signal (between $\delta \sim 4.2$–4.8) is a sensitive probe for the presence of neighboring functional groups.

The substitution pattern of rings A and B in withanone (**22**) was confirmed by crystallographic analysis [11] of Nic-3 (**28**), which has the same functions in this part of the molecule.

Withanolides D, G, and J—Typical 20α-Hydroxywithanolides

The structure assigned [26] to withanolide D (**31**), an isomer of withaferin A, is based on the similarity of the data relevant to the substitution pattern of rings A and B in these compounds as well as on the spectral information concerning the side chain, which points to the presence of a 27-deoxy-δ-lactone and a 20-OH group. The structure was further ascertained by degradation to a pregnan-20-one derivative (**89**) (Fig. 5).

The 17β orientation of the side chain was deduced by CD measurements; the 20α$_F$ stereochemistry in withanolide D (**31**) was determined by chemical

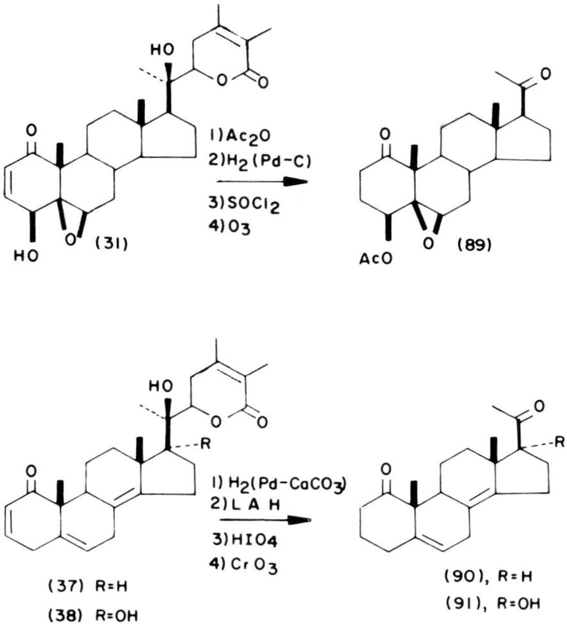

Fig. 5 Structure determination of withanolides D (31), G (37), and J (38). LAH–LiAlH₄,

shift differences, as compared to those found in 20α- and 20β-hydroxycholesterol [43].

The structures assigned [29] to withanolide G (37) and J (38), which are among the least oxygenated withanolides encountered so far, are based as well on degradation to the corresponding pregnan-20-one derivatives (90) and (91) (Fig. 5). The presence of the $\Delta^{8(14)}$-tetrasubstituted double bond is one of the characteristic features of these and several other related compounds (Table 2). So far, there are very few natural $\Delta^{8(14)}$-steroids [44]. The stereochemical assignments at C-17 and C-20 rest on criteria similar to those used for withanolide D (31).

The Structure of Withanolide E (50)

The chemical investigations for the structure of this compound eventually led to the correct positional assignment of the functional groups [45] without obtaining, however, the complete structure. The structure was determined by an X-ray analysis [31] which disclosed that the C-17 side chain is α oriented. Withanolide E (50) is thus the first known natural steroid possessing such a side chain with all its carbon atoms present. The previously known steroids

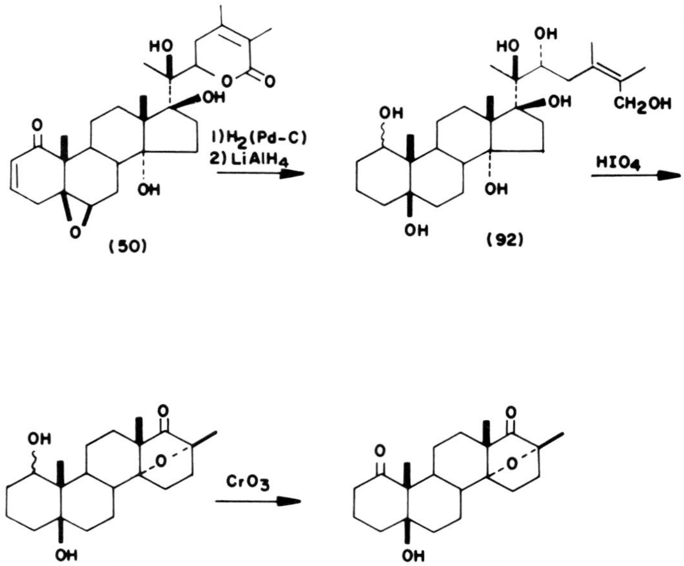

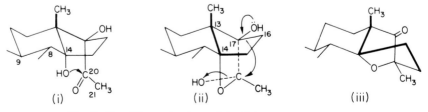

Fig. 6   *D*-Homo rearrangement of withanolide E (**50**).

with a 17α side chain are pregnane derivatives [46]. According to the crystallo-graphic analysis, the rings have the following conformations. Ring A is a boat, ring B is a distorted half-chair, ring C is a chair, and ring D has an envelope shape in which C-14 is the out-of-plane atom. The C-14 and C-20 hydroxy groups are hydrogen bonded.

Originally we attempted to elucidate the structure of this unusual compound by the conventional chemical approach used for withanolides G (**37**) and J (**38**), i.e., degradation to a pregnan-20-one derivative. Such a compound could not, however, be isolated since it underwent a spontaneous *D*-homo rearrange-ment in the exceedingly mild conditions under which the reaction was performed (Fig. 6). This rearrangement, which is triggered by the 14α-OH

Fig. 7   Possible mechanism of the *D*-homo rearrangement.

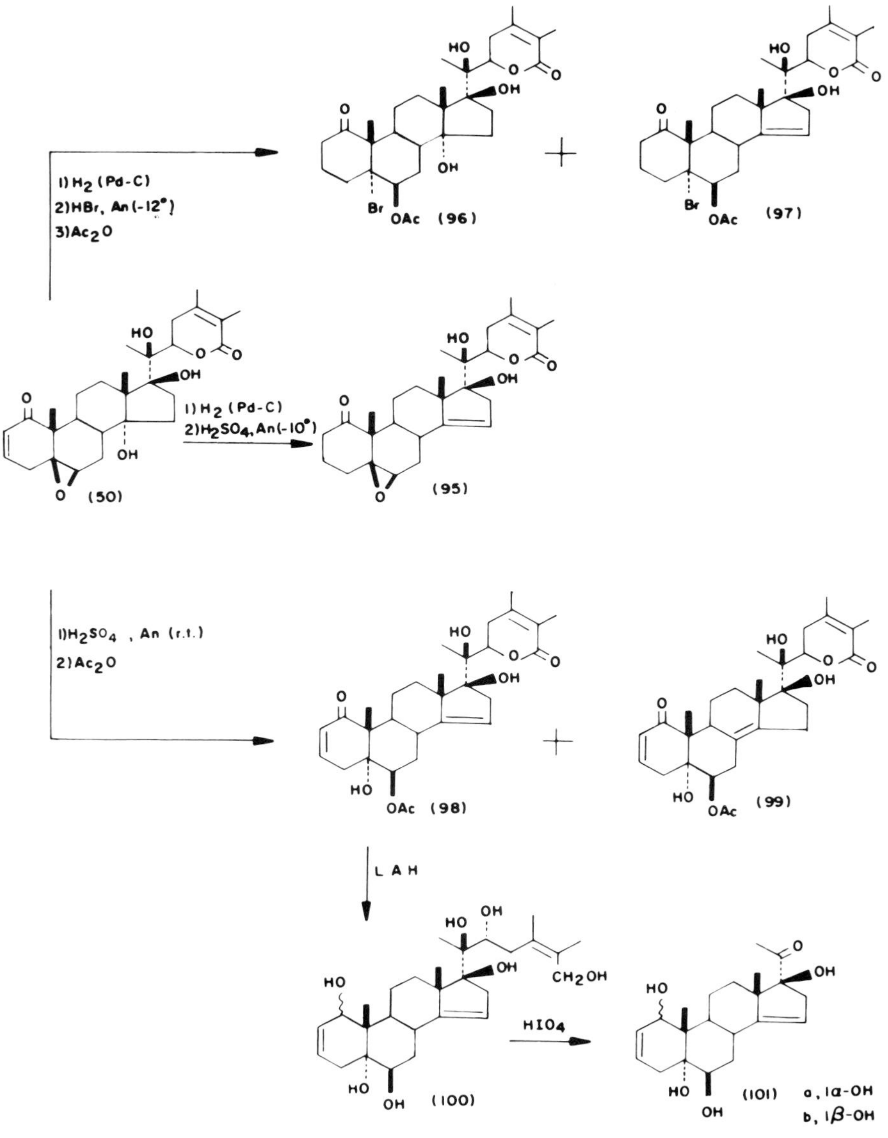

Fig. 8 Chemical transformations of withanolide E (**50**). An (r.t.): acetone, room temperature.

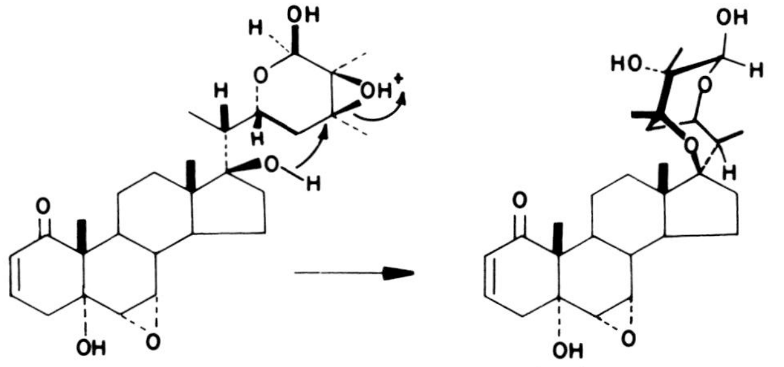

Fig. 9   Possible formation of Nic-11 (**56**).

group, afforded the *D*-homo androstane derivative (**94**), whose structure was also determined by crystallographic analysis.

The mechanism proposed [47] for this reaction involves the transient formation of a 14,20-lactol, as illustrated in Fig. 7.

The main reactions of withanolide E, which are not accompanied by skeletal rearrangement [32], are summarized in Fig. 8.

Nic-2 (**55**) and Nic-11 (**56**) are two biogenetically interrelated compounds with a 17α side chain. Nic-2 was assigned structure (**55**) chiefly by nmr analysis [12a], whereas the assignment for Nic-11 is based on a crystallographic analysis [11b]. It is likely that Nic-11 is formed in nature from Nic-2 through an internal nucleophilic displacement of the 24,25-epoxide by the 17β-hydroxy group (Fig. 9).

The Physalins [(57)–(62)] and the Withaphysalins [(63)–(65)]

The biogenetic pathway from the withanolides to the physalins (Fig. 10) implies a not yet encountered 14,17α,20α-trihydroxy-15-keto-18-carboxy-withanolide as a possible precursor. The substitution pattern of rings A and B could well be a 2,5-dien-1-one as found in physalin B (**59**) and C (**58**), although any of the partial structures found in other withanolides may comply (see Fig. 16), leading to the formation of an array of physalins. In a precursor as postulated above, the 18-carboxy group may react with the 20-OH to form a γ-lactone. The lactone carbonyl at C-18 and the hydroxy group at C-14 bring about a lowering of the electron density in the C-13—C-14 bond which may eventually be cleaved by a suitable oxidative enzyme. Such a cleavage, taking place from the rear of the molecule, should lead to an intermediate 13,14-*seco*-13α,17α-dihydroxy-14,15-diketo structure in which an internal hemiketal may easily be formed between the 17α-hydroxyl and the 14-one. Furthermore, the presence of the 15-one activates the neighboring 16-methylene, which, by an

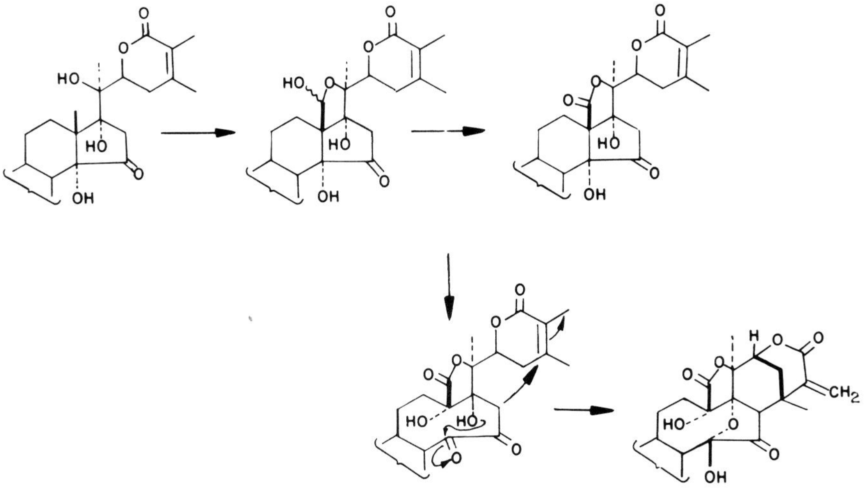

Fig. 10   Biogenetic pathway to the physalins.

intramolecular Michael type of reaction with the 24–25 double bond, affords the final product, in this case physalin C (58). The structures of the other physalins follow the same biogenetic pathway.

The structure of physalin A (57) was established by X-ray analysis of its 5α-acetoxy-6β-bromotetrahydro derivative [9d], whereas the structures of other physalins [9–9c] were obtained by various interrelations involving the reactions shown in Fig. 11.

The few withaphysalins characterized so far [34,35] (Table 5) support this biogenetic scheme (Fig. 10). Although in most of the known withanolides C-18 is at the lowest oxidation level, there exist at least one compound, withacnistin (5) [7b], in which C-18 is a primary alcohol, and two other compounds, withaphysalin B (64) and C (65) [34,35], in which C-18 is at the level of an aldehyde (present as a hemiacetal with 20-OH]. Withaphysalin A (63) is the only known "regular" withanolide in which the C-18 is at the highest oxidation level. Since this compound also possesses a 14α-OH, it can be regarded as a potential candidate for the cleavage of the C-13—C-14 bond; it cannot, however, be considered a precursor for the physalins because it does not have the required 17-hydroxy and 15-keto groups. The isolation of withaphysalin C (65) from the same source (*Physalis minima*) seems to corroborate the biogenetic scheme presented here. In a precursor like withaphysalin A, but with C-18 at the aldehyde level, the electron withdrawal on both sides of the C-13—C-14 bond facilitates its oxidative cleavage. In contrast to the physalins, the attack takes place from the top side of the

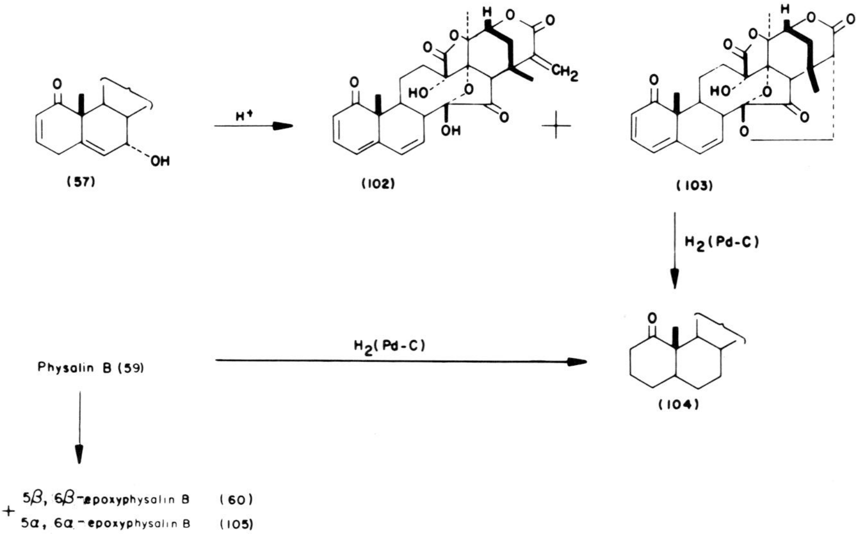

Fig. 11   Interrelation between physalins A and B.

molecule to give a transient $(13R)$-13-hydroxy-14-one structure in a nine-membered ring. In such a structure hemiketalization takes place, leading to the final product (65) [35], whose structure has been recently confirmed by an X-ray analysis performed on 2,3-dihydrowithaphysalin C lactone [48] (Fig. 12).

Withaphysalin C (65) is a good example of the result of a biological reaction in the absence of properly situated functional groups which would have directed the reaction toward the formation of a physalin-like compound.

Compounds with an Aromatic D Ring

The structure of nicandrenone (66), a compound with insect repellent properties isolated in 1951 from *Nicandra physaloides* [49], was determined independently by spectral analysis (without the stereochemistry at several asymmetric centers [12]) and by crystallographic analysis of a related compound [Nic-10 (70)] [11,11a]. The compound was renamed Nic-1. This and

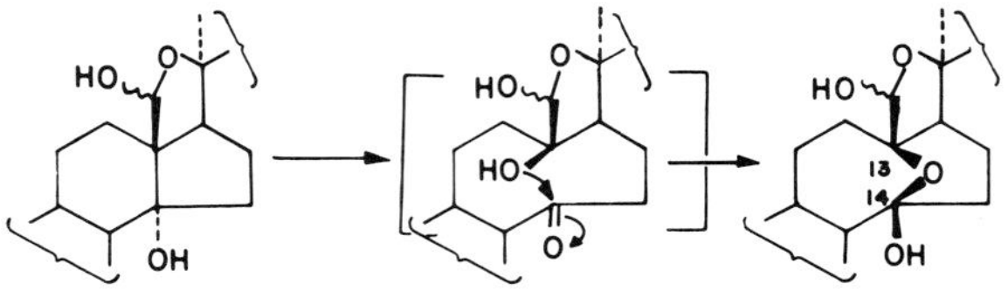

Fig. 12   Possible biogenetic pathway to withaphysalin C (65).

several other related compounds [(67)–(70)] are the first examples of naturally occurring steroids possessing an aromatic six-membered D ring. According to Bates [12], the aromatization of this ring involves a precursor containing a C-18 primary alcohol in which a positive charge generated at this carbon could have triggered its incorporation into the ring system. Such a speculation is supported by the isolation of withacnistin (5) [7b]. With two exceptions, withanicandrin (25) [25] and Nic-1-lactone (67) [36], the other compounds isolated [11–12a] from *N. physaloides* and possessing the entire nine-carbon-atom side chain are epoxylactols (28) and (29). Rings A and B in all these compounds have the same substitution pattern as withanone (22). Another interesting aspect of *N. physaloides* is that it contains several related compounds in which the side chain has been gradually cleaved down to a two-carbon fragment, Nic-10 (70) [11a]. The cleavage of the side chain has also been encountered in androsta-2,5,8(14)-triene-1,17-dione (49), isolated only in minute quantities from *W. somnifera* chemotype III [19].

SYNTHETIC APPROACHES TO THE WITHANOLIDES

Several studies directed toward the synthesis of model compounds possessing rings A and B with the substitution pattern encountered in the withanolides, as well as of compounds possessing their characteristic side chain, have

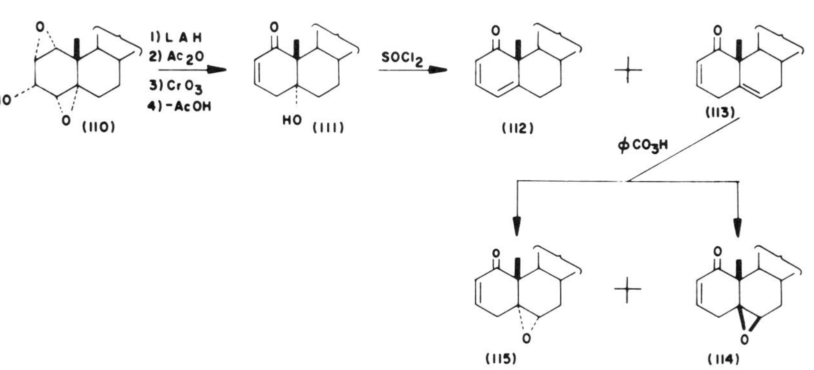

Fig. 13 Synthesis of rings A and B (Rehovot group).

recently been published. According to the scheme developed by the Rehovot group [50], 1α,2α-epoxycholest-4-en-3-one serves as the main intermediate, and the sequence is shown in Fig. 13. The Tokyo group [51] starts from 1α,2α-epoxy-5α-cholestan-6β-ol-3-one acetate, as shown in Fig. 14. The latter

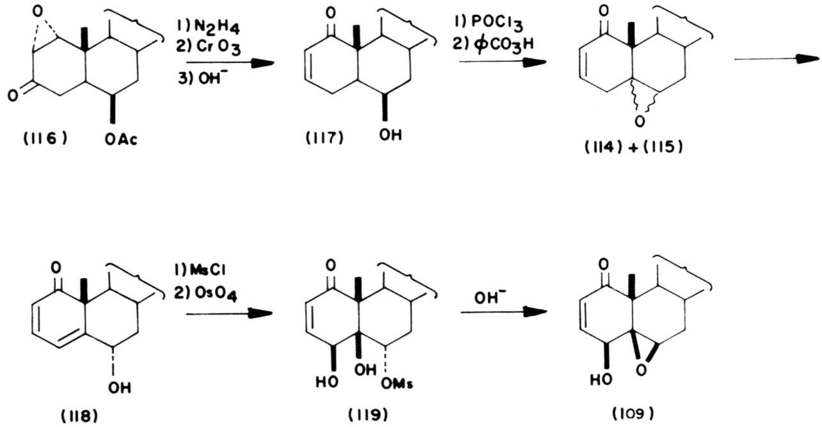

Fig. 14   Synthesis of rings A and B (Tokyo group).

group has also devised a synthesis of the side-chain lactone [52] which resulted, however, in the unnatural $22\alpha_F(22S)$-isomer ($\Delta\varepsilon_{258} - 3.72$). The reaction was done at $-78°C$ using lithium diisopropylamide hexamethyl phosphoramide in order to get the corresponding lithium dienolate (Fig. 15).

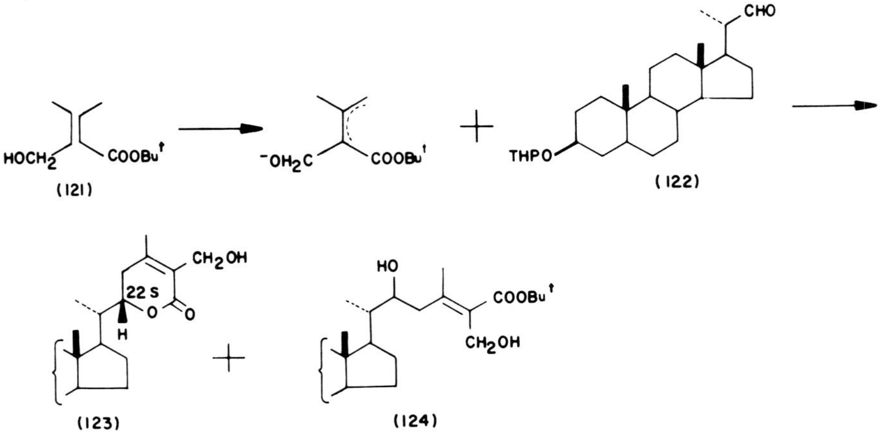

Fig. 15   Synthetic approach to the side chain of withaferin A (**1**).

BIOGENETIC ASPECTS

Although it is almost a truism that acetyl coenzyme A is at the starting point of steroid biosynthesis, the fact that labeled acetate (1-^{14}C and 2-^{14}C) is incorporated by *Physalis minima* to give physalin D (**61**) [34a] demonstrates unequivocally that this complex compound stems indeed from a typical steroid-type skeleton that underwent deep-seated rearrangements.

In a recent investigation [53] on tissue cultures of *W. somnifera*, several simple sterols such as 24-methylene- and 24-ethylidenecholesterol, as well as sitosterol, stigmasterol, and campesterol, were isolated. Surprisingly, no withanolides could be detected in the methanolic extract of these tissue cultures. Concomitantly, an investigation on plants of *W. somnifera* [54] led to the isolation of 24-methylcholesta-5,24-dien-3β-ol along with 24-methylcholesterol and several other sterols. Small amounts of 24-methylenecholesterol were also found. Incorporation experiments performed with 24-methyl-[28-^{3}H]cholesterol and 24-methylene[28-^{3}H]cholesterol led to the conclusion [55] that only the latter is transformed in the leaves of *W. somnifera* to radioactive withaferin A (**1**) and withanolide D (**31**).* The plant material supplied by the Botanical Garden of the University of Liverpool was probably of African origin, since so far this is the only known variety of *W. somnifera* producing (**1**) as well as (**31**) [8]. The ratio between these compounds is similar to that found in the African type (*vide infra*). The eventual intermediacy of 24-methylcholesta-5,24-dien-3β-ol in the biogenetic process could not be clearly established. According to these authors [54], the withanolide biosynthetic pathway may possibly branch from the normal sterol pathway before the 4-demethylsterol stage.

The withanolides are produced in the leaves and do not seem to occur in the roots of *W. somnifera*. In all the experiments done in our laboratory, we never found withanolides in the roots. In order to reach an unambiguous conclusion, *W. somnifera* was grafted on *Nicotiana glauca*, which does not produce any withanolide, and the result was a large branch of *W. somnifera* [45], which grew out of the graft; it contained the usual mixture of withanolides [45]. However, certain species, for instance *W. coagulans* [56], seem to have withanolides in the roots.

In the early stages of this investigation, we became aware that *W. somnifera* exists as distinct populations. Although morphologically indistinguishable from each other, they differ chemically, according to the withanolides they contain. Such populations of plants were designated as chemotypes. In Israel, we succeeded in identifying three chemotypes (I–III) [45,56a,b], and although

---

* We thank Dr. H. Rees for a preprint of this publication.

such a detailed survey on this plant has not yet been done in other parts of the world it is highly probable that the confusion existing in the early literature was due to the fact that the investigations involved various chemotypes. Indeed, our results with *W. somnifera* plants raised from seeds of various origin support this view.

The main constituent of chemotype I [23] is withaferin A (**1**), which is accompanied by small quantities of several other withanolides [(**2**)–(**4**), (**20**)–(**22**), (**54**)]. Although a general characteristic of *W. somnifera* is its ability to introduce hydroxy groups at various sites of the steroid nucleus, none of the constituents of chemotype I are hydroxylated at C-20. The major components are 4β-hydroxy-5β,6β-epoxy derivatives.

All the components of chemotype II [23] are 20-hydroxywithanolides. The main constituent, accounting for more than 90% of the steroid mixture, is withanolide D (**31**), which is accompanied by several other withanolides [(**32**)–(**35**), (**37**)] possessing additional hydroxy groups at various centers of the molecule. With one exception (**37**), these compounds are characterized by the presence of the 4β-hydroxy-5β,6β-epoxy grouping.

Chemotype III [29] also contains only 20-hydroxywithanolides, which are divided into two structural groups, one with a normal, β-oriented side chain, as found in compounds (**37**)–(**44**), and the other with an abnormal, α-oriented side chain, as found in (**50**) and (**51**). None of the compounds isolated from this chemotype possesses a hydroxy group at C-4. With the exception of withanolide E (**50**), which is a 5β,6β-epoxy derivative, all other compounds of this chemotype have a double bond at this position. The rather unusual presence of a $\Delta^{8(14)}$ double bond has already been mentioned.

A common feature of the constituents of *W. somnifera* chemotype Indian I [14] is the lack of a 20-OH group. Several compounds of this type possess rings A and B with the same substitution pattern as (**1**), whereas other compounds [(**22**), (**23**), and (**30**)] have the 5α-hydroxy-6α,7α-epoxide grouping, which had not been previously encountered. This system is also a common feature of the compounds isolated from *Nicandra physaloides*, with a normal [(**28**) and (**29**)] or with a rearranged carbocyclic skeleton [(**66**)–(**70**)]. The isolation of a minor component (**27**) possessing a 6-en-5α-ol group (from the Indian type) and of two other components [(**20**) and (**40**)] with a 5-en-4β-ol group [19,23] provides a clue to the obvious manner of stereospecific introduction of the 6α,7α- and the 5β,6β-epoxide, respectively.

There are two exceptions so far to the general observation that plants which produce 20-hydroxywithanolides do not produce 20-deoxywithanolides as well. Withaferin A (**1**), which accounts for ∼90% of the withanolides of the African type [8], is accompanied by ∼5% of withanolide D (**31**). Most interestingly, this plant also contains small quantities of two withanolides [(**8**) and (**36**)] in which the side-chain lactone is saturated. The second exception is *Lycium chinense* [24], which contains 17-deoxy-20-hydroxywithanone (**47**) [30] as well as 17-deoxywithanone (**24**), a 20-H compound. The relative amounts of these compounds are not specified.

The data presented here suggested that the differences leading to the formation of the various components are of a genetic character. It was attempted, therefore, to confirm this assumption through reciprocal crossings by pollination between the different types of *W. somnifera*. Such an approach would provide an understanding of the mechanisms involved in the biosynthetic pathways from a phytosterol precursor to the withanolides that accumulate in the plant. Crossings would also permit a genetic control, leading to new chemotypes characterized by possible accumulation of withanolides not yet encountered in the natural populations of plants or by transformation of minor into major components.

Although withanolide D (**31**) has never been detected in *W. somnifera* chemotypes I and III, their F_1 offspring contained this compound as the major component. Withaferin A, the major constituent of the parent type I, was absent, whereas withanolide E (**50**) remained in the offspring as an inheritance from the parent type III, although in smaller quantities. The cross of chemotypes I × II again produced mainly withanolide D, whereas that of chemotypes II × III produced a mixture having withanolide D as a dominant product accompanied by withanolide E.

In order to widen the scope of these experiments, the study was extended to cross-breedings between chemotype III (a 20-hydroxywithanolide producer) and the Indian chemotype referred to above (a 20-deoxywithanolide producer). The F_1 offspring of this cross was characterized by significant changes in the main constituents of the parent Indian type, withaferin A (**1**) and withanone (**22**). Whereas the former disappeared and was replaced by withanolide D (**31**), the latter was still present, although in about one-third of the original quantity, and was accompanied by 17-deoxy-20-hydroxy-withanone (**47**) and 20-hydroxywithanone (**48**). Thus, a 20-hydroxy group has been introduced in the withanolides originally present in the Indian type.

Although this work is only in the early stages, the results obtained so far point toward the numerous possibilities of using the chemotypes of the plant in order to induce appropriate transformations in the naturally occurring compounds.

The qualitative and quantitative analysis of the withanolides presented above, in conjunction with the results obtained through cross-breedings (F_1 and F_2 offspring*) permitted the presentation of reasonable biogenetic path-

---

* The F_2 offspring were obtained by selfing the F_1 offspring from the cross-breeding I × III. The work is in progress.

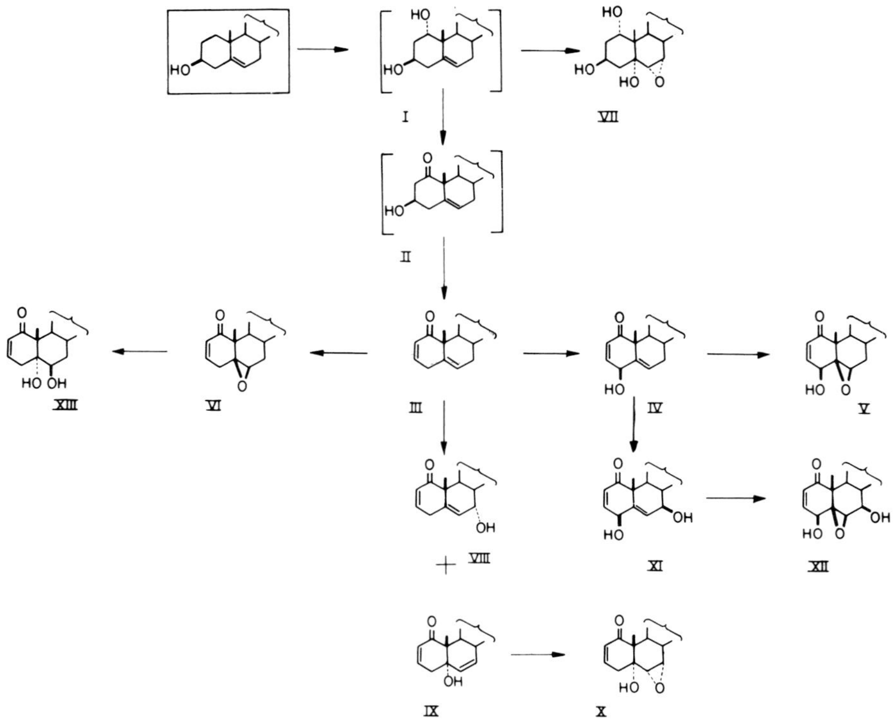

Fig. 16 Routes to the substitution pattern of rings A and B.

ways leading to the substitution pattern of rings A and B in the withanolides, as shown in Fig. 16.*

The first step is presumably hydroxylation at C-1 taking place from the rear, less hindered side of the molecule. The intermediate 5-ene-1α,3β-diol (**I**) is selectively oxidized to the corresponding ketone (**II**), which undergoes β-elimination to give the 2,5-dien-1-one structure **III** encountered in about 18 withanolides. There is only one compound (**30**), a minor constituent of *W. somnifera* Indian type, in which the oxidative processes **III** to **X** preceded the oxidation of the C-1 α-OH. All the compounds possessing structure **III** are rather unstable and have to be reduced to the 2,3-dihydro derivatives in order to prevent their oxidation. It is noteworthy that a saturated 1-one system has rarely been encountered in nature in this series [7,33,35].

The biosynthetic pathway splits at this stage into several directions:

* Structures I–XIII are in Fig. 16.

hydroxylation at C-4, epoxidation at C-5—C-6, or hydroxylation at C-5 (or at C-7). The hydroxylation at C-4 affords the labile 2,5-dien-4β-ol-1-one structure **IV**, which is then rapidly epoxidized to the corresponding 5β,6β-epoxide (**V**). Whereas the former structure (**IV**) is encountered in only 2 compounds [(**20**) and (**40**)] occurring in minute quantities, the latter (**V**) is present in about 16 compounds, among them withanolides (**1**) and (**31**), which are those accumulating in the largest quantities in the plants. It is noteworthy that, with the exception of *Physalis peruviana* [33], the plants which produce withanolides possessing the 4-hydroxy-5,6-epoxy system (**V**) do not produce withanolides having a 5,6-epoxide only (**VI**). A reasonable interpretation is that hydroxylation at C-4 is favored only when this carbon is activated by two neighboring double bonds (Δ^2 and Δ^5).

The allylic hydroxylation at C-7, with or without allylic rearrangement (structures **VIII** and **IX**), accounts for only two compounds [(**17**) and (**27**)]. The stereospecific epoxidation of **IX** affords the 5-hydroxy-6,7-epoxy system **X** encountered in about 15 withanolides, occurring *inter alia* in the Indian type *W. somnifera* as well as in *N. physaloides*. Secondary processes account for all other withanolides; for example, opening of the 5β,6β-epoxide in jaborosalactone A (**9**) leads to the jaborosalactones B–F; allylic hydroxylation of an intermediate like **IV** leads to the dihydroxylated system **XI**, which can undergo further epoxidation to yield system **XII** [cf. compounds (**43**) and (**37**), respectively].

The possible processes taking place in the side chain of the precursor 24-methylenecholesterol and leading to the highly oxidized side chain of the withanolides (in its several variations) are presented in the biogenetic scheme (Fig. 17).*

In a quantitative study [57] of the oxidative pathway from cholesterol to pregnenolone taking place in the adrenal mitochondria of some mammalian species, it was shown that hydroxylation at C-22 precedes hydroxylation at C-20. Both processes are stereospecific, leading to 22β_F(22R)-hydroxycholesterol and 20α_F,22β_F(20R,22R)-dihydroxycholesterol. In the biosynthesis of the latter, 22-hydroxycholesterol is much more rapidly metabolized than 20-hydroxycholesterol. This conclusion is supported by the observation that α-ecdysone (22β_F-OH) is readily metabolized to β-ecdysone (20α_F-OH, 22β_F-OH) [58]. It is noteworthy that there are only 22-hydroxy- and 20,22-dihydroxyecdysterols but no 20-hydroxyecdysterols. The same is true in the withanolide series, in which the stereochemistry at C-20 and C-22 is the same as in the corresponding cholesterols and ecdysterols. Moreover, all the withanolides possess an oxygen function at C-22 (22β_F); however, only some of them are oxygenated at C-20 (20α_F). The C-20 hydroxy group, which is a

* Structures **XIV–XXII** are in Fig. 17.

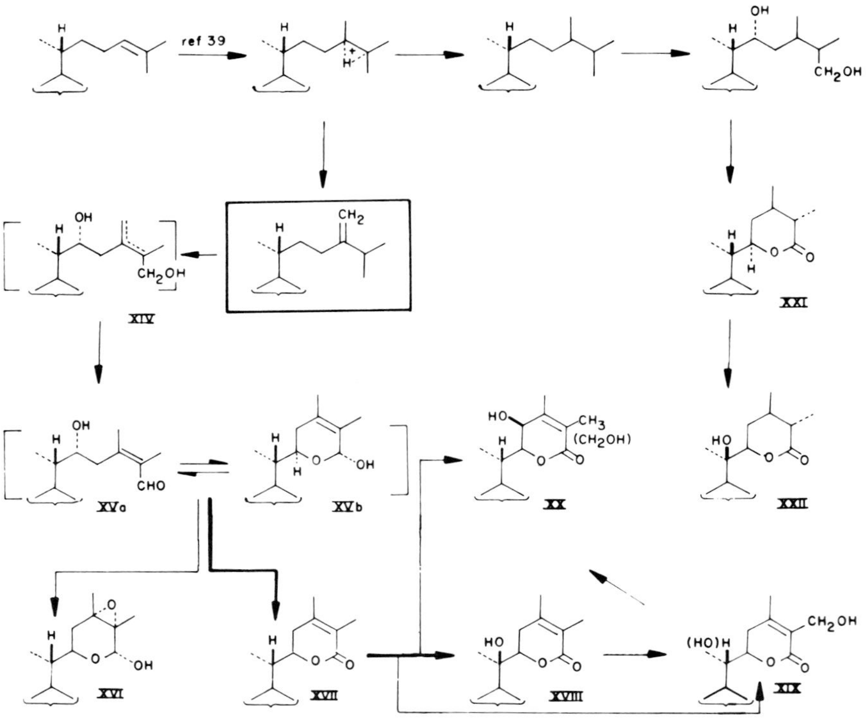

Fig. 17   Routes to the side chain of the withanolides.

dominant feature of the withanolides [45,56a,b], could be introduced by appropriate cross-pollination.

On the basis of the above data, we can assume that the first processes taking place in the side chain of 24-methylenecholesterol are hydroxylation at C-22 and C-26, leading to the intermediate $22\beta_F,26$-diol (**XIV**). The hydroxylation at C-26 is not without precedent, since there are several ecdysterols (inokosterone [59], makysterone B [60], and amarasterone A [61]) possessing such a primary alcohol. The homoallylic position of C-26 in the precursor of the withanolides can only favor such a process. This diol is further oxidized to the corresponding 26-aldehyde **XVa**, in equilibrium with the unsaturated lactol **XVb**. No compound possessing such a structural feature has been isolated so far. However, one can assume that the pathway splits at this stage into two directions: (a) stereospecific epoxidation of the 24,25 double bond leading to the epoxylactol **XVI**, which has been identified so far only in *Nicandra physaloides*; (b) oxidation to the unsaturated δ-lactone **XVII**, which takes place in most of the plants producing withanolides. There is only one compound, Nic-1-lactone (67) [36], in which these two processes seem to converge.

Once the α,β-unsaturated lactone is formed, several secondary oxidative processes may take place, for example, stereospecific hydroxylation at C-20 (XVIII) leading to all the withanolides of *W. somnifera* chemotypes II and III and of several other species; allylic hydroxylation at C-27 (XIX) leading to withaferin A (1), as well as to the withanolides of *Jaborosa integrifolia* [(9), (11)–(15)]; and allylic hydroxylation at C-23 (XX) leading only to withanolides Q (18) and R (26) [22],* which are minor components in the plants in which they occur.

When the enzymes responsible for hydroxylation at either C-20 or C-27 coexist in the same plant (for example, in the F_1 offspring of *W. somnifera* I × III), withanolide D (31) accumulates since the rate of hydroxylation at C-20 is faster than that at C-27. In contrast to the biological hydroxylations at C-20 and C-22, which are always stereospecific, the hydroxylation at C-23 may assume both possible directions. In withanolides Q and R it is $23\alpha_F$, the same as in the tetracyclic triterpene cyclograndisolide [62], whereas in the fungal sexual hormone antheridiol [63] it is $23\beta_F$.

It is quite possible that the two withanolides (8) and (36) possessing a saturated lactone ring (XXI and XXII in Fig. 17) are formed from a precursor preceding the formation of 24-methylenecholesterol [54].

The hydroxy groups introduced in the withanolide skeleton at C-14 and C-17 have the same orientation as the original hydrogen atom, thus pointing toward an insertion mechanism. The only exception, albeit important, is that of the withanolides with a 17α-oriented side chain (Table 3), all of which possess a 17β-OH group. The biogenetic implications of this problem are now being investigated.

BIOLOGICAL PROPERTIES OF THE WITHANOLIDES

The antibacterial properties of withaferin A (1) are mentioned in the Introduction.

Withaferin A (1) and withacnistin (5) isolated from *Acnistus arborescens* showed cytotoxicity *in vitro* against KB cell cultures derived from human carcinoma of the nasopharynx [7a,b]. Subsequently, it was shown that withaferin A (1) produced significant growth retardation in a number of tumor systems in mice. Growth of Ehrlich ascites carcinoma was completely inhibited in more than half of the mice, which survived for 100 days without

---

* The stereochemistry of these compounds is 22S,23S (22β_F,23α_F) and not 22S,23R as erroneously reported [22].

evidence of tumor growth [64]. It has been further demonstrated that (1) produces mitotic arrest in the metaphase of the same tumor system and that the compound also induces vacuolization of the cytoplasm [65,66]. It was found that, in the experiments in which the tumor growth disappeared, the mice were resistant to rechallenge with Ehrlich ascites tumor cells. Such behavior can be explained as a combined effect of withaferin A and an immune response [65].

Withaferin A could suppress secondary lesions in adjuvant-induced arthritis in rats. The locally induced graft (lymphocytes) versus host reaction in chicks was also strongly inhibited [67]. It seems that, besides antiproliferative activity, withaferin A has a complex influence on inflammation and immune response.

Withanolide E (50) has recently been studied to some extent. Several tumor systems used in cancer chemotherapy screening have responded to treatment with this compound. Forty-eight percent of mice bearing P388 leukemia survived compared to controls, whereas with the L1210 system a definite increase in survival was achieved; compound (50) was found to be active against B-16 mouse melanoma and Lewis lung carcinoma [68].

Several other withanolides now under investigation show a definite activity of the type described above.

ACKNOWLEDGMENTS

We thank the National Cancer Institute, National Institutes of Health, and the United States–Israel Binational Science Foundation, Jerusalem, Israel, for financial support during different phases of the work performed in Rehovot.

REFERENCES

1. P. A. Kurup, *Curr. Sci.* **25**, 57 (1956); *Antibiot. Chemother.* (*Washington, D.C.*) **8**, 511 (1958).
2. Cited by W. Kopaczewski, *Therapie* **3**, 98 (1948).
3. J. M. Watt and M. G. Breyer-Brandwijk, "The Medicinal and Poisonous Plants of Southern and Eastern Africa." Livingstone, Edinburgh, 1962.
4. N. S. Dhalla, M. S. Sastry, and C. L. Malhotra, *J. Pharm. Sci.* (India) **50**, 876 (1961).
5. F. B. Power and A. H. Salway, *J. Chem. Soc.* **99**, 490 (1911).
6. S. Ben-Efraim and A. Yarden, *Antibiot. Chemother.* (*Washington, D.C.*) **12**, 575 (1962); S. Kohlmuenzer and J. Krupinska, *Diss. Pharm.* **14**, 501 (1963).
7. D. Lavie, E. Glotter, and Y. Shvo, *J. Org. Chem.* **30**, 1774 (1965); *J. Chem. Soc.* p. 7517 (1965).
7a. S. M. Kupchan, R. W. Doskotch, P. Bollinger, A. T. McPhail, G. A. Sim, and J. A. Saenz-Renauld, *J. Am. Chem. Soc.* **87**, 5805 (1965).

7b. S. M. Kupchan, W. K. Anderson, P. Bollinger, R. W. Doskotch, R. M. Smith, J. A. Saenz-Renauld, H. K. Schnoes, A. L. Burlingame, and D. H. Smith, *J. Org. Chem.* **34**, 3858 (1969).

8. I. Kirson, E. Glotter, A. Abraham, and D. Lavie, *Tetrahedron* **26**, 2209 (1970).

9. T. Matsuura, M. Kawai, R. Nakashima, and Y. Butsugan, *Tetrahedron Lett.* p. 1083 (1969); *J. Chem. Soc. C* p. 664 (1970).

9a. T. Matsuura and M. Kawai, *Tetrahedron Lett.* p. 1765 (1969).

9b. M. Kawai and T. Matsuura, *Tetrahedron* **26**, 1743 (1970).

9c. M. Kawai, T. Taga, K. Osaki, and T. Matsuura, *Tetrahedron Lett.* p. 1087 (1969).

9d. M. Kawai, T. Matsuura, T. Taga, and K. Osaki, *J. Chem. Soc. B* p. 812 (1970).

9e. T. Matsuura, personal communication.

10. N. B. Mulchandani, S. S. Iyer, and L. P. Badheka, cf. *Chem. Abstr.* **82**, 171286 (1975).

11. M. J. Begley, L. Crombie, P. J. Ham, and D. A. Whiting, *Chem. Commun.* p. 1108 (1972).

11a. M. J. Begley, L. Crombie, P. J. Ham, and D. A. Whiting, *Chem. Commun.* p. 1250 (1972).

11b. M. J. Begley, L. Crombie, P. J. Ham, and D. A. Whiting, *Chem. Commun.* p. 821 (1973).

11c. M. J. Begley, L. Crombie, P. J. Ham, and D. A. Whiting, *J. Chem. Soc., Perkin Trans.* **1** pp. 296 and 304 (1976).

12. R. B. Bates and D. J. Eckert, *J. Am. Chem. Soc.* **94**, 8258 (1972).

12a. R. B. Bates and S. R. Morehead, *Chem. Commun.* p. 125 (1974).

13. E. Glotter, R. Waitman, and D. Lavie, *J. Chem. Soc. C* p. 1765 (1966).

14. I. Kirson, E. Glotter, D. Lavie, and A. Abraham, *J. Chem. Soc. C* p. 2032 (1971).

15. A. G. Gonzalez, J. L. Breton, and J. Trujillo, *An. Quim.* **68**, 107 (1972).

16. D. Lavie, I. Kirson, E. Glotter, and G. Snatzke, *Tetrahedron* **26**, 2221 (1970).

17. R. Tschesche, H. Schwang, and G. Legler, *Tetrahedron* **22**, 1121 (1966).

18. R. Tschesche, H. Schwang, H. W. Fehlhaber, and G. Snatzke, *Tetrahedron* **22**, 1129 (1966).

19. I. Kirson, unpublished results.

20. R. Tschesche, M. Baumgarth, and P. Welzel, *Tetrahedron* **24**, 5169 (1968).

21. R. Tschesche, K. Annen, and P. Welzel, *Tetrahedron* **28**, 1909 (1972).

22. I. Kirson, A. Cohen, and A. Abraham, *J. Chem. Soc., Perkin Trans. 1* p. 2136 (1975).

23. A. Abraham, I. Kirson, D. Lavie, and E. Glotter, *Phytochemistry* **14**, 189 (1975)

24. R. Hänsel, J. T. Huang, and D. Rosenberg, *Arch. Pharm. (Weinheim, Ger.)* **308**, 653 (1975).

25. I. Kirson, D. Lavie, S. S. Subramanian, P. D. Sethi, and E. Glotter, *J. Chem. Soc. Perkin Trans. 1* p. 2109 (1972).

26. D. Lavie, I. Kirson, and E. Glotter, *Israel J. Chem.* **6**, 671 (1968).

27. I. Kirson, D. Lavie, S. M. Albonico, and H. R. Juliani, *Tetrahedron* **26**, 5063 (1970).

28. L. Barata, W. B. Mors, I. Kirson, and D. Lavie, *An. Acad. Bras. Cienc.* **42**, Suppl. 401 (1970).

29. E. Glotter, I. Kirson, A. Abraham, and D. Lavie, *Tetrahedron* **29**, 1353 (1973).

29a. G. Adam and M. Hesse, *Tetrahedron Lett.* p. 1199 (1971); *Tetrahedron* **28**, 3527 (1972),

30. S. S. Subramanian, P. D. Sethi, E. Glotter, I. Kirson, and D. Lavie, *Phytochemistry* **10**, 685 (1971).

31. D. Lavie, I. Kirson, E. Glotter, D. Rabinovich, and Z. Shakked, *Chem. Commun.* p. 877 (1972).

32. E. Glotter, A. Abraham, G. Günzberg, and I. Kirson, *J. Chem. Soc., Perkin Trans. 1*, 341 (1977).

33. I. Kirson, A. Abraham, P. D. Sethi, S. S. Subramanian, and E. Glotter, *Phytochemistry* **15**, 340 (1976).
34. E. Glotter, I. Kirson, A. Abraham, S. S. Subramanian, and P. D. Sethi, *J. Chem. Soc., Perkin Trans. 1* p. 1370 (1975).
34a. N. B. Mulchandani, S. S. Iyer, and L. P. Badheka, *Chem. Abstr.* **82**, 135841 (1975).
35. I. Kirson, Z. Zaretskii, and E. Glotter, *J. Chem. Soc., Perkin Trans. 1* p. 1244 (1976)
36. E. Glotter, I. Kirson, A. Abraham, and P. Krinsky, *Phytochemistry* **15**, 1317 (1976).
37. D. Lavie, S. Greenfield, and E. Glotter, *J. Chem. Soc. C* p. 1753 (1966).
38. E. Glotter and D. Lavie, *J. Chem. Soc. C* p. 2298 (1967).
39. A. T. McPhail and G. A. Sim. *J. Chem. Soc. B* p. 1962 (1968).
40. D. Lavie, Y. Kashman, E. Glotter, and N. Danieli, *J. Chem. Soc. C* p. 1757 (1966).
41. S. Greenfield, E. Glotter, D. Lavie, and Y. Kashman, *J. Chem. Soc. C* p. 1460 (1967).
42. R. Tschesche, K. Annen, and P. Welzel, *Chem. Ber.* **104**, 3556 (1971).
43. A. Mijares, D. S. Cargill, J. A. Glasel, and S. Lieberman, *J. Org. Chem.* **32**, 810 (1967).
44. See, for instance, L. H. Zalkow, G. A. Cabat, C. L. Chetty, M. Gohsal, and G. Keen. *Tetrahedron Lett.* p. 5727 (1968).
45. A. Abraham, I. Kirson, E. Glotter, and D. Lavie, *Phytochemistry* **7**, 957 (1968).
46. See references cited in Bates and Morehead [12a].
47. D. Rabinovich, Z. Shakked, I. Kirson, G. Günzberg, and E. Glotter, *Chem. Commun.* p. 461 (1976).
48. D. Rabinovich and F. Frolow, unpublished results.
49. F. V. Gizycki and G. Kotitsche, *Arch. Pharm. (Weinheim, Ger.)* **284**, 129 (1951); D. Nalbandov, R. T. Yamamoto, and G. S. Fraenkel, *Agric. Food Chem.* **12**, 55 (1964).
50. M. Weissenberg, E. Glotter, and D. Lavie, *Tetrahedron Lett.* p. 3063 (1974); *J. Chem. Soc., Perkin Trans. 1* (in press) (1977).
51. M. Ishiguro, A. Kajikawa, T. Haruyama, M. Morisaki, and N. Ikekawa, *Tetrahedron Lett.* p. 1421 (1974); M. Ishiguro, A. Kajikawa, T. Haruyama, Y. Ogura, M. Okubayashi, M. Morisaki, and N. Ikekawa, *J. Chem. Soc., Perkin Trans. 1* p. 2295 (1975).
52. A. Kajikawa, M. Morisaki, and N. Ikekawa, *Tetrahedron Lett.* p. 4135 (1975).
53. P. L. C. Yu, M. M. El-Olemy, and S. J. Stohs, *Lloydia* **37**, 593 (1974).
54. W. J. S. Lockley, D. P. Roberts, H. H. Rees, and T. W. Goodwin, *Tetrahedron Lett.* p. 3773 (1974).
55. W. J. S. Lockley, H. H. Rees, and T. W. Goodwin, *Phytochemistry* **15**, 937 (1976).
56. S. S. Subramanian and P. D. Sethi, *Curr. Sci.* **38**, 267 (1969).
56a. D. Lavie, I. Kirson, and A. Abraham, *Israel. J. Chem.* **14**, 60 (1975), paper presented at *25th IUPAC Congr., 1975.*
56b. D. Lavie, *in* "Chemistry in Botanical Classification" (G. Bendz and J. Santesson, eds.), pp. 181–188. Academic Press, New York, 1974.
57. S. Burstein, H. L. Kimball, and M. Gut, *Steroids* **15**, 809 (1970).
58. D. S. King and J. B. Sidall, *Nature (London)* **221**, 955 (1969); H. Moriyama, K. Nakanishi, D. S. King, T. Okauchi, J. B. Sidall, and W. Hafferl, *Gen. Comp. Endocrinol.* **15**, 80 (1970); L. Cherbas and P. Cherbas, *Biol. Bull.* **138**, 115 (1970).
59. T. Takemoto, Y. Hikino, S. Arihara, H. Hikino, S. Ogawa, and N. Nishimoto, *Tetrahedron Lett.* p. 475 (1968).
60. S. Imai, S. Fujioka, E. Murata, Y. Sasakawa, and K. Nakanishi, *Tetrahedron Lett.* p. 3887 (1968).
61. T. Takemoto, K. Nomoto, and H. Hikino, *Tetrahedron Lett.* p. 4953 (1968).
62. F. H. Allen, J. P. Kutney, J. Trotter, and N. D. Westcott, *Tetrahedron Lett.* p. 2629 (1971).

63. G. P. Arsenault, K. Biemann, A. W. Barksdale, and T. C. McMorris, *J. Am. Chem. Soc.* **90**, 5635 (1968); J. A. Edwards, J. Sundeen, W. Salmond, I. Iwadare, and J. H. Fried, *Tetrahedron Lett.* p. 791 (1972).
64. B. Shohat, S. Gitter, A. Abraham, and D. Lavie, *Cancer Chemother. Rep.* **51**, 271 (1967).
65. B. Shohat, S. Gitter, and D. Lavie, *Int. J. Cancer* **5**, 244 (1970).
66. B. Shohat, *Z. Krebsforsch.* **80**, 97 (1973).
67. A. Fügner, *Arzneim.-Forsch.* **23**, 932 (1973).
68. B. Shohat, unpublished results.

4

Novel Piperidine Alkaloids from the Fungus *Rhizoctonia leguminicola*: Characterization, Biosynthesis, Bioactivation, and Related Studies

F. Peter Guengerich and Harry P. Broquist

INTRODUCTION

The fungus *Rhizoctonia leguminicola* produces two rather novel piperidine alkaloids: 1-acetoxy-6-aminooctahydroindolizine (**I**, Fig. 1) and 3,4,5-tri-hydroxyoctahydro-1-pyrindine (**III**, Fig. 1). The former compound, termed slaframine, is a parasympathomimetic toxin of which both the bioactivation and mode of action appear to be unique. Both alkaloids are derived from lysine via a novel biosynthetic pathway. Studies of this pathway have led to a better understanding of lysine anabolism and catabolism in fungi.

The work described in this chapter was started in response to an agricultural problem. In the late 1950's, cattle and other domestic animals were reported to be afflicted with a syndrome characterized by uncontrollable salivation, refusal to eat, and occasionally death after the consumption of certain forages [1,2]. It was found that these forages were infected with *R. leguminicola* [3]; the mold was isolated, and conditions for growth and toxin production in stationary culture were established [1,4–7].

The toxin slaframine was isolated using salivation in guinea pigs as an assay [1,2,6–9]. The elucidation of the structure of slaframine has been reviewed [1].

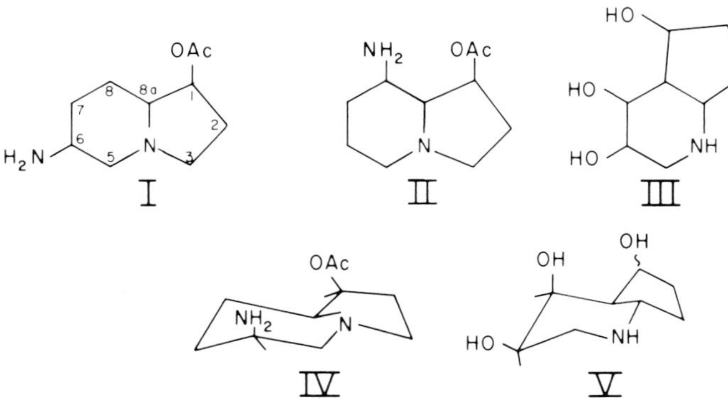

Fig. 1   Structures of slaframine and 3,4,5-trihydroxyoctahydro-1-pyrindine.

Slaframine was originally assigned the structure 1-acetoxy-8-aminoocta-hydroindolizine [8,10,11] (**II**, Fig. 1); proton magnetic resonance (pmr) spin decoupling led to a subsequent reassignment of the structure as 1-acetoxy-6-aminooctahydroindolizine [1,12]. The absolute configuration (1S,6S,8aS; **IV**, Fig. 1) was deduced by a combination of pmr and chemical methods [1,12].

More recently, Rinehart and his associates have achieved the total synthesis of slaframine [13]. 2-Hydroxy-5-nitropyridine was converted in several steps to the key intermediate, ethyl 5-acetamidopipecolate, which was condensed with ethyl acrylate; Dieckmann cyclization established the carbon skeleton. Hydrolysis, decarboxylation, reduction, and acetylation gave a mixture of the isomers of 1-acetoxy-6-acetamidooctahydroindolizine, which were separated by chromatography. Only one pair of enantiomers appeared to be identical with N-acetylslaframine; after removal of the N-acetyl group, the synthetic material was identical chromatographically and spectrally with authentic slaframine. 1-Acetoxy-8-aminooctahydroindolizine was also synthesized via a similar route but was distinguished from slaframine [13]. This synthetic work further establishes the assigned structure of slaframine; however, the procedure is not yet practical in terms of providing adequate quantities of slaframine for further physiological studies.

3,4,5-TRIHYDROXYOCTAHYDRO-1-PYRINDINE

During studies on the biosynthesis of slaframine (*vide infra*), labels from the precursors of slaframine were found to be incorporated into a previously unrecognized compound. Although this compound did not prove to be a precursor or metabolite of slaframine, it was characterized and found to possess a novel structure, namely, a 1-pyrindine nucleus, previously unrecognized in biological systems [14,15].

The compound was labeled *in vivo* by incubation with [*ring*-^{3}H]pipecolic acid to aid in purification. Since the alkaloid did not partition well into organic solvents, ion-exchange chromatography and thin-layer chromatography (tlc) were utilized in the isolation procedure. An identical compound was obtained when the ion-exchange systems, which utilize both strongly acidic and basic conditions, were replaced with a Dowex-50 system utilizing sodium citrate buffer (0.38 *N*, pH 4.25), indicating that the characterized alkaloid is probably not an artifact of the isolation procedure.

The empirical formula $C_8H_{15}NO_3$ was obtained by high-resolution mass spectrometry; the infrared (ir) spectrum indicated strong hydroxyl absorption. The pmr and ir spectra provided no evidence for unsaturation; sodium borohydride treatment and attempted acidic and basic hydrolysis were without effect on the compound. The presence of a secondary amine was consistent with positive reactions with both ninhydrin and nitroprusside and a negative test with Dragendorff's reagent. Migration on tlc was retarded by borate, and the compound reacted with periodate. Thus, a bicyclic eight-carbon structure containing as the only functional groups a secondary amine and three hydroxyls (at least two of which are vicinyl) was implicated.

The structure was assigned [15] on the basis of pmr spin decoupling; most of the stereochemistry could be assigned on the basis of the observed coupling constants. However, the configuration at C-5 could not be definitely assigned from available data. The absolute configuration is either that of structure **V** (Fig. 1) or its enantiomer. The major fragments observed in the high-resolution mass spectrum [15] are in accord with the assigned structure. Acetylation of the alkaloid (pyridine–acetic anhydride) gave rise to the 1,3,5-triacetyl derivative in quantitative yield. The remaining hydroxyl group (C-4) resisted acetylation and chromic oxide–pyridine oxidation (probably due to steric hindrance by acetoxy groups at C-3 and C-5).

In preliminary experiments, the trihydroxy compound was without observable effect when injected intravenously into female mice at a concentration of 10 mg/kg; thus, a role in the "slobber forage" syndrome is not indicated. More recently, a group in Japan has isolated the compound 5-(*trans*-1'-butenyl)-3-methyl-3,4,6,7-tetrahydro-1(2*H*)-pyrindine ("pyrindicine") from *Streptomyces* and has found narcoantagonistic and analgesic properties for the compound. Moreover, the alkaloid relaxes intestine, inhibits collagen-induced platelet aggregation, and exhibits weak antimicrobial activity [16,17].

BIOSYNTHESIS OF PIPERIDINE AND PYRIDINE ALKALOIDS: GENERAL FEATURES

Piperidine and pyridine alkaloids are formed in nature via several biosynthetic pathways (Fig. 2). The Δ^1-piperideine ring of betanin is formed from dihydroxyphenylalanine by a sequence of reactions involving the opening of the catechol ring, as shown in Fig. 2 [18]. Several alkaloids, e.g., anabasine (Fig. 2), are believed to be derived from Δ^1-piperideine, which is postulated to arise from lysine via decarboxylation to cadaverine, subsequent transamination (or oxidation), and ring closure [19].

Leete and his co-workers were unable to effect the incorporation of lysine or some of its metabolites into coniine; however, acetate was incorporated well into the molecule. Octanoic acid, 5-ketooctanoic acid, and 5-ketooctanal are also efficiently incorporated into coniine [19]. Acetate has also been shown to be a more efficient precursor than lysine in the biosynthesis of the piperidine alkaloids pinidine [20], carpaine [21], and nigrifactin [22]. Leete postulates that these piperidine alkaloids arise via poly-β-keto intermediates [19].

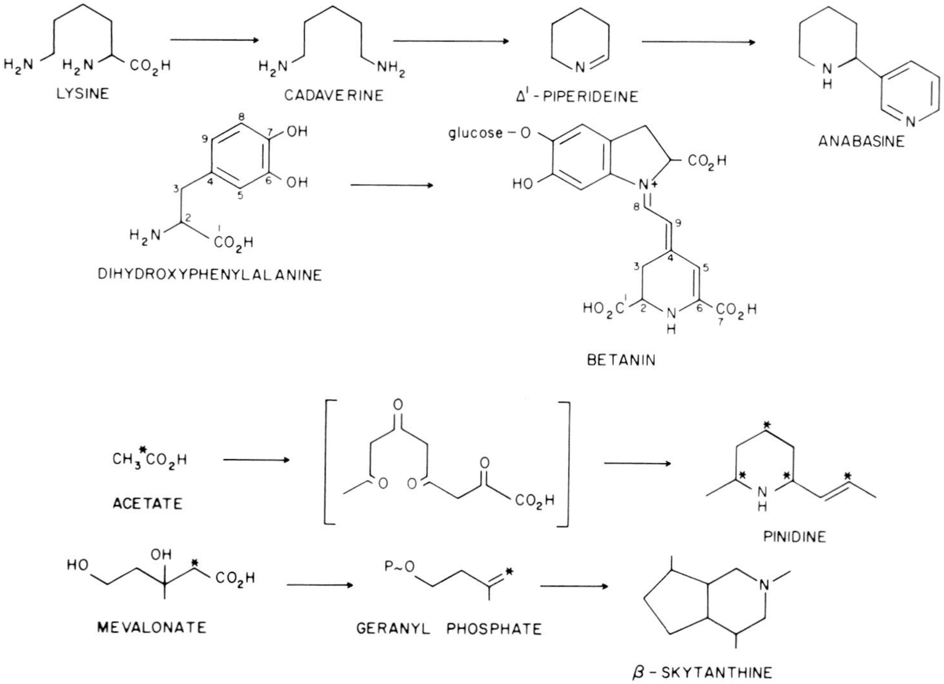

Fig. 2 Pathways for biosynthesis of piperidine and pyridine alkaloids.

Studies by Waller and his associates indicate that lysine is not a precursor of the *Skytanthus* alkaloids β-skytanthine and actinidine. Incorporation of acetate, mevalonate, and geranyl phosphate into actinidine and of mevalonate into skytanthine suggests an isoprenoid mode of biosynthesis [23,24].

BIOSYNTHESIS OF PIPECOLIC ACID
AND ITS ROLE IN ALKALOID BIOSYNTHESIS

Preliminary experiments indicated that both carbons 1 and 6 of lysine were incorporated into slaframine [1,8,25,26]. Subsequent experiments showed that label from [2-^{14}C]- and [4,5-^{3}H]lysines were also well incorporated [25,27], suggesting strongly that the lysine skeleton is incorporated into slaframine *in toto*, in contrast to other piperidine alkaloids in which the carboxyl group is lost (cf. Fig. 2). The most obvious pathway would involve cyclization of lysine to form pipecolic acid (Fig. 3), which would then form the piperidine ring of slaframine. Carrier pipecolic acid diluted the label incorporated from lysine into slaframine [25,26]. Moreover, DL-[1-^{14}C]-, DL-[6-^{14}C]-, DL-[4,5-^{3}H]-, DL-[*ring*-^{3}H]-, and L-[2-^{3}H]pipecolate were all found to label slaframine

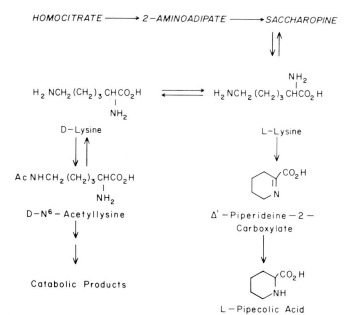

Fig. 3   Biosynthesis and catabolism of lysine in *Rhizoctonia leguminicola*. [Reprinted with permission from H. P. Broquist, *Biochemistry* 12, 4270 (1973). Copyright by the American Chemical Society.]

[25–27]. 2-Aminoadipic acid was found to label lysine, pipecolic acid, and slaframine [25], indicating that *R. leguminicola* utilizes the homocitric–aminoadipic acid pathway of lysine biosynthesis [28]. Moreover, the fungus was found to have saccharopine dehydrogenase activity [27], previously shown to be a marker for this pathway [28]. Labels from variously labeled lysines were found to be incorporated into pipecolic acid [29]. 2-Aminoadipic acid, lysine, and pipecolic acid were found to label 3,4,5-trihydroxyoctahydro-1-pyrindine; as in the case of slaframine, all carbon atoms are apparently utilized in the biotransformations [14,15]. Thus, slaframine and 3,4,5-tri-hydroxyoctahydro-1-pyrindine probably derive from a branch point after pipecolic acid.

In experiments designed to find which isomer of lysine is utilized in the biosynthesis of pipecolic acid, the incorporation of label from DL-lysine into pipecolic acid was diluted by both carrier D- and L-lysines [29]. These experiments suggested that (a) both isomers of lysine are metabolized to pipecolic acid and/or (b) lysine racemase activity is present. In support of the latter explanation, the conversions of L- to D-lysine and D- to L-lysine were demonstrated in cell-free extracts [27,29]. Surprisingly, the conversion of L- to D-lysine was found to occur significantly more rapidly than the reverse reaction under the conditions used [27,29].

The *in vitro* racemizations of both D- and L-lysines were almost completely inhibited by hydroxylamine [29], as was found to be the case with the lysine racemase of *Pseudomonas putida* [30]. When resting cultures of *R. leguminicola* were incubated with the respective labeled lysine isomers in the presence of the inhibitor, it was found that L-lysine was utilized largely for the biosynthesis of pipecolic acid and D-lysine was utilized for production of N^ε-acetyllysine [29] (Table 1). The pipecolic acid was found to be exclusively of the L- configuration, and the N^ε-acetyllysine was almost exclusively of the D- configuration, regardless of the stereochemistry of the proffered lysine label [29]. Further experiments with both resting cultures and cell-free extracts of

TABLE 1

Stereospecificity of Lysine Catabolism in *Rhizoctonia leguminicola*[a]

Radiotracer	Incorporation (%)		N^ε-Acetyllysine/pipecolate
	N^ε-Acetyllysine	Pipecolate	
DL-[2-^{14}C]Lysine	10.6	2.9	3.7
D-[1-^{14}C]Lysine	10.5	1.0	10.5
L-[U-^{14}C]Lysine	2.9	3.7	0.8

[a] Reproduced with permission from H. P. Broquist, *Biochemistry* **12**, 4270 (1973). Copyright by the American Chemical Society.

TABLE 2

Incorporation of the Nitrogen Atoms of Lysine into *Rhizoctonia leguminicola* Metabolites[a]

Added lysine	Excess ^{15}N (%)	
	N^ε-Acetyllysine	Pipecolate
DL-[^{15}N$^\alpha$], 60.7% excess	62.8	3.0
DL-[^{15}N$^\varepsilon$], 89.7% excess	76.1	40.3

[a] Reproduced with permission from H. P. Broquist, *Biochemistry* **12**, 4270 (1973). Copyright by the American Chemical Society.

R. leguminicola have provided evidence that the major pathway of lysine catabolism involves the sequence lysine → N^ε-acetyllysine → 2-keto-6-acetamidohexanoic acid → 5-acetamidovaleric acid → 5-aminovaleric acid → → glutaric acid → → → short-chain organic acids [31].

Experiments with [^{15}N$^\varepsilon$]- and [^{15}N$^\alpha$]lysines indicated that the nitrogen atom of pipecolic acid is derived exclusively from the ε-nitrogen of lysine (Table 2) [29], consistent with the role of Δ^1-piperideine-2-carboxylic acid as an intermediate in the conversion of lysine to pipecolic acid. Moreover, cell-free extracts of *R. leguminicola* reduced Δ^1-piperideine-2-carboxylic acid, but not its isomer Δ^1-piperideine-6-carboxylic acid (the expected product of transamination or oxidation of the ε-amino group of lysine), to pipecolic acid in the presence of NADPH (neither isomer was reduced with NADH [29]).

The pathway 2-aminoadipic acid → lysine → pipecolic acid is not reversible in *R. leguminicola*; incubation with labeled lysine did not produce label in aminoadipic acid, and no label in either lysine or aminoadipic acid was found after incubation with labeled pipecolic acid [27,31].

Studies with lysine auxotrophs of another mold, *Neurospora crassa*, showed that lysine was converted to L-pipecolic acid and to both isomers of N^ε-acetyllysine. As in the case of *R. leguminicola*, the L-isomer of lysine is preferentially used in forming pipecolic acid [31]. [^{15}N]Lysine experiments also demonstrate that the nitrogen atom of pipecolic acid derives from the ε-nitrogen of lysine [31].

The relationship of the D- and L-isomers of lysine to anabolic and catabolic events of lysine catabolism in *R. leguminicola* is summarized in Fig. 3. L-Lysine is oxidized to give rise to pipecolic acid, which is utilized in alkaloid production, while D-lysine is acetylated prior to catabolism to acidic metabolites. The pathways for the two isomers are probably not completely stereospecific [31]; however, studies with other yeasts and molds suggest that, with certain qualifications, these pathways are shared by these organisms [31].

LATER STEPS OF SLAFRAMINE BIOGENESIS

In the conversion of pipecolic acid to slaframine, the question arises as to whether functionalization of the C-5 position of pipecolic acid or condensation to form the octahydroindolizine nucleus occurs first. Because of difficulty in preparing labeled 5-aminopipecolic acid, experiments were done with the hydroxy analog, which would be expected to be on the pathway to the amino derivative. Label from neither 5-hydroxylysine [25] nor either isomer of L-5-hydroxypipecolic acid [27] was incorporated into slaframine; moreover, the hydroxypipecolate labels were not incorporated into 3,4,5-trihydroxy-octahydro-1-pyrindine [15].

Attention was turned to derivatives of the octahydroindolizine nucleus as possible precursors of slaframine. 1-Ketooctahydroindolizine and both the *cis*- and *trans*-isomers of 1-hydroxyoctahydroindolizine were found to dilute the level of labeled pipecolic acid incorporated into slaframine; moreover, radioactive labels from all three compounds were incorporated into the alkaloid (Table 3) [25–27]. However, the incorporation of *trans*-1-hydroxy-octahydroindolizine is not consistent with the *cis* configuration of slaframine about carbons 1 and 8a [12,13]. It was suspected that the *trans*-alcohol was being oxidized to the ketone. The reduction of the ketone in cell-free extracts was found to yield almost exclusively the *cis*-alcohol in the presence of reduced pyridine nucleotides; the reductase was found to prefer NADPH to NADH, and its activity was dependent on the addition of reduced thiols to the enzyme preparations [26,27].

The conversions of 1,6-dihydroxyoctahydroindolizine (Table 3) [26] and

TABLE 3

Incorporation of Substituted Octahydroindolizines into Slaframine[a]

Added tracer	^{3}H incorporated into slaframine (%)	Specific activity of slaframine/specific activity of added tracer
1-[Piperidine ring-^{3}H]ketooctahydro-indolizine, 0.22 mM	2.29	0.25
1-[1-^{3}H]Hydroxyoctahydroindolizine, 0.22 mM	0.73	0.20
1-[6,7-^{3}H]Hydroxyoctahydroindolizine		
cis, 0.20 mM	0.83	0.13
trans, 0.20 mM	0.04	0.08
1,6-[^{3}H]Dihydroxyoctahydroindolizine, 2.75 μM	13.85	0.015

[a] Reproduced with permission from H. P. Broquist, *Biochemistry* **12**, 4624 (1973). Copyright by the American Chemical Society.

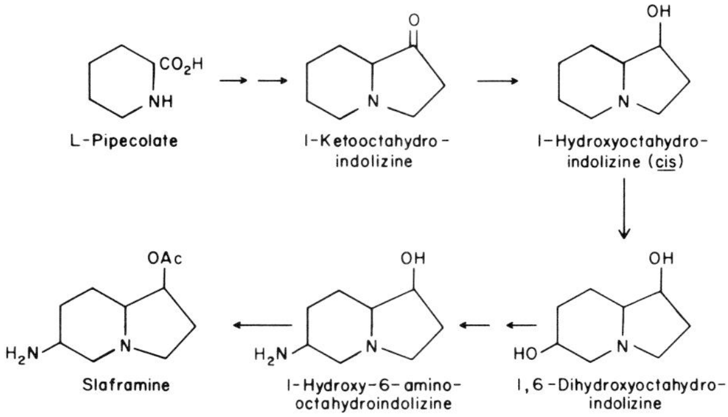

Fig. 4 A postulated scheme for slaframine biogenesis in *Rhizoctonia leguminicola*. [Reproduced with permission from H. P. Broquist, *Biochemistry* **12**, 4624 (1973). Copyright by the American Chemical Society.]

1-hydroxy-6-aminooctahydroincolizine (deacetylslaframine) [26] to slaframine were also demonstrated to occur in *R. leguminicola* cultures. Attempts to trap labeled intermediates from labeled pipecolic acid were not particularly successful; however, the accumulation of deacetylslaframine was demonstrated. Cell-free extracts of *R. leguminicola* converted deacetylslaframine to slaframine in the presence of acetyl-CoA [26].

These results were used to postulate the pathway depicted in Fig. 4 for the biosynthesis of slaframine [26]. Studies of the origin of carbons 2 and 3 of slaframine are in progress in this laboratory. Radioactivity from serine, a possible source of the two-carbon fragment, was not incorporated into slaframine [25,27] (or into 3,4,5-trihydroxyoctahydro-1-pyrindine) [15]. Recent experiments indicate that a form of malonic acid may possibly be involved, since either [1-^{14}C]- or [2-^{14}C]malonate is more efficient than acetate in labeling both alkaloids in resting cultures of *R. leguminicola* [32]. The specific activity of the slaframine isolated after [2-^{14}C]malonic acid incubation was twice that isolated from a [1-^{14}C]malonic acid incubation, consistent with the loss of one carboxyl group; moreover, pipecolic acid-dependent malonic acid decarboxylation in the presence of coenzyme A and ATP has been demonstrated *in vitro*. The labels are incorporated into the octahydroindolizine nucleus of slaframine, as no label was found in the acetoxy moiety [32].

BIOACTIVATION OF SLAFRAMINE

A time lag in the onset of the physiological response to administered slaframine first suggested that activation of this alkaloid is required for its parasympathomimetic activity [1,8,33]. Surgical experiments demonstrated that the liver is the principal site of bioactivation [1,33]. After such bioactivation, slaframine appears to bind to and directly stimulate acetylcholine receptors [8,33,34]. This form of slaframine (and slaframine itself) are not substrates for acetylcholinesterase [1,8]; thus, a prolonged stimulation of nerves is observed both *in vivo* and *in vitro* in isolated organ preparations [8,33,34] (slaframine does not inhibit the esterase [8]). Target organs of activated slaframine are the exocrine glands, principally the pancreas and salivary glands [8,35]. Activated slaframine apparently binds to muscarinic acetylcholine receptors, as atropine will compete for binding of the activated form both *in vivo* and *in vitro* [34].

The resemblance of slaframine to acetylcholine should be noted, i.e., an alkylated amine and an acetoxy moiety separated by two carbons. The acetoxy group is absolutely essential for activity. Neither *N*-acetylslaframine, 1-acetoxyoctahydroindolizine, 1-acetoxy-8-aminooctahydroindolizine, 3,4,5-trihydroxyoctahydro-1-pyrindine, nor several other compounds exhibit *in vivo* or *in vitro* activity, even after attempted bioactivation (*vide infra*).

Liver microsomes activate slaframine as ·judged by bioassay utilizing contraction of the isolated guinea pig ileum [34]. Activation can also be achieved with a model system using flavins in the presence of light; under anaerobic conditions, flavin is reduced concomitantly with the production of the active metabolite [34]. Since the flavin is reduced, it was suspected that slaframine is oxidized to form the active metabolite. However, electron paramagnetic resonance measurements showed no evidence for the production of radicals in the reaction, arguing against the role of either a free radical of slaframine or an *N*-oxide as an active species [36]. No detectable products could be extracted with organic solvents. However, small amounts of sodium borohydride destroy the activity of the activated metabolite; after such reduction, three compounds can be extracted into chloroform. These have been identified as 1-acetoxyoctahydroindolizine, slaframine, and 1-acetoxy-6-hydroxyoctahydrolizine [27,37,37a]. Label from [³H]sodium borohydride was incorporated into all three compounds; after sodium borodeuteride reduction, the first compound contained one deuterium and the latter compound contained two deuteriums (because of excess unreacted slaframine, the excess deuterium in this fraction could not be calculated) [27]. Two reactions are postulated to occur (Fig. 5), one involving a four-electron oxidation and subsequent deamination, and another involving nonoxidative deamination and rearrangement [27,37,37a]. Ammonia was found to be released in stoichiometric yields [27].

The photochemical flavin–slaframine reaction was chromatographed on columns of Dowex-50 and BioRex-70 to separate the metabolites. Apparently only the ketoimine (**VII**, Fig. 5) resulting from the oxidation of slaframine is capable of stimulating the isolated guinea pig ileum [27,37,37a,38]. The

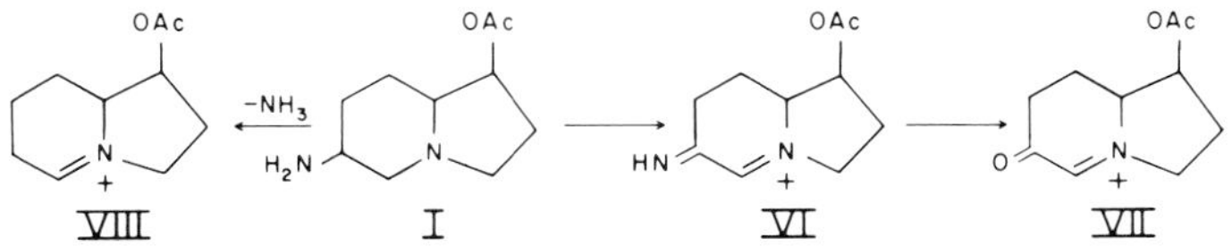

Fig. 5 Postulated steps in the bioactivation of slaframine.

compound was also found to produce salivation when separated and injected intravenously into male rats [27].

The hepatic enzymes responsible for slaframine bioactivation are localized in the endoplasmic reticulum [34]. Although *in vivo* experiments showed that the time lag observed for slaframine activation was reduced by treatment with phenobarbital and increased by treatment with 2-ethylaminoethyl-2,2-diphenylvalerate (SKF-525A) [33], *in vitro* results to date suggest that mixed-function oxidation is not responsible for the bioactivation of slaframine. Microsomal activation is unaffected by carbon monoxide or anaerobiosis, as judged by bioassay of slaframine activation using the guinea pig ileum [34]. A new bioactivation assay has been developed using the incorporation of label from [³H]sodium borohydride into compounds **VI** and **VII** (Fig. 5) [27,37, 37a], permitting better quantitation. Contrary to previous work [34], slaframine activation was found not to be NADPH dependent. Moreover, the reaction is unaffected by the presence of carbon monoxide, cyanide, SKF-5254A NADH, octylamine, or *N,N*-dimethylanaline. Pretreatment of male rats with either phenobarbital or 3-methylcholanthrene was without effect on the reaction. Recently, the microsomal activation of slaframine has been attributed to a flavoprotein oxidase [37a].

CONCLUSION

The foregoing work illustrates, among other things, the futility of drawing hard and fast lines between pure and applied research. This project had its origin in a stable, where a farmer observed that cattle fed certain leguminous fodder slobbered profusely and refused further feed. The finding that such fodder was infected with *Rhozoctonia leguminicola* led to the realization that the "slobber hay" problem was a manifestation of mycotoxin action. Subsequent biosynthetic studies led to the discovery of a role of lysine in the fashioning of such unique nitrogen heterocycles as the octahydroindolizine and pyrindine ring systems. These piperidine alkaloids may prove to be of interest pharmacologically. The opportunity to utilize a fungus, rather than the traditional higher plant, for studies of alkaloid biosynthesis offered important technical advantages. The complete elucidation of the biosynthesis of these piperidine alkaloids should add to existing knowledge of secondary pathways of amino acid metabolism.

ACKNOWLEDGMENTS

This work was supported in part by USPHS NIH Grants AM 3156, AM 05441, AM 14338, and ES 00267.

REFERENCES

1. H. P. Broquist and J. J. Snyder, *Microb. Toxins* **7**, 319 (1971).
2. B. L. O'Dell, W. O. Regan, and T. J. Beach, *Mo.*, *Agric. Exp. Stn.*, *Res. Bull.* **702**, 12 (1959).
3. F. J. Gough and E. S. Elliot, *W. Va.*, *Agric. Exp. Stn.*, *Bull.* **387T** (1947).
4. J. H. Byers and H. P. Broquist, *J. Dairy Sci.* **43**, 873 (1960).
5. J. H. Byers and H. P. Broquist, *J. Dairy Sci.* **44**, 1179 (1961).
6. S. D. Aust and H. P. Broquist, *Nature (London)* **205**, 204 (1965).
7. D. P. Rainey, E. B. Smalley, M. H. Crump, and F. M. Strong, *Nature (London)* **205**, 203 (1965).
8. S. D. Aust, Doctoral Thesis, University of Illinois, Urbana (1965).
9. S. D. Aust, H. P. Broquist, and K. L. Rinehart, Jr., *Biotechnol. Bioeng.* **10**, 403 (1968).
10. S. D. Aust, H. P. Broquist, and K. L. Rinehart, Jr., *J. Am. Chem. Soc.* **88**, 2879 (1966).
11. B. J. Whitlock, D. P. Rainey, N. V. Riggs, and F. M. Strong, *Tetrahedron Lett.* p. 3819 (1966).
12. R. A. Gardiner, K. L. Rinehart, Jr., J. J. Snyder, and H. P. Broquist, *J. Am. Chem. Soc.* **90**, 5638 (1968).
13. D. Cartwright, R. A. Gardiner, and K. L. Rinehart, Jr., *J. Am. Chem. Soc.* **92**, 7615 (1970).
14. F. P. Guengerich, S. J. DiMari, and H. P. Broquist, *Abstr.*, *164th Natl. Meet.*, *Am. Chem. Soc.*, *New York*, *1972* No. Biol. 226 (1972).
15. F. P. Guengerich, S. J. DiMari, and H. P. Broquist, *J. Am. Chem. Soc.* **95**, 2055 (1973).
16. M. Onda, Y. Konda, Y. Narimatsu, S. Omura, and T. Hata, *Chem. Pharm. Bull.* **21**, 2048 (1973).
17. S. Omura, H. Tanaka, J. Awaya, Y. Narimatsu, Y. Konda, and T. Hata, *Agric. Biol. Chem.* **38**, 899 (1974).
18. H. E. Miller, H. Rösler, A. Wohlpart, H. Wyler, M. E. Wilcox, H. Frohofer, T. J. Mabry, and A. S. Dreiding, *Helv. Chim. Acta* **51**, 1470 (1968).
19. E. Leete, *Acc. Chem. Res.* **4**, 100 (1971).
20. E. Leete and K. N. Juneau, *J. Am. Chem. Soc.* **91**, 5614 (1969).
21. C. W. L. Bevan and A. U. Ogan, *Phytochemistry* **3**, 591 (1964).
22. T. Terashima, Y. Kuroda, and Y. Kanelo, *Tetrahedron Lett.* p. 2535 (1969).
23. H. Auda, H. R. Juneja, E. J. Eisenbraun, G. R. Waller, W. R. Kays, and H. H. Appel, *J. Am. Chem. Soc.* **89**, 2476 (1967).

24. H. Auda, G. R. Waller, and E. J. Eisenbraun, *J. Biol. Chem.* **242**, 4157 (1967).
25. J. J. Snyder, Doctoral Thesis, University of Illinois, Urbana (1969).
26. F. P. Guengerich, J. J. Snyder, and H. P. Broquist, *Biochemistry* **12**, 4264 (1973).
27. F. P. Guengerich, Doctoral Thesis, Vanderbilt University, Nashville, Tennessee (1973).
28. H. P. Broquist and J. S. Trupin, *Annu. Rev. Biochem.* **35**, 231 (1966).
29. F. P. Guengerich and H. P. Broquist, *Biochemistry* **12**, 4270 (1973).
30. D. L. Miller and V. W. Rodwell, *J. Biol. Chem.* **246**, 2758 (1971).
31. F. P. Guengerich and H. P. Broquist, *J. Bacteriol.* **125**, 338 (1976).
32. E. C. Clevenstine and H. P. Broquist, *Fed. Proc., Fed. Am. Soc. Exp. Biol.* **35**, 1478 (1976).
33. S. D. Aust, *Biochem. Pharmacol.* **18**, 929 (1969).
34. T. E. Spike and S. D. Aust, *Biochem. Pharmacol.* **20**, 721 (1971).
35. S. D. Aust, *Biochem. Pharmacol.* **19**, 427 (1970).
36. T. E. Spike, Master's Thesis, Michigan State University, East Lansing (1969).
37. F. P. Guengerich, S. D. Aust, and K. L. Rinehart, Jr., *Abstr., 166th Natl. Meet., Am. Chem. Soc., Chicago, 1973* No. Biol. 221 (1973).
37a. F. P. Guengerich and S. D. Aust, *Mol. Pharmacol.* (in press).
38. J. A. Buege, F. P. Guengerich, and S. D. Aust, *Pharmacologist* **15**, 191 (1973).

CHAPTER

5

Enzymatic Stereospecificity at Prochiral Centers of Amino Acids

Richard K. Hill

INTRODUCTION

Biological reactions are stereospecific; since the time of Pasteur chemists have recognized that, when living organisms deal with chiral molecules, they generally synthesize or metabolize a single stereoisomer. Consequently, we are accustomed to finding that most natural products are optically active as a result of having been constructed by the catalytic action of three-dimensional enzymes, which distinguish chiral sites.

In 1948 Ogston added a new dimension to this subject with his recognition [1] that enzyme stereospecificity can extend to certain molecules normally thought of as symmetrical, specifically that in compounds of the type Cxxyz an enzyme can discriminate between the like groups x. This discrimination is frequently explained in terms of Ogston's hypothesis of a three-point attachment of the substrate to the enzyme, but the principle is fundamentally that the two transition states for attack of any chiral reagent on x and x′ are diastereomers and that consequently the rates of reaction proceeding through them will differ [2].

Ogston's hypothesis was quickly verified in the citric acid cycle, in which the two carboxymethyl groups of citric acid are differentiated [3] and was soon followed by demonstrations of enzymatic discrimination between like groups in ethanol [4] and glycerol [5]. Examples have multiplied in recent years, and enzyme stereospecificity is the theme of two recent monographs

[6,7], several reviews [8–12a], and the 1975 Nobel Prize in Chemistry awarded to J. W. Cornforth and V. Prelog [13].

The significance of these stereochemical studies lies not only in revealing the subtle yet powerful specificity with which enzymes carry out their tasks, but more importantly in providing clues to the mechanisms by which they operate. At one level, comparing the stereochemical course of enzymatic eliminations, addition to double bonds, displacements, and rearrangements with well-established *in vitro* analogs allows plausible mechanisms to be suggested, as in the carbon–carbon bond-forming steps of terpene synthesis from mevalonate [13], or perhaps calls attention to the exceptional, such as the vitamin B_{12}-catalyzed rearrangements of propanediol [14] and methyl-malonyl-CoA [15], for which unprecedented mechanisms are necessary. At a deeper level, as Hanson and Rose have pointed out [16], comparative results among families and classes of enzymes, revealing those that display and those that lack stereochemical consistency, furnish valuable clues to enzyme evolution and the development of basic principles which guide their catalytic action.

This survey is confined to recent examples of enzymatic discrimination at prochiral centers in the biosynthesis and transformations of amino acids. Most of the common amino acids contain prochiral centers, and enzymatic attack at these centers is common.

The nomenclature of Mislow and Raban [17] is used to describe the relationship of like groups; thus, the methylene hydrogens of glycine (1) and the hydroxymethyl groups of hydroxymethylserine (2) are *enantiotopic* (mirror-image relationship), while the methylene hydrogens of tyrosine (3) and the methyl groups of valine (4) are *diastereotopic* (nonequivalent, non-mirror-image relationship). The reader will recall that in principle both kinds of groups can be distinguished by chiral reagents such as enzymes, while diastereotopic groups can be differentiated in addition by achiral reagents and nuclear magnetic resonance (nmr) spectrometers. The terminology of Hanson [18] is employed for specification of configuration at prochiral centers and faces; H_R in (1) and (3) refers to the pro-*R* hydrogen and H_S to the pro-*S* hydrogen. A characteristic of a prochiral carbon is that if one of the like groups is replaced by its isotopic counterpart, such as hydrogen by deuterium or tritium, or $^{12}CH_3$ by $^{13}CH_3$, the carbon becomes asymmetric and the R/S configurational notation can be applied. The rapid advance in our understanding of enzyme specificity at prochiral centers has come about through the widespread use of specific isotopic labeling, requiring imaginative solutions to the problem of labeling in a chiral rather than stereorandom pattern. This review focuses on methods used to prepare chirally labeled amino acids and their use in elucidating enzyme stereospecificity and suggests additional possible applications.

$$
\begin{array}{cccc}
\underset{H_2N}{\overset{H_S}{\diagup}}\underset{\overset{}{COOH}}{\overset{H_R}{\diagdown}} & \underset{H_2N}{\overset{HOCH_2}{\diagup}}\underset{\overset{}{COOH}}{\overset{CH_2OH}{\diagdown}} & & \underset{H_3C}{\overset{H}{\diagup}}\underset{\overset{}{COOH}}{\overset{NH_2}{\diagdown}}
\end{array}
$$

(1)   (2)   (3)   (4)

GLYCINE

Chiral Labeling

Two related enzymatic methods and a chemical route have been used to prepare samples of glycine in which one of the enantiotopic methylene hydrogens has been replaced by an isotope. Both of the biochemical methods involve enzymes dependent on pyridoxal phosphate and so undoubtedly take place by the mechanism shown in Scheme 1. Reaction of glycine with the pyridoxal phosphate–lysine complex forms the Schiff base (5), in which the methylene protons are much more acidic than in glycine itself. Loss of a proton affords the resonance-stabilized anion (6); anions of this type are intermediates in most of the pyridoxal-dependent reactions of amino acids such as racemization, β-elimination, hydroxymethyl transfer, and decarboxylation [19]. If no other substrate is present, anion (6) is simply protonated, presumably stereospecifically from one face or the other of the planar π system; if the protonation is stereospecific and carried out in isotopically labeled water, then the glycine liberated by hydrolysis of the Schiff base will be chirally labeled.

An early clue came from the finding of Schirch and Jenkins [20] that, in the presence of serine transhydroxymethylase, pyridoxal phosphate, and tetrahydrofolate in ^3H_2O, both D-alanine and glycine exchanged one α-hydrogen for tritium. This enzyme normally catalyzes the interconversion of glycine and formaldehyde with serine by way of intermediate (6), but if no formaldehyde is present (6) is enzymatically protonated. This reaction was investigated more thoroughly by Wellner [20a] and by Akhtar and Jordan [21], who not only repeated the preparation of [^3H]glycine by this method, but also prepared an enantiomeric [^3H]glycine by exchanging [2-$^3H_2$]glycine with the enzyme in ordinary water. Proof that these samples were chiral and enantiomeric, and assignment of their configurations, came from a study of their reaction with D-amino-acid oxidase. This enzyme catalyzes the conversion of D-alanine to pyruvate but also accepts glycine as a substrate. This oxidase converted the labeled glycines to glyoxylate with 85% loss of tritium from one isomer but only 17% loss of tritium from the enantiomer, indicating that

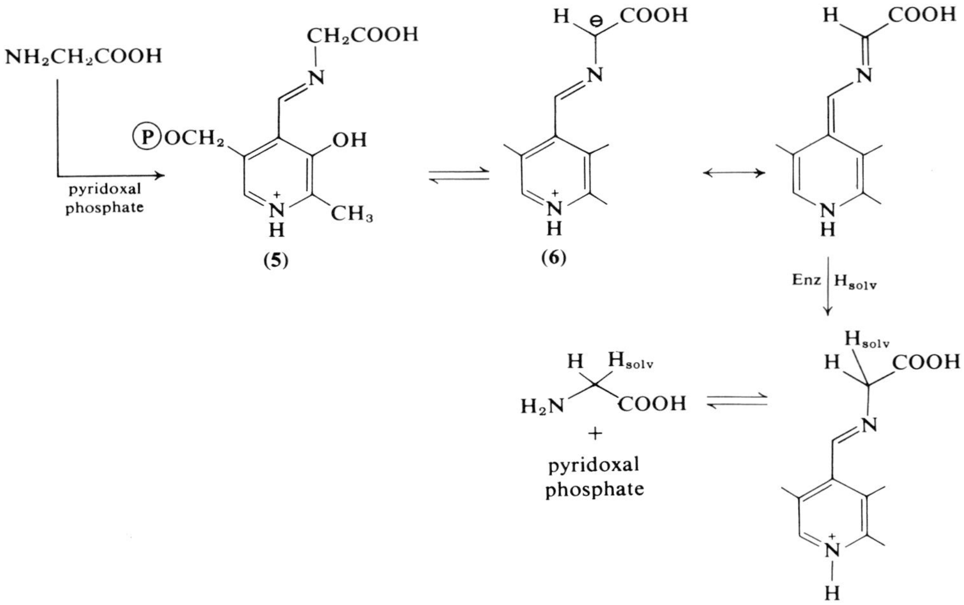

Scheme 1 Chiral labeling of glycine.

tritium had entered the samples with an optical purity of about 85%. On the reasonable assumption that both serine transhydroxymethylase and D-amino-acid oxidase react with glycine in the same stereochemical sense as they do with D-alanine, configurations can be assigned to the tritiated glycines as shown in Scheme 2. These results show that, when anion (6) is protonated by

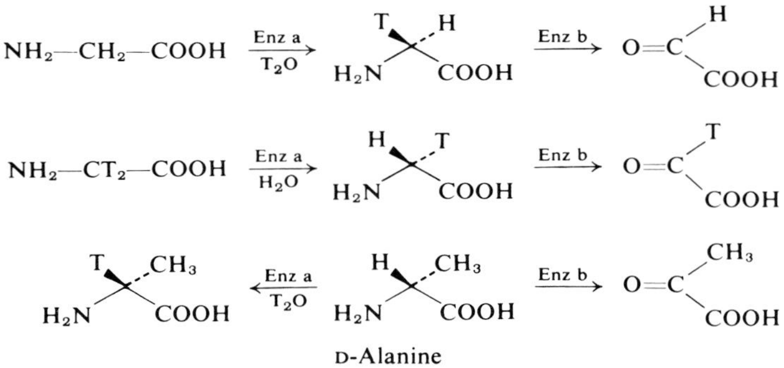

D-Alanine

Scheme 2 Preparation of chirally tritiated glycines. Enz a, serine transhydroxymethylase; Enz b, D-amino-acid oxidase.

serine transhydroxymethylase, the proton added occupies the pro-S position of glycine.

Besmer and Arigoni independently developed a similar method using L-alanine aminotransferase, a transaminase enzyme [22]. In the presence of D_2O or T_2O and pyridoxal phosphate, one of the methylene hydrogens of glycine was stereospecifically replaced by the isotope. The [^{2}H]glycine was optically active, giving a plain positive optical rotatory dispersion (ORD) curve. Nitrous acid deamination, known to occur with retention of configuration, gave (R)-$(+)$-[^{2}H]glycolic acid, proving that the [^{2}H]glycine had the (R) configuration. Confirmation came from a chemical synthesis of [^{2}H]-glycine. (S)-[α-^{2}H]Benzylamine [23], after protecting the amino group, was degraded by ozonolysis to (S)-$(-)$-[^{2}H]glycine. A final proof of configuration was provided by enzymatic oxidation of the tritiated glycolic acid with glycolic acid oxidase, which is known to remove the pro-R hydrogen; the isotope was lost during this oxidation (Scheme 3). These results demonstrate that the transaminase labilizes the pro-R hydrogen in glycine, corresponding to the hydrogen removed in L-alanine. Wellner [20a] independently obtained (R)-[^{3}H]glycine by transamination of glyoxylate with L-aspartate in T_2O and proved the configuration of the product by oxidation with D-amino-acid oxidase. The complementary results of these two enzymatic methods, one exchanging H_S and one exchanging H_R, makes this approach particularly useful for preparing either enantiomer.

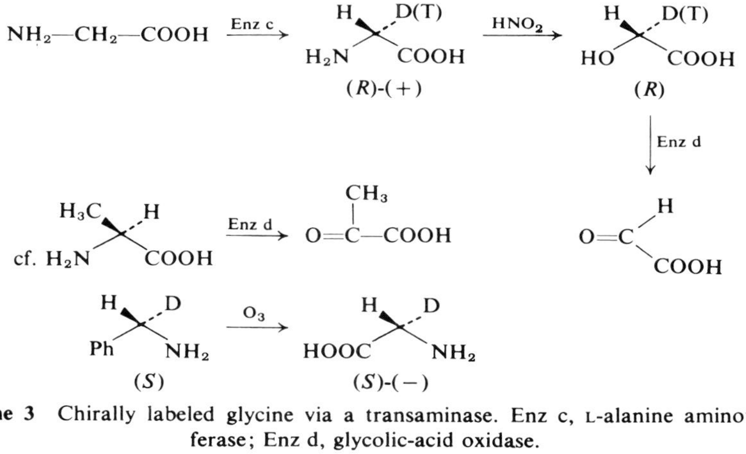

Scheme 3 Chirally labeled glycine via a transaminase. Enz c, L-alanine aminotrans-ferase; Enz d, glycolic-acid oxidase.

Golding *et al.* [23a] have devised an ingenious approach to the synthesis of chirally labeled glycine using stereoselective exchange of glycine coordinated to an optically active Co^{III} complex. A similar method has been used by

McClarin *et al.* [23b] for stereospecific exchange at C-3 in chelated aspartic acid.

Stereochemistry of the Serine Hydroxymethyltransferase Reaction

The stereochemistry of the reversible process interconverting glycine and serine has been studied from both directions. Jordan and Akhtar [21] used their tritiated glycines to determine which of the enantiotopic protons is replaced by formaldehyde. (R)-[^{3}H]Glycine was converted to serine with 13.5% loss of tritium, while the (S)-somer lost 87% of its isotope. Consequently, it is the pro-S hydrogen that is lost in forming anion (6), just as it is in the exchange reaction with this enzyme, and the overall replacement of hydrogen by —CH$_2$OH takes place with retention of configuration. A similar result was found for threonine synthesis from glycine, in which acetaldehyde takes the place of formaldehyde [21] and for the cleavage of L-allothreonine into acetaldehyde and glycine by L-allothreonine aldolase [23c].

Both Besmer and Arigoni [22] and Wellner [20a] carried out the reverse reaction in T$_2$O, isolating the [^{3}H]glycine formed from L-serine. It was found to be more than 90% optically pure and of the (S) configuration, confirming that replacement of —CH$_2$OH by H occurs with retention and that anion (6) is attacked by both H$^+$ and CH$_2$O at the same face.

These results not only reveal the stereospecificity of the glycine–serine interconversion, but also exclude some plausible mechanisms. One, for example, involves dehydration of intermediate (7) to (8), to which tetrahydrofolate then adds [21,22]. This mechanism, however, requires that both methylene hydrogens of glycine be lost during conversion to serine and is incompatible with the results.

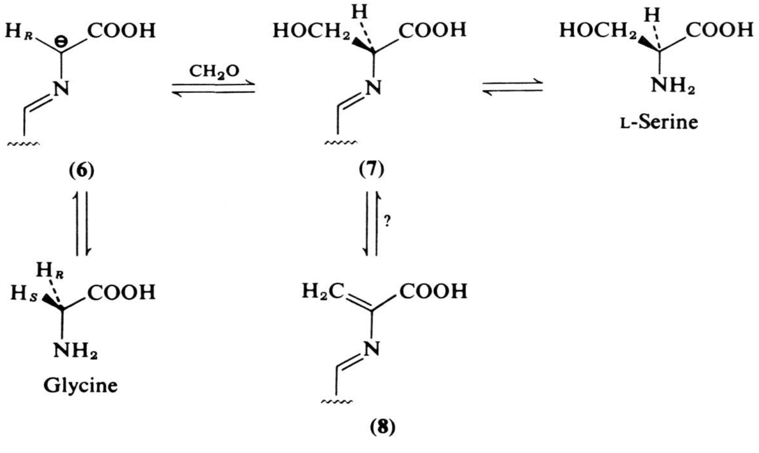

Wilson and Snell [24] isolated from soil bacteria an inducible enzyme, α-methylserine hydroxymethyltransferase, which acted on serine analogs containing no α-hydrogens; two examples are shown below:

(i)

$$\underset{\substack{\\ (S)\text{-}(+)\text{-}\alpha\text{-Methylserine}}}{\underset{\text{H}_2\text{N}\qquad\text{COOH}}{\overset{\text{HOCH}_2\qquad\text{CH}_3}{\diagup\!\!\!\diagdown}}} \rightleftharpoons \underset{\substack{\\ \text{D-Alanine}}}{\underset{\text{H}_2\text{N}\qquad\text{COOH}}{\overset{\text{H}\qquad\text{CH}_3}{\diagup\!\!\!\diagdown}}} + \text{CH}_2\text{O}$$

(ii)

$$\underset{\substack{\\ \alpha\text{-Hydroxymethylserine}}}{\underset{\text{H}_2\text{N}\qquad\text{COOH}}{\overset{\text{HO}^{14}\text{CH}_2\qquad\text{CH}_2\text{OH}}{\diagup\!\!\!\diagdown}}} \rightleftharpoons \underset{\substack{\\ \text{D-Serine}}}{\underset{\text{H}_2\text{N}\qquad\text{COOH}}{\overset{\text{H}\qquad\text{CH}_2\text{OH}}{\diagup\!\!\!\diagdown}}} + {}^{14}\text{CH}_2\text{O}$$

From the (S) configuration of $(+)$-α-methylserine [25] it can be seen that the stereochemical course of reaction (i) parallels exactly that of the serine–glycine exchange. Reaction (ii) involves an achiral substrate but was shown to be stereospecific as well. When α-hydroxymethylserine was prepared enzymatically from D-serine and $^{14}\text{CH}_2\text{O}$ and then resubjected to the transferase reaction, only the labeled —CH$_2$OH group was lost. Determination of the absolute configuration of the labeled hydroxymethylserine will reveal whether this reaction follows the same stereochemical pathway [26].

Stereochemistry of δ-Aminolevulinic Acid Biosynthesis

An early step in porphyrin biosynthesis is the condensation of glycine with succinyl-CoA to yield δ-aminolevulinic acid (ALA). The synthetase that accomplishes this requires pyridoxal phosphate, so presumably carbanion (6) is again involved, this time in a Claisen-like condensation (Scheme 4). Although the methylene group of glycine is still present as C-5 in ALA, its configuration has been affected as a consequence of three steps: (1) removal of a proton to form (6), (2) attachment of the succinyl moiety to form (9), and (3) decarboxylation of the glycine carboxyl to replace —COOH by a proton from the solvent. Since intermediate (9) has not been isolated, it has not been possible to elucidate the stereochemistry of each of these steps, but the overall configurational change has been investigated by Akhtar [27], proving that each step is stereospecific.

First, NH$_2$CT$_2$COOH was incorporated into ALA, which was then reduced with borohydride and cleaved with periodate to liberate CH$_2$O from C-5. The formaldehyde had lost half of the original radioactivity, consistent with the mechanism in Scheme 4. (R)-[^{3}H]Glycine lost 97% of its tritium in the same experiment, while the (S)-enantiomer retained its tritium. Consequently,

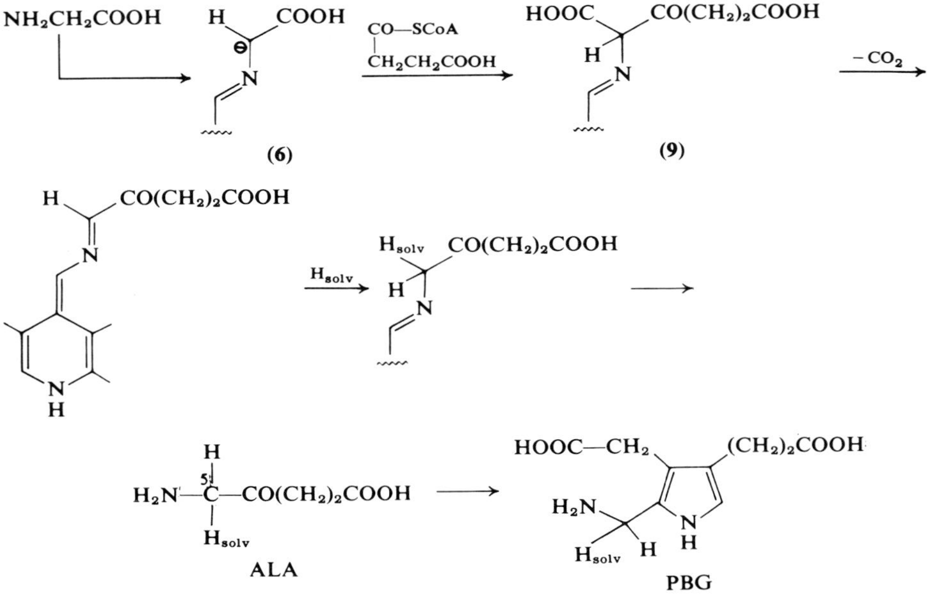

Scheme 4 δ-Aminolevulinic acid biosynthesis.

the ALA synthetase is totally stereospecific in removing only the pro-*R* hydrogen from glycine.

In order to determine the configuration of the labeled ALA formed from (*S*)-[³H]glycine, it was carried one step further in the biosynthetic pathway to porphobilinogen (PBG). After protecting the amino group the pyrrole ring was destroyed by permanganate oxidation, and the [³H]glycine formed was found on assay with serine transhydroxymethylase to be at least 87% (*S*).

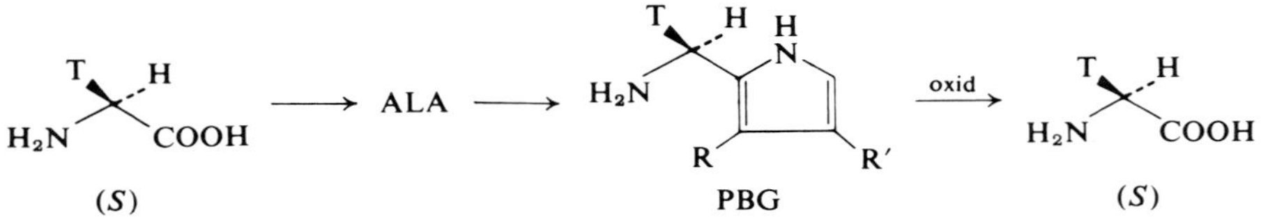

Thus, the pro-*S* hydrogen is retained during this sequence and occupies the pro-*S* configuration in ALA. Since, however, the carboxyl of the glycine obtained by degradation is not the original glycine carboxyl, an inversion must have occurred; one of the bond-forming steps must take place with inversion and one with retention:

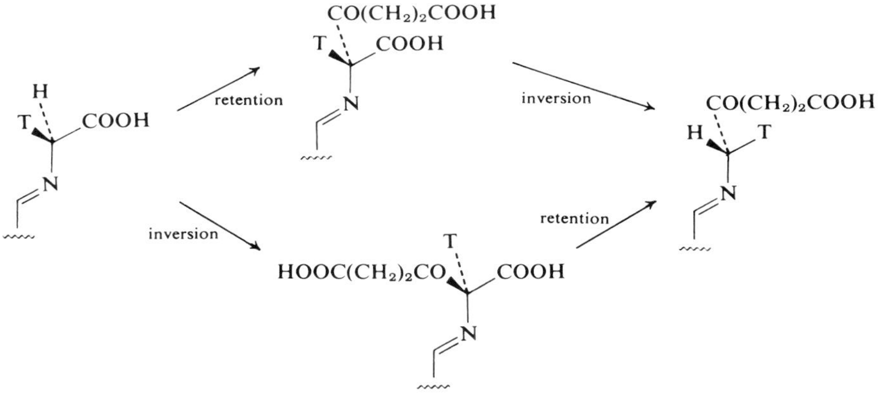

In either case, the hydrogen added in the last stage occupies the same configuration as the hydrogen initially removed from glycine, suggesting that the same group on the enzyme may be involved in both processes.

Stereochemistry of Transamination

Although it is not prochirality but rather the chirality of the α-carbon of amino acids that is established during transamination of α-keto acids, the simultaneous interchange of pyridoxal with pyridoxamine involves a prochiral center in the latter, and the stereospecificity at this center has been investigated with the aid of chirally labeled glycine. Dunathan and co-workers [28] have demonstrated that the protonation of intermediate (10) (Scheme 5) is stereospecific by preparing two enantiomeric forms of [α-²H]pyridoxamine. Using several different transaminases in D_2O, one of the methylene hydrogens was exchanged by deuterium, and conversely one of the deuterium atoms of [α-²H₂]pyridoxamine was exchanged by protium in ordinary water. A tentative assignment of configuration was made by chemical synthesis of one enantiomer: The L-threonine–pyridoxal–Mn complex (11) was reduced with NaBD₄ and the product oxidized with periodate. Assuming that the deuterium is delivered from the less hindered face, the [²H]pyridoxamine formed would have the (S) configuration.

This assignment was corroborated by Arigoni and Besmer [22,29], who converted pyridoxal to [α-³H]pyridoxamine by aspartate aminotransferase in ³H₂O. The pyridine ring of pyridoxamine was then removed by ozonolysis, leaving (S)-[³H]glycine. All the transaminases studied so far follow the same stereochemical pattern: Intermediate (10) is protonated from the *si* face in generating pyridoxamine, the same face from which protonation occurs in forming L-amino acids. For a more detailed discussion of the stereochemical

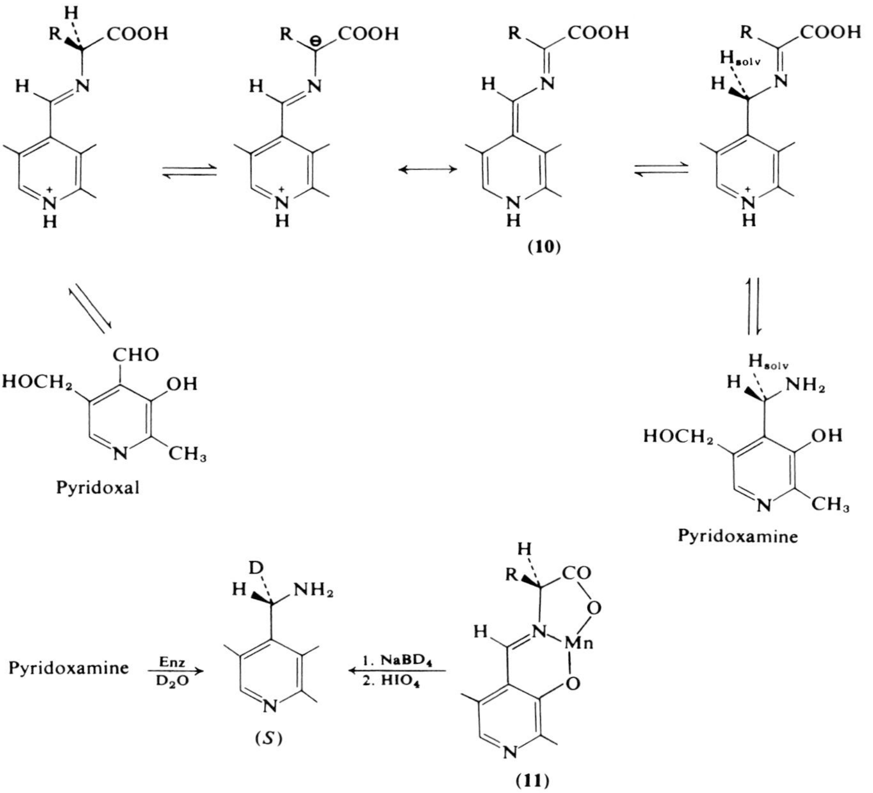

Scheme 5 Chirally labeled pyridoxamine from transamination.

aspects of transamination and the mechanistic inferences that can be drawn from them, the reader is referred to the papers of Dunathan [9,28].

Stereochemistry of Glycine Reductase

A phosphorylation reaction used in *Clostridium sticklandii* involves conversion of glycine to acetate by a glycine reductase:

$$NH_2CH_2COOH + ADP + P_i + 2H \xrightarrow[\text{reductase}]{\text{glycine}} H—CH_2COOH + NH_3 + ATP$$

Akhtar has shown [30] that the methylene group of glycine does not undergo exchange during this process, allowing the investigation of stereochemistry through the generation of "chiral acetate." For this purpose, glycine-d_2 was enzymatically labeled with tritium, and the resulting $NH_2CDTCOOH$ was converted to acetate by glycine reductase in water. The chiral acetate,

analyzed by established methods [31], proved to be (S). When the enantiomeric NH₂CDTCOOH was used as substrate, the acetate formed was (R).
These results show that the replacement of the glycine amino group by
hydrogen occurs with inversion of configuration.

$$NH_2-CD_2-COOH \xrightarrow[\text{T}_2\text{O}]{\text{Enz a}} \quad \begin{array}{c} \text{T} \quad \text{D} \\ \diagdown / \\ \diagup \diagdown \\ \text{H}_2\text{N} \quad \text{COOH} \end{array} \xrightarrow{\text{reductase}} \quad \begin{array}{c} \text{T} \quad \text{D} \\ \diagdown / \\ \diagup \diagdown \\ \text{HOOC} \quad \text{H} \end{array}$$

(S)

$$NH_2-CT_2-COOH \xrightarrow[\text{D}_2\text{O}]{\text{Enz a}} \quad \begin{array}{c} \text{D} \quad \text{T} \\ \diagdown / \\ \diagup \diagdown \\ \text{H}_2\text{N} \quad \text{COOH} \end{array} \xrightarrow{\text{reductase}} \quad \begin{array}{c} \text{D} \quad \text{T} \\ \diagdown / \\ \diagup \diagdown \\ \text{HOOC} \quad \text{H} \end{array}$$

(R)

Stereochemistry of Aminomalonate Decarboxylation

Although aminomalonate does not appear to be a normal biological
substrate, Palekar *et al.* [31a] found that it can be decarboxylated to glycine
by aspartate β-decarboxylase. When the enzymatic reaction is carried out in
³H₂O, the [³H]glycine formed has the (S) configuration, as shown by release
of the tritium by D-amino-acid oxidase. Palekar and his co-workers further
prepared enantiomeric chirally ¹⁴COOH-labeled aminomalonates by oxidation of the corresponding labeled L-serines, and showed that the decarboxylase
discriminated between the enantiotopic carboxyl groups, expelling CO₂ from
the pro-*R* carboxyl. The replacement of —COOH by H thus occurs with
retention of configuration. This display of specificity confirms in a satisfying
way the prediction of the specific example used by Ogston in his original
discussion [1] of enzymatic discrimination at prochiral centers.

$$\begin{array}{c} \text{COOH} \\ | \\ \text{H}_2\text{N}-\!\!\!-\!\text{H} \\ | \\ \text{COOH} \end{array} \xrightarrow[\text{T}_2\text{O}]{\text{aspartate } \beta\text{-decarboxylase}} \quad \begin{array}{c} \text{T} \quad \text{H} \\ \diagdown / \\ \diagup \diagdown \\ \text{H}_2\text{N} \quad \text{COOH} \end{array}$$

(S)

$$\begin{array}{c} \text{HOCH}_2 \quad \text{H} \\ \diagdown / \\ \diagup \diagdown \\ \text{H}_2\text{N} \quad \text{COOH} \end{array} \xrightarrow{\text{oxid}} \begin{array}{c} \text{HOOC} \quad \text{H} \\ \diagdown / \\ \diagup \diagdown \\ \text{H}_2\text{N} \quad \overset{\bullet}{\text{C}}\text{OOH} \end{array} \xrightarrow{\text{Enz}} \text{H}_2\text{NCH}_2\overset{\bullet}{\text{C}}\text{OOH} + \text{CO}_2$$

$$\begin{array}{c} \text{HO}\overset{\bullet}{\text{C}}\text{H}_2 \quad \text{H} \\ \diagdown / \\ \diagup \diagdown \\ \text{H}_2\text{N} \quad \text{COOH} \end{array} \xrightarrow{\text{oxid}} \begin{array}{c} \text{HOO}\overset{\bullet}{\text{C}} \quad \text{H} \\ \diagdown / \\ \diagup \diagdown \\ \text{H}_2\text{N} \quad \text{COOH} \end{array} \xrightarrow{\text{Enz}} \text{H}_2\text{NCH}_2\text{COOH} + \overset{\bullet}{\text{C}}\text{O}_2$$

When aminomalonate decarboxylation is effected by a rat liver decarboxylase, apparently identical with cytoplasmic serine hydroxymethylase, however,
the reaction is *non*stereospecific with respect to the carboxyls and to introduction of hydrogen [23c].

Other

The enzymatic incorporation of the methylene group of glycine into methylenetetrahydrofolate, known to involve pyridoxal and lipoic acid and to leave the methylene hydrogens intact [32], would seem to present a reasonable case for study; use of chirally labeled glycine might show whether the glycine is first broken down to formaldehyde or whether the methylene is incorporated while chirality is still present.

SERINE

Preparation of Chirally Labeled Serine

The first preparation of chirally labeled serine was that of Biellmann [33], who prepared it from glycine in rat liver slices using ^{3}HCOOH as the source of the extra carbon. In this system formate provides the carbon and one of the hydrogens of methylenetetrahydrofolate (13), the other coming from NADPH:

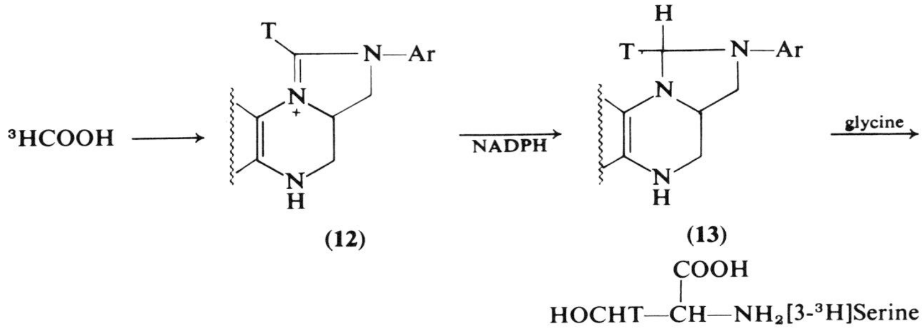

$[3\text{-}{}^3\text{H}]$Serine

The serine was chemically degraded to CH_3CHTOH and analyzed for enantiomeric purity by oxidation with yeast alcohol dehydrogenase [4]; the acetaldehyde formed retained about 70% of the radioactivity. These results showed that formate furnishes the pro-*S* hydrogen at C-3 of serine with about 40% optical purity. The important conclusion is not only that the NADPH reduction of (12) is stereospecific, leading to a chiral methylene carbon in (13), but that the methylene is transferred to glycine with appreciable retention of chirality [33a].

An enzymatic synthesis of tritiated serine proceeding with complete optical purity has been effected by Floss [34], starting with (2*S*,3*S*)- and (2*S*,3*R*)-3-phospho[3-^{3}H]glyceric acids (14) and (15), which are available from [1-^{3}H]glucose and [1-^{3}H]mannose, respectively, by the successive action of isomerases and glycolytic enzymes [35]. The labeled phosphoglyceric acids were then used for enzymatic conversion to phosphoserines in *Escherichia coli*, without affecting the asymmetric center at C-3, and hydrolyzed to serine (Scheme 6). Subsequent conversion to tryptophan (see below) showed these serine samples to be optically pure.

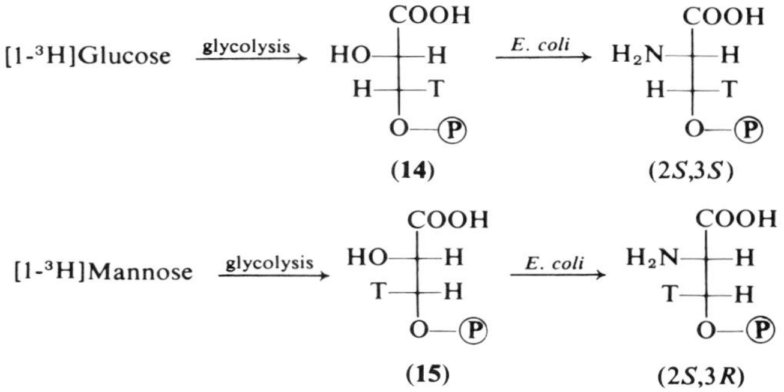

Scheme 6 Enzymatic synthesis of [3-³H]serine.

In addition to these enzymatic syntheses, three chemical routes to chirally labeled serines have been reported (Scheme 7):

(i) Fuganti *et al.* [36] converted (*S*)-2-[1-²H]phenylethanol to (2*RS*,3*S*)-[3-²H]serine in 15% yield by introduction of an amino group and oxidation of phenyl to carboxyl.

(ii) Without furnishing experimental details, Kainosho and Ajisaka [37] reported the synthesis of (2*S*,3*S*;2*R*,3*R*)-[3-²H]serine by hydrogenating the condensation product of ethyl hippurate and deuterated ethyl formate.

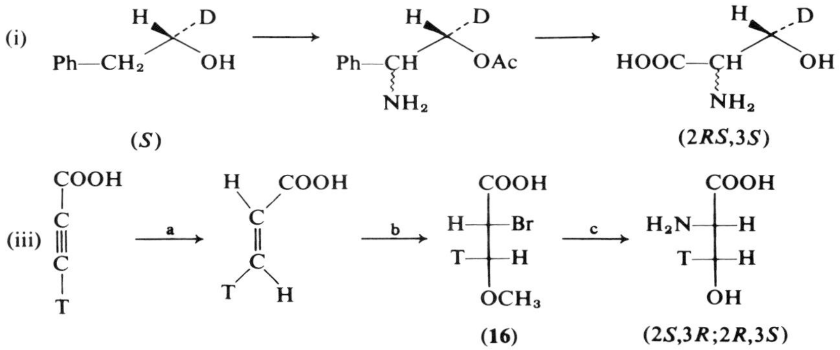

Scheme 7 Chemical syntheses of chirally labeled serine.
Reagents: (a) Cr²⁺ in H₂O; (b) Br₂, AgOAc, CH₃OH; (c) NH₃, then HBr.

(iii) Cheung and Walsh [38] used a stereospecific *trans* addition to tritiated acrylic acid to generate the two asymmetric centers in (16) in known relative configuration and then converted (16) to serine. The stereochemical purity at C-3 was shown to be 80% by oxidation to glycolic acid with D-amino-acid oxidase followed by peroxide, and then assay of the glycolate with glycolate oxidase (see the section on glycine, chiral labeling). This method can be used to prepare gram quantities of [³H]- or [²H]serine, and the serine can be easily resolved by hydrolysis of the *N*-acetate with hog kidney acylase.

Stereochemistry of β-Replacement Reactions of Serine

Among the most intriguing reactions that affect the prochiral center of serine are the enzymatic conversions to tyrosine and tryptophan, in which the hydroxyl is replaced by an indole or phenol ring:

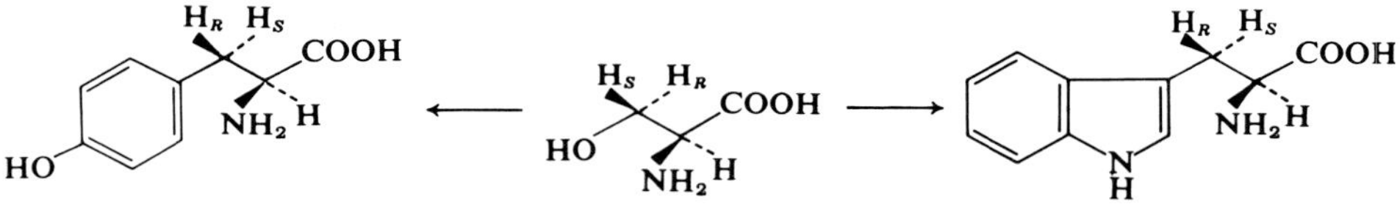

L-Tyrosine ← L-Serine → L-Tryptophan

These reactions formally represent nucleophilic substitution at a primary carbon, and consequently the stereochemistry can be studied only through the use of chirally labeled substrates.

The conversion to tyrosine has been studied by Fuganti *et al.* [39] using both (2*RS*,3*R*)- and (2*RS*,3*S*)-[³H]serine with phenol and tyrosine-phenol lyase; the configuration at C-3 of tyrosine was determined as outlined in the section on aromatic amino acids. The stereochemistry of the conversion of serine to tryptophan has been elucidated independently by Fuganti [36] and Floss [34,40] using both tryptophan synthetase from *Neurospora crassa* and tryptophanase from *E. coli*. Floss determined the product configuration at C-3 in two ways: (1) by incorporating the biosynthesized tryptophan into the antibiotic indolmycin, in which a methyl group is known to specifically replace the pro-*R* hydrogen [41], and (2) by oxidizing the tryptophan to [³H]aspartic acid and determining its configuration by enzymatic methods.

In all cases, the serine hydroxyl is replaced with complete. retention of configuration. This result, of course, rules out an S_N2 bimolecular displacement; the most likely mechanism is that shown in Scheme 8 [42]. Both enzymatic syntheses require pyridoxal phosphate, so the initial step is likely

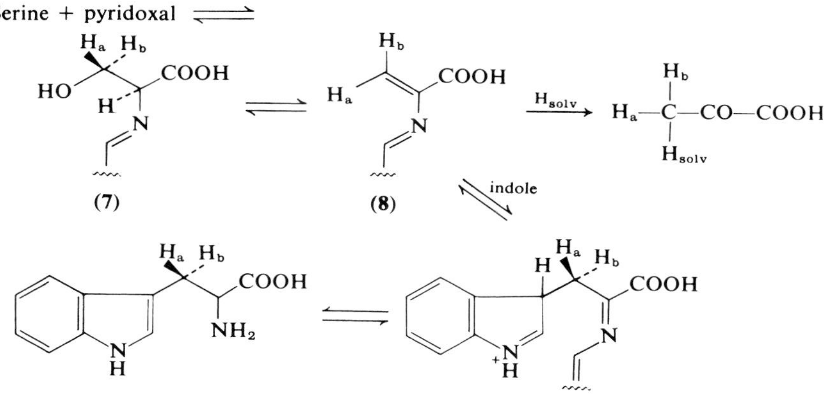

L-Serine + pyridoxal ⇌

Scheme 8 β-Replacement reactions of serine.

to be formation of the usual Schiff base (7). β-Elimination leads to an enzyme-bound aminoacrylate (8), in which the methylene protons remain diastereo-topic. Nucleophilic addition of indole or phenol to the extended Michael acceptor (8) generates tryptophan or tyrosine. The experimental results prove that both the β-elimination and Michael addition are stereospecific and that the incoming nucleophile adds to (8) from the same face (re) from which the hydroxyl departs. These results reinforce the concept [9] that reactions of pyridoxal enzymes take place on only one face of the amino acid–pyridoxal complex.

If these enzymes are allowed to act on serine in the absence of any nucleo-philic substrate other than water, intermediate (8) is hydrolyzed to pyruvate and ammonia. The possibility that protonation of the enamine double bond in (8) might also be enzyme controlled and stereospecific has been confirmed by generation of chiral pyruvate. Cheung and Walsh [38] used both (3S)- and (3R)-[³H]serine as substrates for serine dehydrase in D_2O. The doubly labeled pyruvate formed was analyzed by the method of Rose [43] and found to have nearly the calculated optical purity. The solvent proton attacks (8) from the same face as do nucleophiles, with overall retention of configuration. Floss et al. [40] have found the same result using either [³H]serine or [³H]-tryptophan as substrates with the enzyme tryptophanase in D_2O, although with the tryptophan synthetase β_2 protein the pyruvate is racemic, suggesting that protonation in this system is nonenzymatic.

Addition of sulfur nucleophiles to (8) also appears to be feasible, since serine is converted to cysteine by serine sulfhydrase and H_2S, and to S-methyl-cysteine by methylmercaptan in yeast extracts [44]. The stereochemistry of these reactions has not been investigated with chirally labeled serine, which would be a revealing experiment.

CYSTEINE AND CYSTINE

Chiral Labeling

Spurred by the interest in the mechanism of incorporation of cysteine into penicillin, two groups have devised syntheses of stereospecifically tritiated cysteine or cystine (Scheme 9):

(i) Morecombe and Young have used the approach of generating two adjacent asymmetric centers simultaneously by reduction of a double bond [45]. Catalytic hydrogenation of thiazoline (17) delivers hydrogens *cis* so that, after resolution of (18) and hydrolysis, the L-[3-³H]cysteine formed has the (2R,3S) configuration. If unlabeled pyruvic acid is used as the starting material and the thiazoline is reduced with tritium gas, the product is (2R,3R)-[2,3-³H₂]cysteine.

(ii) An alternative synthesis in which the relationship between adjacent

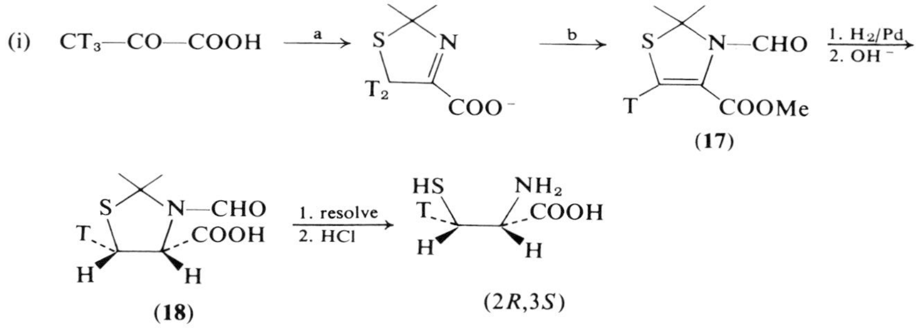

[Reagents: (a) Br₂; NH₄SH; Me₂CO, NH₃; (b) Me₂SO₄; H—CO—O—CO—CH₃]

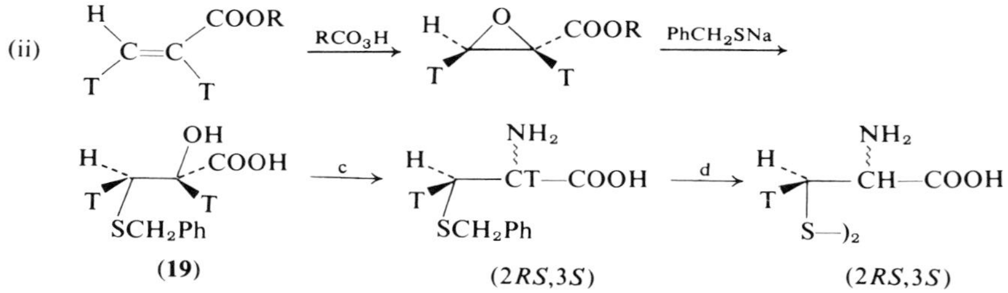

[Reagents: (c) resolution; PBr₃; NH₃; (d) Ac₂O, AcOH; HCl; Na/NH₃]
Scheme 9   Synthesis of chirally labeled cysteine and cystine.

asymmetric centers was achieved by epoxide opening was reported by Aberhart *et al.* [46]. (*E*)-*n*-Butyl [2,3-³H₂]acrylate, prepared following a published procedure [47], was epoxidized and the epoxide opened with benzylmercaptan; S$_N$2 displacement gave (2*R*,3*S*;2*S*,3*R*)-(**19**) [48]. Resolution gave the enantiomers of established configuration. Each was converted to *S*-benzylcysteine as shown; the racemization that accompanies the amination step is irrelevant. Heating with acetic anhydride in acetic acid removed the tritium at C-2, and after reductive removal of the benzyl protecting group the cysteine was allowed to oxidize to cystine.

Stereochemistry of Cystine and Cysteine Incorporation into Penicillin

The bicyclic skeleton of the penicillin antibiotics is known to be constructed enzymatically from cysteine and valine [49], although the exact mechanism of its formation is unknown; the Arnstein tripeptide, δ-(L-aminoadipyl)-L-cysteinyl-D-valine, may be involved [50]. The incorporation of L-cysteine without loss of the α-hydrogen [51] rules out an α,β-dehydrocysteine moiety in any intermediate. A hydrogen is lost from C-3, however, perhaps during oxidation to a thioaldehyde, in bonding C-3 to the valine nitrogen. The labeled cysteines described above have been used to probe the specificity of hydrogen loss.

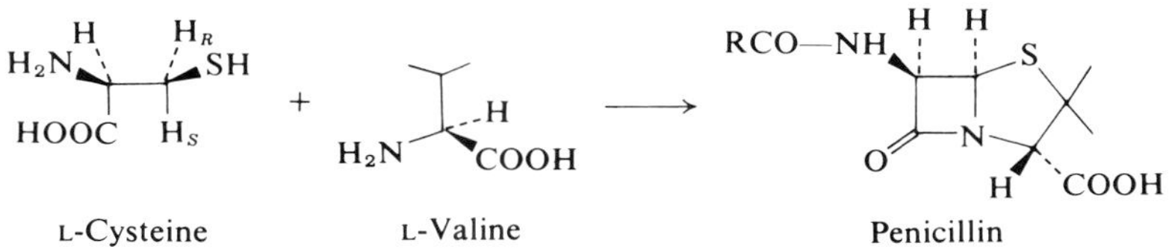

L-Cysteine   L-Valine   Penicillin

Both groups [45,46] fed their synthetic labeled samples (cystine is reduced to cysteine under the conditions of the experiment) to cultures of *Penicillium chrysogenum* and analyzed the penicillin G for radioactivity. In both experiments the penicillin derived from (3*R*)-[3-³H]cysteine retained most of the radioactivity, while that derived from the (3*S*)-isomer lost most of the tritium. Loss of the 3-pro-*S* hydrogen from cysteine shows that the C—N bond is formed with predominant retention of configuration.

Other

The enzymatic conversion of cysteine and cystine to pyruvate, ammonia, and hydrogen sulfide by cysteine desulfhydrase, tryptophanase, and related enzymes is dependent on pyridoxal phosphate in many systems and thus of the type discussed in the section on the stereochemistry of β-replacement reactions of serine, with sulfur as the leaving group [52]. Consequently, it is likely that chiral pyruvate could be generated from chirally labeled cysteine in the same way as from serine.

VALINE

Valine differs from the other amino acids considered here in containing two methyls that are diastereotopic rather than hydrogens. The nonequivalence of the methyls can be clearly seen in the nmr spectrum, which shows two separate three-proton doublets. While relatively few metabolic reactions directly affect the methyl groups (incorporation into cephalosporin and dehydrogenation to methacrylate are two examples), both the biosynthesis of valine and its incorporation into pantoic acid and penicillin involve reactions at the prochiral carbon. To investigate the stereochemistry of these reactions, a number of syntheses of specifically methyl-labeled valines have been devised, using both carbon and hydrogen isotopes.

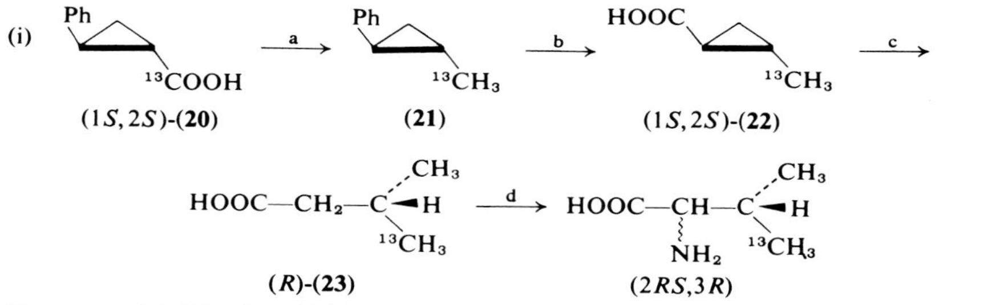

(i)

^{13}COOH
(1S,2S)-(**20**)
→ a →
(**21**) ^{13}CH$_3$
→ b →
(1S,2S)-(**22**) ^{13}CH$_3$
→ c →

HOOC—CH$_2$—C$\overset{CH_3}{\underset{^{13}CH_3}{\overset{|}{-}}}$H
(*R*)-(**23**)
→ d →
HOOC—CH—C$\overset{CH_3}{\underset{^{13}CH_3}{\overset{|}{-}}}$H
$\overset{|}{NH_2}$
(2*RS*,3*R*)

[Reagents: (a) CH$_2$N$_2$; LiAlH$_4$; MeSO$_2$Cl; LiAlH$_4$; (b) O$_3$; (c) CH$_3$CHN$_2$; Li/NH$_3$; NaOH; (d) Br$_2$, PCl$_3$; NH$_3$]

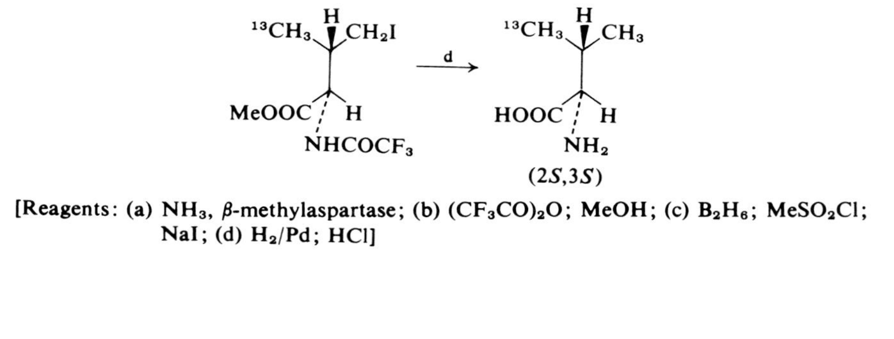

(ii)

^{13}CH$_3$ and HOOC on alkene, COOH and H
→ a →
(2*S*,3*R*) ^{13}CH$_3$, COOH, HOOC, H, NH$_2$
→ b →
^{13}CH$_3$, COOH, MeOOC, H, NHCOCF$_3$
→ c →

^{13}CH$_3$, CH$_2$I, MeOOC, H, NHCOCF$_3$
→ d →
^{13}CH$_3$, CH$_3$, HOOC, H, NH$_2$
(2*S*,3*S*)

[Reagents: (a) NH$_3$, β-methylaspartase; (b) (CF$_3$CO)$_2$O; MeOH; (c) B$_2$H$_6$; MeSO$_2$Cl; NaI; (d) H$_2$/Pd; HCl]

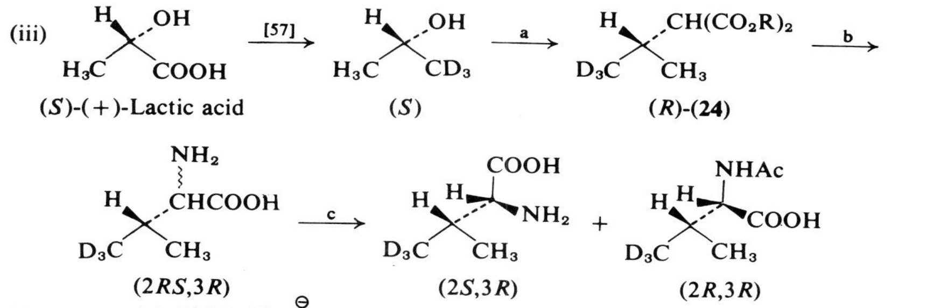

(iii)

(S)-(+)-Lactic acid → [57] → (S) → a → (R)-(24) → b →

(2RS,3R) → c → (2S,3R) + (2R,3R)

[Reagents: (a) PhSO$_2$Cl; $\overset{\ominus}{C}$H(CO$_2$R)$_2$; (b) NaOH; Br$_2$, heat; NH$_3$; (c) Ac$_2$O; hog kidney acylase]

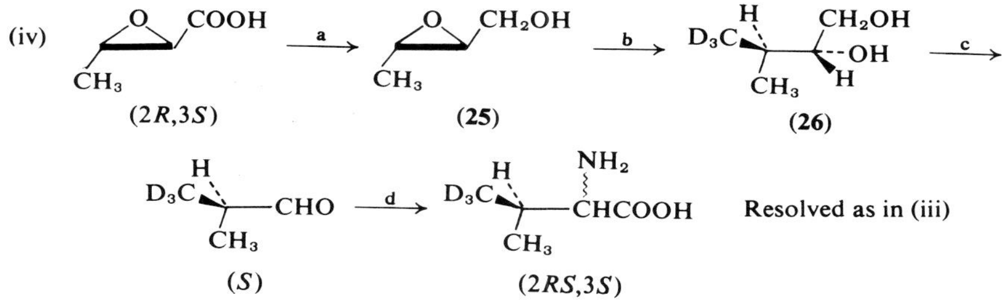

(iv)

(2R,3S) → a → (25) → b → (26) → c →

(S) → d → (2RS,3S)   Resolved as in (iii)

[Reagents: (a) LiAlH$_4$; (b) CD$_3$Li; (c) NaIO$_4$; (d) NH$_4$OH, NaCN; HCl]

Scheme 10 Preparation of chirally labeled valine.

Chemical Syntheses of Chirally Labeled Valine (Scheme 10)

BY RING OPENING OF A CYCLOPROPANE

A route developed by Baldwin et al. [53] for a study of penicillin biosynthesis began with the optically active trans-2-phenylcyclopropanecarboxylic acid (20); carbon-13 was introduced into the carboxyl group by Grignard synthesis with ^{13}CO$_2$. A standard hydride reduction sequence converted the carboxyl to methyl, and the benzene ring of hydrocarbon (21) was ozonized to afford 2-methylcyclopropanecarboxylic acid (22). Its ethyl ester was reduced with lithium in liquid ammonia to afford (3R)-[4-^{13}C]isovaleric acid (23), which was finally converted to (2RS,3R)-[4-^{13}C]valine [48] by bromination and displacement of the bromine with ammonia.

FROM β-METHYLASPARTIC ACID

Taking advantage of the known enzymatic trans addition of ammonia to mesaconic acid, Sih and his co-workers [54] used ^{13}CH$_3$-labeled mesaconic acid to prepare labeled β-methylaspartic acid and then employed a series of selective chemical reductions (Scheme 10, ii) to convert the 4-carboxyl to methyl. Subsequently, the two diastereotopically CD$_3$-labeled valines were prepared by this route [55], either using CD$_3$I to prepare mesaconic acid or carrying out the reduction of the carboxyl group with deuterated reagents. This approach has the particular advantage of leading directly to the natural L-valine configuration without a resolution step.

FROM LACTIC ACID

A straightforward route to labeled valine used by Hill *et al.* [56] in a study of valine biosynthesis started with (S)-$(+)$-lactic acid, converted by a published procedure [57] to (S)-2-[1-$^2H_3$]propanol. Malonic ester displacement (S_N2 with inversion) of the benzenesulfonate led to diester (**24**), converted to a diastereomeric mixture of trideuterated valines by standard methods. Enzymatic resolution of the *N*-acetyl derivative with hog kidney acylase gave the natural $(2S,3R)$-[4-$^2H_3$]valine along with the acetyl derivative of the unnatural $(2R,3R)$-isomer.

BY EPOXIDE RING OPENING

An approach published by Aberhart and Lin [58] used the stereospecific opening of an epoxide to label the prochiral center. Epoxide (**25**), prepared by reduction of the optically active epoxide of *trans*-crotonic acid, was opened with CD_3Li to diol (**26**). It was then found that the chirally labeled isobutyraldehyde formed by periodate cleavage underwent Strecker synthesis without racemization under mild conditions. The diastereotopic valine-d_3 samples thus prepared were again separated by enzymatic hydrolysis of the *N*-acetates. Substitution of $^{13}CH_3Li$ in the ring-opening step gave a valine chirally labeled with ^{13}C.

In the last three syntheses described the stereochemical purity at C-3 could be accurately measured from the proton or ^{13}C nmr spectra since, if the samples are diastereomerically pure, only one of the methyl doublets appears in the proton spectrum and only one of the methyl signals in the ^{13}C nmr spectrum is enhanced over natural abundance. These syntheses provide labeled valine samples of rigorously established configuration and agree in showing that the upfield doublet in the nmr spectrum of L-valine is due to the pro-*S* methyl [59].

Stereochemistry of Valine Incorporation into Penicillin and Cephalosporin

As pointed out in the section on the stereochemistry of cystine and cysteine incorporation into penicillin, the heterocyclic skeleton of the penicillin and cephalosporin antibiotics is derived from cysteine and valine. The diastereotopic methyls of valine become the diastereotopic methyls of penicillin, and the stereochemical question of concern is whether formation of the bond from sulfur to C-3 of valine is stereospecific. During incorporation into cephalosporin, on the other hand, the valine methyls become functionally differentiated, one becoming the ring methylene C-2, the other the side-chain methylene C-17, and the question is one of enzymatic discrimination between the diastereotopic methyls.

Three different groups have answered these questions with the labeled valines described above. Analysis of the ^{13}C incorporation pattern in the antibiotics is straightforward by nmr, since C-2 and C-17 are easily distinguished in cephalosporin, and the α- and β-methyls of penicillin give rise to discrete signals in the ^{13}C nmr spectrum of the sulfoxides which have been assigned to the respective methyls [60]. When $(2RS,3R)$-[4-^{13}C]valine is supplied as substrate for growth of *P. chrysogenum*, penicillin labeled only in the β-methyl is obtained, while in *C. acremonium*, the pro-*R* methyl provides C-2 of cephalosporin C [61]. Complementarily, $(2S,3S)$-[4-^{13}C]valine labels the α-methyl of penicillin and C-17 of cephalosporin [54,58]. Thus, the incorporation of valine into both antibiotics is stereospecific at C-3, the penicillin C—S bond being formed with retention of configuration.

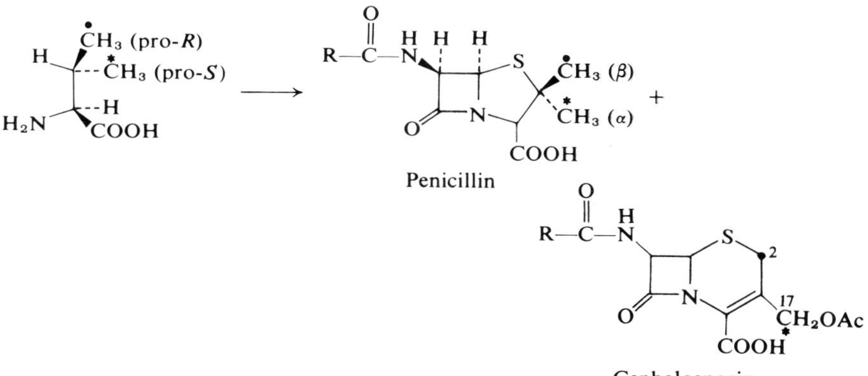

Stereochemistry of Valine Biosynthesis

The biosynthetic pathway to valine in bacteria [62] is outlined in Scheme 11. The first intermediate, α-acetolactic acid (**27**), formed from pyruvate by acetohydroxy-acid synthetase, is then converted by an isomeroreductase to the dihydroxyisovaleric acid (**28**). Dehydration to α-ketoisovaleric acid (**30**) has been shown to proceed via enol (**29**), since a proton is incorporated from the solvent [63]. The final step is transamination of (**30**) to valine. The enzymes of this pathway also catalyze the corresponding reactions in isoleucine biosynthesis.

The stereochemistry of the dehydration step, in which the hydroxyl at the prochiral carbon (C-3) of (**28**) is replaced by hydrogen, has been investigated

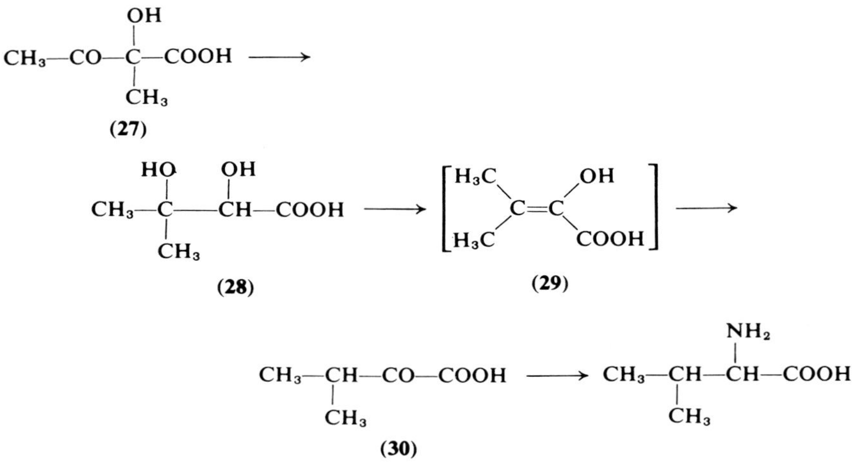

$$CH_3-CO-\underset{\underset{CH_3}{|}}{\overset{\overset{OH}{|}}{C}}-COOH \longrightarrow$$

(27)

$$CH_3-\underset{\underset{CH_3}{|}}{\overset{\overset{HO}{|}}{C}}-\overset{\overset{OH}{|}}{CH}-COOH \longrightarrow \left[\underset{H_3C}{\overset{H_3C}{\diagdown}}C=C\underset{\diagup}{\overset{\diagup}{\underset{COOH}{\overset{OH}{}}}}\right] \longrightarrow$$

(28)   **(29)**

$$CH_3-\underset{\underset{CH_3}{|}}{CH}-CO-COOH \longrightarrow CH_3-\underset{\underset{CH_3}{|}}{CH}-\overset{\overset{NH_2}{|}}{CH}-COOH$$

(30)

Scheme 11   Valine biosynthesis.

by Hill and co-workers [56] using chirally labeled samples of **(28)** as substrates. The preparation of the diastereomeric d_3 derivatives is shown in Scheme 12. The unsaturated ester **(31)**, prepared by *cis* addition of dimethyllithium cuprate to ethyl tetrolate-d_3, was *cis* hydroxylated to (2R,3R;2S,3S)-**(28)**-d_3 and *trans* hydroxylated to (2R,3S;2S,3R)-**(28)**-d_3. It had earlier been shown that only the (2R)-enantiomer of **(28)** is converted to valine in *E. coli* [64,65], so the racemic deuterated dihydroxy acids could be used for conversion to deuterated valines by purified dehydratase and transaminase. Examination of the biosynthesized valine-d_3 by nmr showed that the enzymatic conversion is at least 95% stereospecific; each valine-d_3 displayed only one of the methyl

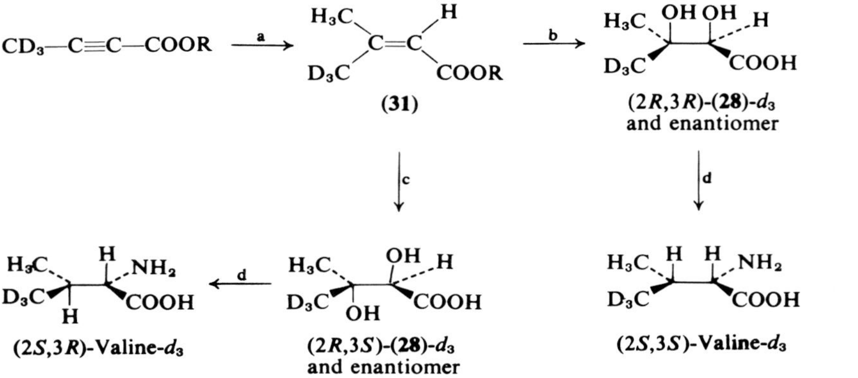

$$CD_3-C\equiv C-COOR \xrightarrow{\text{a}} \underset{D_3C}{\overset{H_3C}{\diagdown}}C=C\underset{\diagup}{\overset{\diagup}{\underset{COOR}{\overset{H}{}}}} \xrightarrow{\text{b}} \underset{D_3C}{\overset{H_3C}{\diagdown}}\overset{OH\ OH}{\underset{COOH}{}}H$$

                                 (31)   (2R,3R)-**(28)**-d_3
                                                                 and enantiomer

                                 ↓ c   ↓ d

$$\underset{D_3C}{\overset{H_3C}{\diagdown}}\overset{H}{\underset{H}{\diagup}}\underset{COOH}{\overset{NH_2}{}} \xleftarrow{\text{d}} \underset{D_3C}{\overset{H_3C}{\diagdown}}\overset{OH}{\underset{OH}{}}\underset{COOH}{\overset{H}{}} \qquad \underset{D_3C}{\overset{H_3C}{\diagdown}}\overset{H\ H}{\underset{COOH}{}}NH_2$$

(2S,3R)-Valine-d_3 (2R,3S)-**(28)**-d_3 (2S,3S)-Valine-d_3
                                    and enantiomer

Scheme 12   Stereochemistry of dehydration step in valine biosynthesis. Reagents: (a) $(CH_3)_2CuLi$; (b) OsO_4, $Ba(ClO_3)_2$; (c) MCPBA; H_3O^+; (d) dihydroxy-acid dehydratase plus transaminase.

doublets. Comparison with the chemically synthesized valine-d_3 (Scheme 10, iii) showed that the pro-R methyl of (28) becomes the pro-S methyl of valine and vice versa; i.e., the replacement of hydroxyl by hydrogen occurs with retention of configuration, paralleling the finding deduced earlier [64] for isoleucine biosynthesis. The mechanistic significance of this result is that (1) enzymatic dehydration of (28) to enol (29) must be stereospecific, and (2) protonation of the enol must be enzyme mediated, since the proton is delivered from only one face of (29).

Another step in valine biosynthesis, the migration of a methyl group as (27) rearranges to (28), creates a prochiral center. The determination of which of the chemically distinct methyls of (27) provides the pro-R and which the pro-S methyl of (28) can now be undertaken with labeled samples of (27) [66].

Other Reactions at the Valine Prochiral Center

Metabolic degradation of valine proceeds initially by transamination to (30) and oxidative decarboxylation to isobutyryl-CoA (32). This has been shown [67] to be enzymatically dehydrogenated to methacrylyl-CoA (33). The dehydrogenase is likely to be stereospecific in attacking only one of the diastereotopic methyls of (32); an attempt to study the stereochemistry with labeled valine gave inconclusive results, but better success is anticipated with chirally labeled isobutyric acid [68].

Valine, by condensation of the derived α-keto acid (30) with formaldehyde, provides the carbon skeleton of pantoic acid (34), a constituent of pantothenic acid and coenzyme A [69]. The stereochemistry of the condensation at the prochiral center of (30) is unknown but would be susceptible to a chiral labeling study.

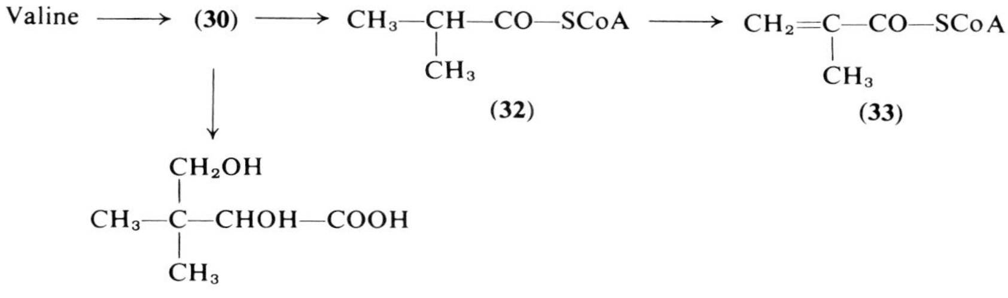

$$\text{Valine} \longrightarrow (30) \longrightarrow \underset{(32)}{\underset{\underset{CH_3}{|}}{CH_3-CH-CO-SCoA}} \longrightarrow \underset{(33)}{\underset{\underset{CH_3}{|}}{CH_2=C-CO-SCoA}}$$

$$\underset{(34)}{\underset{\underset{CH_3}{|}}{\overset{\overset{CH_2OH}{|}}{CH_3-C-CHOH-COOH}}}$$

ISOLEUCINE

The prochiral center at C-4 of isoleucine is involved in metabolic degradation to angelic and tiglic acid via a pathway analogous to the degradation of

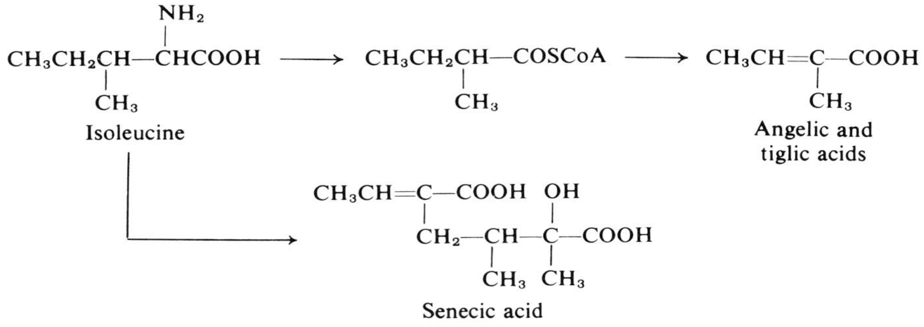

$$CH_3CH_2CH\overset{\overset{\displaystyle NH_2}{|}}{-}CHCOOH \longrightarrow CH_3CH_2CH\overset{\overset{\displaystyle}{}}{-}COSCoA \longrightarrow CH_3CH=C-COOH$$

(under first structure: CH₃ / Isoleucine)
(under second: CH₃)
(under third: CH₃ / Angelic and tiglic acids)

$$CH_3CH=C-COOH \quad OH$$
$$CH_2-CH-C-COOH$$
$$CH_3 \quad CH_3$$

Senecic acid

valine (see the section on other reactions at the valine prochiral center), as well as in conversion to senecic acid in *Senecio* species [70].

For investigation of the stereospecificity of hydrogen removal at C-4 several routes to chirally labeled isoleucine have been developed (Scheme 13). Crout [71] has used deuterated or tritiated diimide to effect *cis* reduction of olefin (35); isoleucine was separated from alloisoleucine and resolved enzymatically [72]. Hill and Rhee [73] have used a route analogous to the valine synthesis in Scheme 10, iii, employing malonic ester displacement of the tosylate of 2-[3-²H]butanol; the *threo*-alcohol is prepared by reduction of *cis*-2-butene oxide with LiAlD₄, while the *erythro*-isomer is the product of hydroboration of *cis*-2-butene with B₂D₆ [74].

No results have been reported on the stereochemistry of conversion of isoleucine to tiglic acid, but the incorporation into senecic acid appears to be stereospecific at C-4, since all the tritium is lost from (4S)-[4-³H]isoleucine [71].

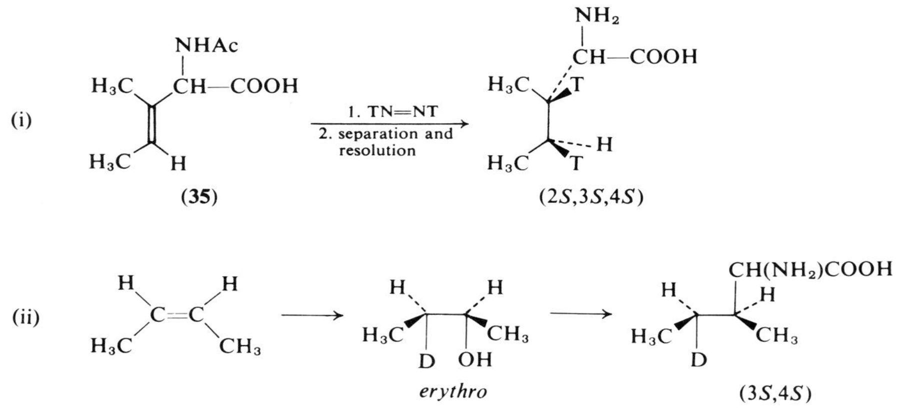

(i)

(35)   (2S,3S,4S)

(ii)

erythro   (3S,4S)

Scheme 13   Preparation of chirally labeled isoleucine.

AROMATIC AMINO ACIDS

In the aromatic amino acids phenylalanine, tyrosine, tryptophan, and histidine the methylene carbon at C-3 is prochiral. Attack at one of the diastereotopic hydrogens occurs in the course of enzymatic hydroxylation, dehydrogenation, alkylation, and elimination of ammonia, and the aromatic ring itself may be displaced from the prochiral carbon. Both chemical and enzymatic methods have been used to prepare samples stereospecifically labeled with ^2H or ^3H at C-3 to investigate these enzymatic processes.

Chiral Labeling

1. The most general method for the specific introduction of isotope at C-3 is that developed independently by Battersby and Hanson [75] and Kirby [76] and their co-workers, involving hydrogenation of a labeled acylamino-cinnamic acid; the route to phenylalanine is shown in Scheme 14. Reaction of [^2H]- or [^3H]benzaldehyde with hippuric acid gives the oxazolone (36), opened by alkali to the benzoylaminocinnamic acid (37). Catalytic hydrogenation then creates contiguous asymmetric centers simultaneously. Since acid (37) has been shown by X-ray analysis to have the (Z) configuration [77], and catalytic hydrogenation can be assumed to proceed cis (see below), the labeled amino acid should be the (2S,3R;2R,3S) racemate. If the synthesis is carried out using unlabeled aldehyde and reducing with deuterium gas, the amino acid is the (2S,3S;2R,3R) dideuterated racemate [78].

For any experiments in which the biological system is known to be specific for only the natural (2S)-amino acid, the synthetic racemate may be employed as the equivalent. The racemate can easily be resolved, however; phenylalanine is usually resolved by enzymatic hydrolysis of the N-chloroacetyl derivative [75,79] or hydrolysis of the N-acetyl methyl ester with chymotrypsin [80], while labeled tyrosine has been resolved via the N-chloroacetate with hog kidney acylase [76], by hydrolysis of the ethyl ester with chymotrypsin [78], or by decarboxylation of the L-isomer with tyrosine carboxylyase [81]. A commercial synthesis of L-[2,3-$^3H_2$]phenylalanine involves opening the unlabeled oxazolone (36) with L-glutamic acid to yield (38), followed by catalytic reduction with tritium. One diastereomer can be obtained pure by crystallization and hydrolyzed to the (2S,3S)-amino acid [82].

The same route has been applied to the synthesis of labeled tryptophan, beginning with deuterated or tritiated 3-formylindole [83]. The racemic tryptophan was resolved with α-phenylethylamine.

Several methods have been used to verify the configuration at C-3

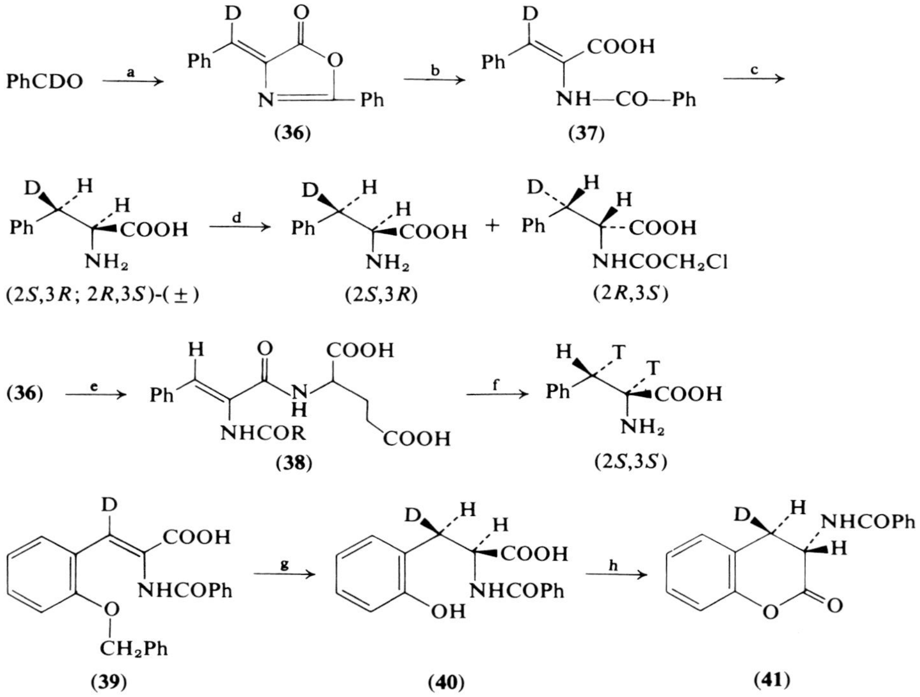

Scheme 14 Preparation of phenylalanine labeled at C-3. Reagents: (a) PhCONHCH₂-COOH, Ac₂O; (b) NaOH; (c) H₂/Pd; H₃O⁺; (d) ClCH₂COCl; acylase; (e) L-glutamic acid; (f) T₂/Pd; H₃O⁺; (g) H₂/Pd; (h) DCC.

and to assess the optical purity of the samples synthesized by this method:

a. The C—H protons of the side chains of phenylalanine, tyrosine, and tryptophan form an ABX pattern in the nmr, with values for J_{AX} and J_{BX} of about 8.5 and 4.5 Hz. In the (2S,3R) or corresponding racemic mono-deuterated amino acids only one doublet, $J = 4.5$ Hz, appears in the AB region. This result supports the (3R) configuration (assuming the staggered conformation) and reveals that the hydrogenation is at least 95% cis [75,76, 79,83]. Moreover, if the azlactone synthesis is carried out on o-benzyloxy-benzaldehyde-d and the cinnamic acid (39) is hydrogenated, the product (40) can be cyclized to a lactone (41) which shows a large trans coupling constant (14.8 Hz), confirming cis hydrogenation [84].

b. Ozonolysis of the synthetic (2S,3R;2R,3S)-tyrosine-d gave aspartic acid-d, shown by nmr to have the threo configuration [76]. Alternatively, the aspartic acid derived from ozonolysis of tyrosine-t was resolved and successively converted to malic and fumaric acids by reactions of known stereo-chemistry [81].

c. The (2S,3R)-[3-²H]phenylalanine was converted, by treatment with HNO_2–HBr followed by hydrogenation, to β-[²H]phenylpropionic acid, which was then ozonized to [²H]succinic acid; ORD analysis showed that this was the (S)-enantiomer [75]. More accurate measurements on the corre-sponding tritiated phenylalanine showed that the racemic samples synthesized by the azlactone route have an optical purity of at least 99% at C-3 but that, when the samples are enzymatically resolved, about 7% epimerization takes place during chloroacetylation, so that the resulting optically active amino acids are about 93% optically pure at C-3 (still ample for most investigations) [75,76].

2. A second approach begins with enzymatic reduction of ArCDO, using either an actively fermenting yeast [85] or a purified alcohol dehydrogenase [75,85a] to provide optically pure (S)-ArCHDOH (Scheme 15). Its tosylate undergoes S_N2 displacement with acetamidomalonic ester, followed by hydrolysis, to give (3R)-[3-²H]phenylalanine; a variation is displacement with malonic ester, followed by bromination, hydrolysis, and amination [75]. If the (S)-benzyl alcohol-d is converted to the chloride with inversion (PCl_3 and pyridine) and the chloride is used in the acetamidomalonate displacement, (3S)-phenylalanine-d results. Degradation to succinic acid-d showed that the optical purity at C-3 was at least 95%.

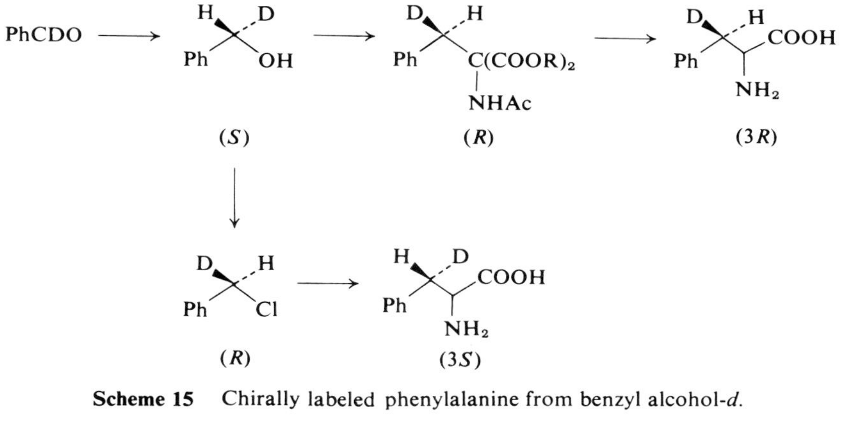

Scheme 15 Chirally labeled phenylalanine from benzyl alcohol-d.

3. In addition to these chemical methods, several of the aromatic amino acids can be prepared in labeled form by enzymatic routes. As discussed in the section on the stereochemistry of β-replacement reactions of serine, biosyn-thesis of tyrosine and tryptophan from chirally labeled serine preserves the chirality at C-3. Preparation of chirally labeled histidine has been achieved by

allowing L-histidine to undergo partial deamination in the presence of histidine ammonia-lyase in tritiated water; the recovered histidine has incorporated tritium at C-3 [86]. Degradation of the histidine-*t* to aspartic acid-*t* showed that the tritium introduced occupied the pro-*R* configuration [48,87]. This result was independently confirmed by chemical degradation of the histidine-*t* to succinic acid-*t* followed by dehydrogenation with succinate dehydrogenase [88].

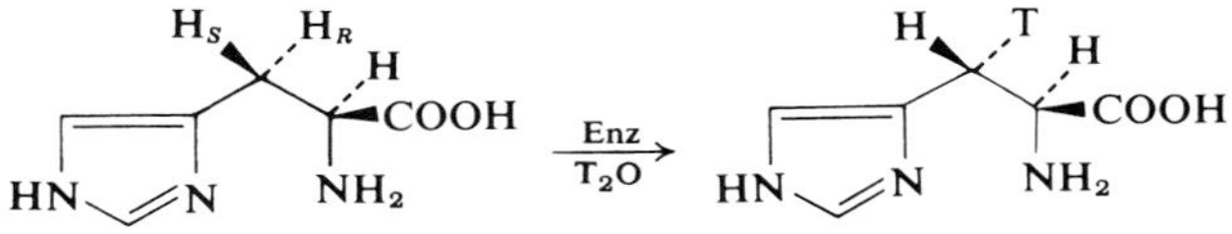

Stereospecificity of Enzymatic Attack at C-3

The stereospecificity of an appreciable number of enzymatic processes involving the prochiral center of the aromatic amino acids has been investigated with the aid of the labeled substrates described above; most of these can be classified in the following five groups:

AMMONIA LYASES

Phenylalanine, tyrosine, and histidine each undergoes enzymatic elimination of ammonia to an unsaturated acid, catalyzed by an ammonia-lyase:

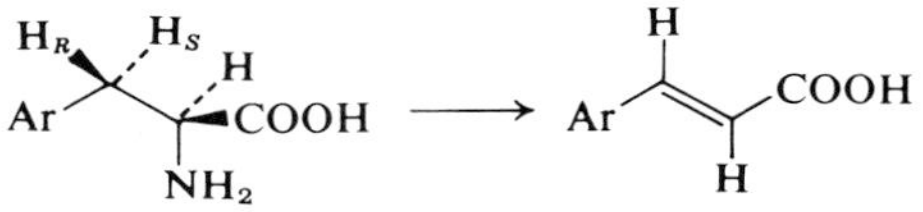

Use of chirally labeled substrates has shown that the elimination is *anti*-periplanar in each case, with the specific loss of the pro-*S* hydrogen from phenylalanine and tyrosine and the corresponding (but pro-*R*) hydrogen from histidine. The enzymes investigated include phenylalanine ammonia-lyase from potatoes [75,79] and bacteria [78], tyrosine ammonia-lyase [81,89], and histidine ammonia-lyase [87,88]. The stereochemistry parallels that found earlier for aspartate ammonia-lyase and β-methylaspartase; the reaction has been reviewed in depth by Hanson and Havir [90].

Phenylalanine is incorporated into colchicine (**42**) with loss of the C-3 pro-*S* hydrogen, consistent with initial deamination to cinnamic acid, although in the biosynthesis of haemanthamine (**43**), where phenylalanine is again first converted to cinnamic acid, both (3*R*)- and (3*S*)-[3-³H]phenylalanine lose all the radioisotope [75].

DEHYDROGENATION

In the biosynthesis [91] of the alkaloid securinine (**44**) and also of mycelian-amide (**45**), a metabolite of *P. griseofulvum* [92], tyrosine is incorporated with dehydrogenation at C-3; studies with specifically labeled substrates have shown that in both cases H_S is lost.

HYDROXYLATION

During the incorporation of tyrosine into haemanthamine (**43**) in daffodils, C-3 is hydroxylated; use of stereospecifically tritiated tyrosine showed that the hydroxyl selectively replaces H_R with predominant retention of configuration, consistent with an electrophilic substitution [76]. This result has been con-

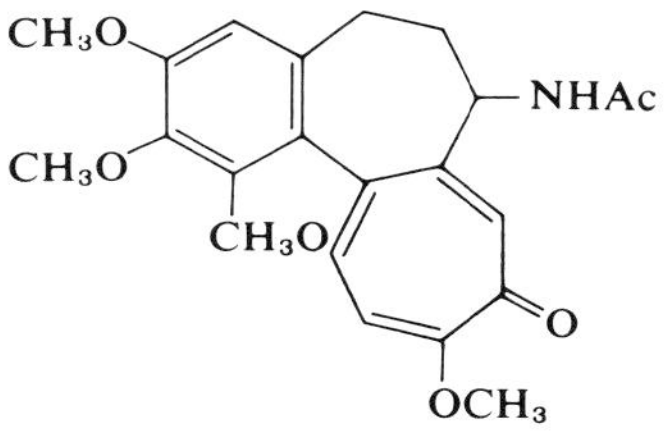

(**42**)

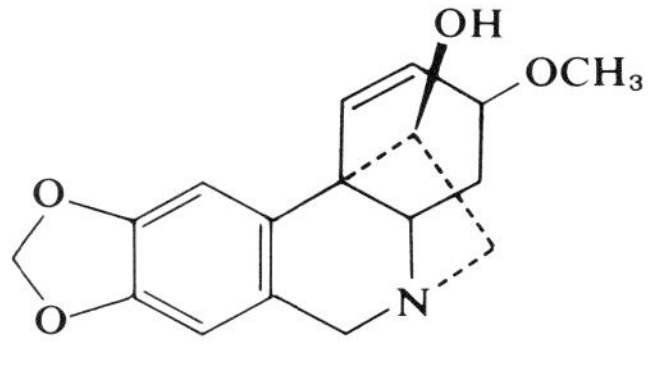

(**43**)

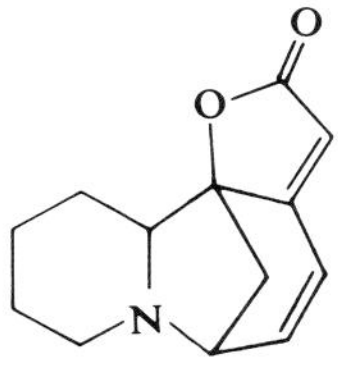

(**44**)

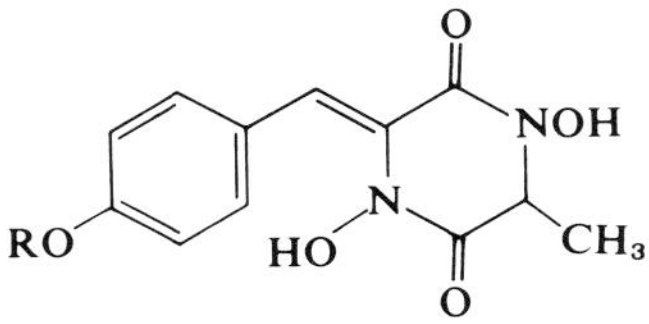

(**45**)

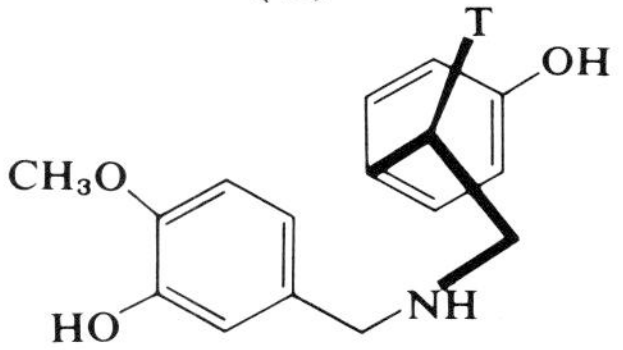

(**46**)

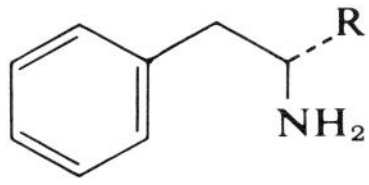

(**47**), R = H
(**48**), R = CH₃

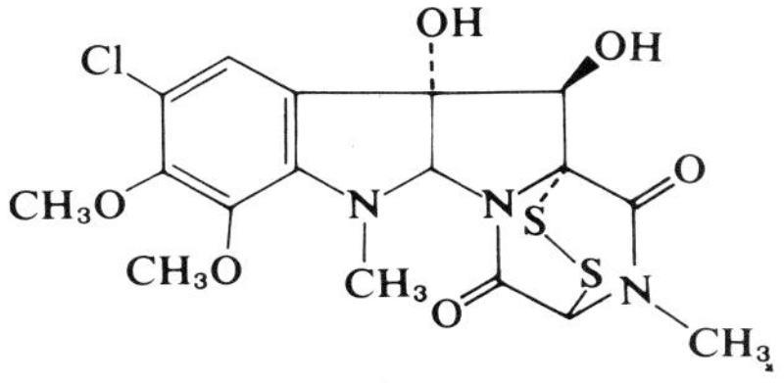

(**49**)

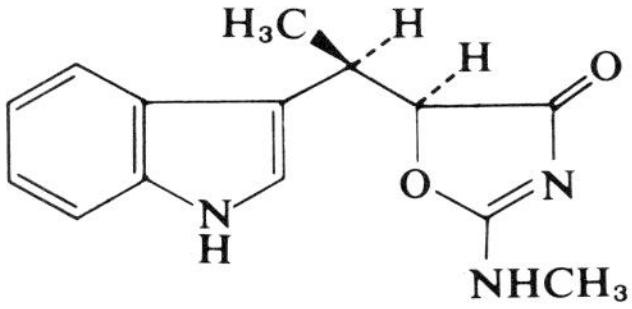

(**50**)

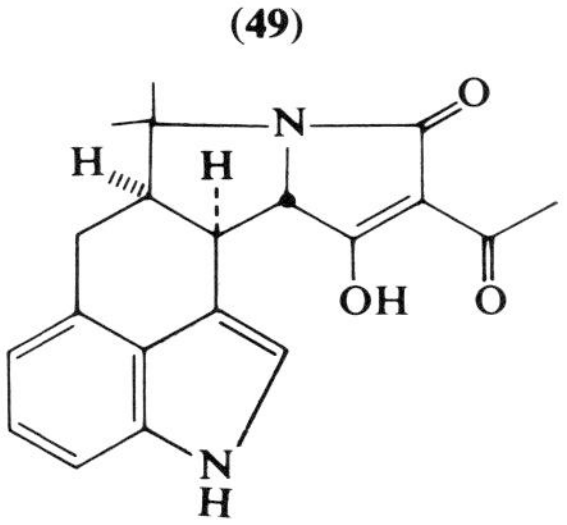

(**51**)

firmed and information added about the timing of the hydroxylation step through use of the specifically labeled later precursor (46) [93]. Hydroxylation of the simpler amines phenylethylamine (47), amphetamine (48), and dopamine by dopamine-β-hydroxylase also replaces the pro-R hydrogen with retention of configuration [94–95a].

Hydroxylation at C-3 of tryptophan accompanies biosynthesis of the fungal metabolite sporidesmin A (49). Here again H_R is replaced with retention of configuration [83].

ALKYLATION

Reference has already been made to the specific replacement of H_R at C-3 of tryptophan by a methyl group in indolmycin (50) biosynthesis [41]; the substitution occurs with retention of configuration. Another natural product in which one of the diastereotopic hydrogens of tryptophan is replaced by an alkyl group is α-cyclopiazonic acid (51), a metabolite of *P. cyclopium*. In this case it has been shown by Steyn *et al.* [96] that the pro-S hydrogen is lost with inversion of configuration.

SUBSTITUTION

Tyrosine-phenol lyase catalyses, as well as the synthesis and cleavage of tyrosine (see the section on the stereochemistry of β-replacement reactions of serine), the replacement of the phenol ring by substituted phenols such as catechol or resorcinol; the mechanism undoubtedly involves Michael addition to the aminoacrylate intermediate (8) (cf. Scheme 8). The stereochemistry of the resorcinol substitution was investigated by Sawada *et al.* [97] using tyrosine deuterated at the prochiral center. The configuration at C-3 of the product was proved by stereospecific synthesis using the azlactone method (Scheme 14). The replacement was found to occur with complete retention of configuration, paralleling other reactions of this pyridoxal-catalyzed reaction type discussed earlier.

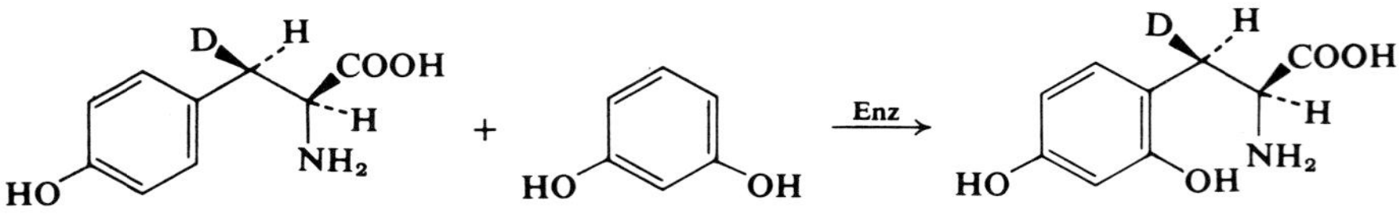

Several notes of caution in using aromatic amino acids labeled at C-3 have been sounded. Johns *et al.* [98] found that the incorporation of phenylalanine into gliotoxin was accompanied by loss of the pro-R hydrogen. Loss of the C-3 proton is not an obligatory step in the biosynthetic pathway, however, and is apparently caused by the presence of pyridoxal phosphate enzymes which catalyze a rapid exchange. Vederas and Tamm [98a] showed that incorporation of [3-³H]phenylalanine into cytochalasin D occurs with appreciable loss of isotope from both 3R and 3S samples. Since, in addition, both D- and L-phenylalanine are incorporated with loss of tritium from C-2, phenylpyruvic acid was implicated as an otherwise unsuspected intermediate. These results emphasize the need for consideration of biochemical equilibria in biosynthetic tracer studies.

Stereochemistry of the Shikimic Acid Pathway

The biosynthetic pathway by which the benzene rings of aromatic amino acids and many other aromatic natural products are constructed from sugars is replete with reactions at prochiral sites. The shikimate pathway [99] is

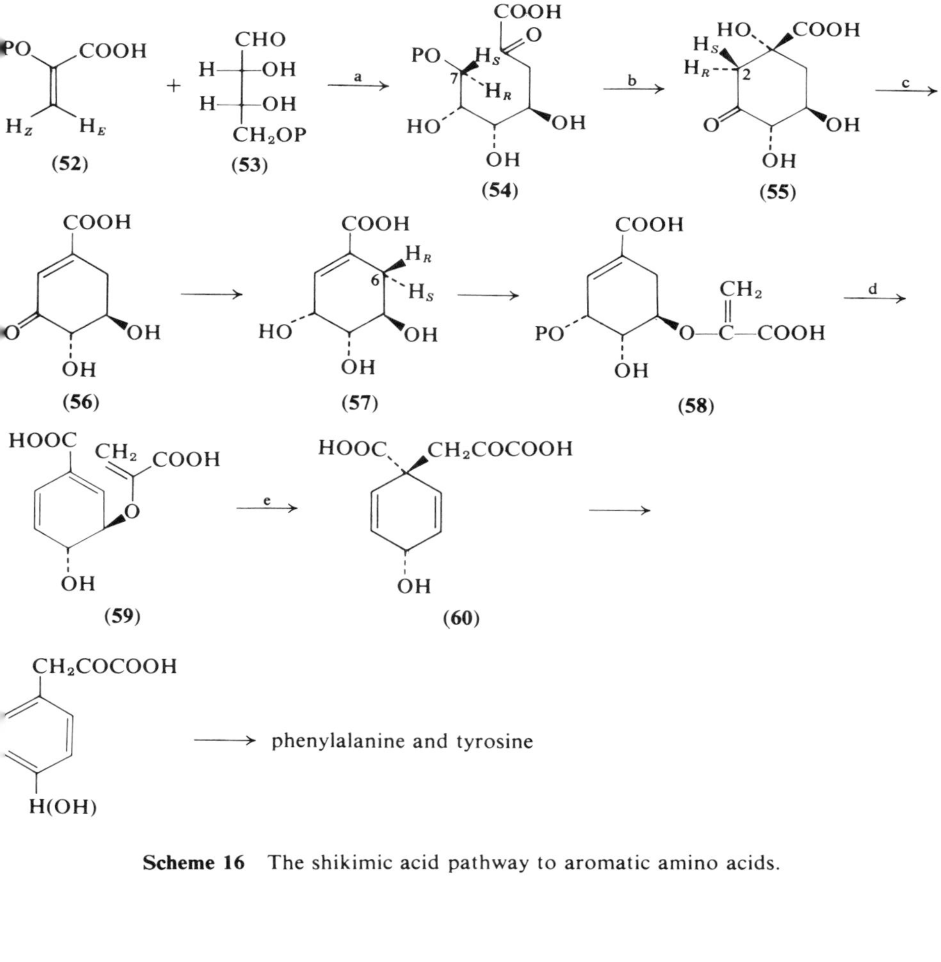

Scheme 16 The shikimic acid pathway to aromatic amino acids.

outlined in Scheme 16, and the stereochemistry of steps a–e is summarized below; for a more detailed discussion of some of these steps see Bentley [6].

a. DAHP synthetase assembles the seven-carbon sugar 3-deoxy-3-arabinoheptulosonate 7-phosphate (54) from phosphoenolpyruvate (52) and erythrose 4-phosphate (53), using an aldol-like condensation to form the new C—C bond. The stereochemistry of addition to the prochiral carbon of (52) has been studied by Floss [35] using stereospecifically tritiated samples of (52) derived from the enolase-catalyzed *anti* elimination of water from 3-phosphoglyceric acid [100]. The label eventually appears at C-6 of shikimic acid (57); its configuration was deduced by chemical degradation to malic acid followed by dehydration with fumarase. The finding that (Z)-[^{3}H](52) leads to $(6R)$-[^{3}H](57) shows that the condensation is indeed stereospecific, with attack of the aldehyde at the *si* face of (52).

b. The mechanism of the cyclization of (54) to dehydroquinic acid (55) is still unknown, but the stereochemistry at C-7 of (54) has been studied by Haslam *et al.* [101,102]. Tritium was introduced stereospecifically at C-7 from labeled glucose or mannose (cf. Scheme 6) and, using an *E. coli* mutant that accumulated dehydroquinic acid, the (55) formed was then converted to shikimic acid (57). Knowing the stereochemistry of dehydration of (55) to (56) (see next section), it could be deduced that the 2-pro-*R* hydrogen of (55) was derived from the 7-pro-*R* hydrogen of (54).

c. The stereochemistry of dehydration of (55) to (56) has been studied from both directions. Hanson and Rose [103] carried out the enzymatic hydration of dehydroshikimic acid (56) in T_2O, degraded the quinic acid formed by enzymatic reduction to citric acid, and determined the configuration of the tritium by dehydration with aconitate hydratase. The addition of water was shown to be *cis*, with the hydrogen added in the pro-*R* configuration.

This result has been confirmed by Haslam in the forward direction [101, 102]. Dehydroquinic acid (55) was found to undergo a slow stereospecific exchange with D_2O at pH 7; the pro-*S* (axial) hydrogen was 90% exchanged [104]. When this material was used as substrate for the enzymatic dehydration deuterium was retained and, when the isomeric $(2R)$-[^{2}H]dehydroquinic acid was used, the deuterium was lost, again evidencing a *cis* elimination.

It is instructive to compare this result with the course of nonenzymatic dehydration. Haslam showed, using these deuterated dehydroquinic acids, that *in vitro* dehydration with either acid or base was predominantly *anti*, the normal stereochemistry of 1,2-eliminations. It has been suggested [105] that the enzyme forces the ring into a skew-boat conformation, reversing the usual stereochemistry of elimination.

d. The second double bond is introduced by 1,4-conjugate elimination of phosphate from enolshikimylpyruvate phosphate (58), during which one of the diastereotopic hydrogens at C-6 is lost. Two groups have shown independently that the enzymatic process is an *anti* elimination. Floss and co-workers [35] used the (6R)- and (6S)-[6-^{3}H]shikimate samples (described above in part a) for conversion to chorismic acid (59); the (6R)-[^{3}H]shikimate lost 82–88% of its tritium, while the (6S)-isomer retained the same amount of its tritium. Hill and Newkome [106] synthesized the specifically deuterated shikimic acids by a chemical route, using the Diels–Alder reaction to introduce the isotope stereospecifically (Scheme 17). The substrates were converted by *E. coli* mutants directly to phenylalanine and tyrosine; H_R was totally lost and H_S was totally retained in the amino acids.

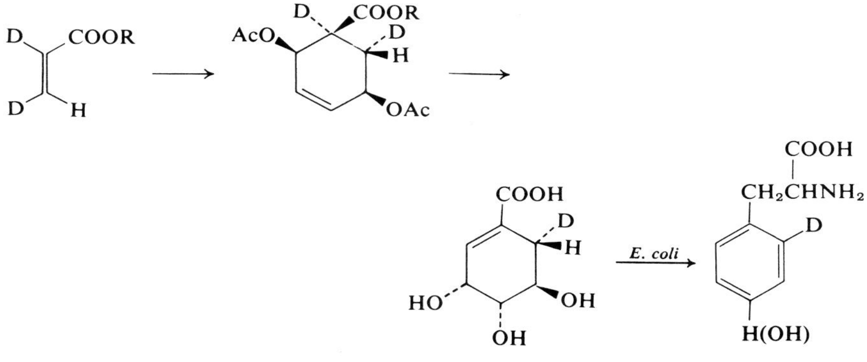

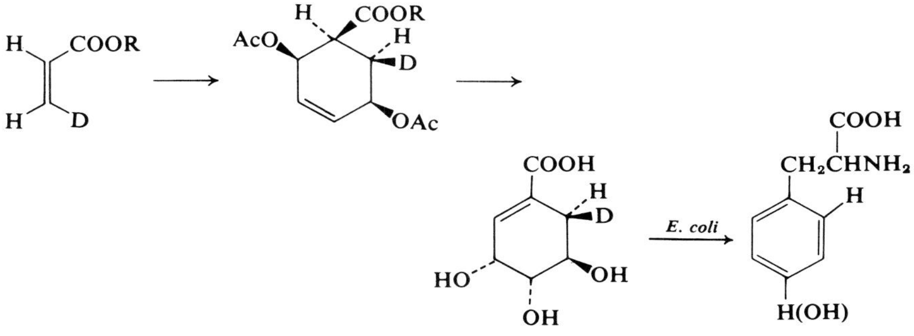

Scheme 17   Stereochemistry of chorismic acid biosynthesis.

This enzymatic elimination, as well as the earlier one in step c, takes a stereochemical course opposite that of the nonenzymatic counterpart. Model studies with simple cyclohexenol derivatives show that base-catalyzed or thermal 1,4-eliminations, both presumably concerted, are almost exclusively *syn*, while carbonium ion type of eliminations are stereorandom [107]. Floss [35] has suggested that chorismate synthesis may involve an initial *syn* S_N2' displacement of phosphate by an enzymatic nucleophile followed by *anti* 1,2-elimination, while Rose has argued [10] that the *anti* elimination may simply be due to maximum separation of catalytic sites on the enzyme.

e. The rearrangement of chorismic acid (**59**) to prephenic acid (**60**) is a remarkable biochemical Claisen rearrangement; since these are not normally catalyzed, the role of the enzyme must be to hold the molecule in the optimum conformation. The rearrangement is suprafacial, as are model Claisen rearrangements in cyclic systems [108]. Haslam [79] has pointed out that the diastereotopic *exo*-methylene hydrogens of (**59**) remain stereochemically distinct in (**60**) and that, if rearrangement occurs through the usual chair-like conformation, the (*Z*)-hydrogen of (**59**) becomes the pro-*S* hydrogen of (**60**). This hypothesis should be amenable to experimental test.

STEREOCHEMISTRY OF ENZYMATIC DECARBOXYLATION OF AMINO ACIDS

When α-amino acids are decarboxylated, the replacement of —COOH by H creates a new prochiral center in the amine formed. Mandeles *et al.* [109] showed early that enzymatic decarboxylation of several amino acids in D_2O was stereospecific, leading to chiral monodeuterated amines.

The stereochemistry of tyrosine decarboxylation was elucidated by Belleau and co-workers [110,111], who prepared one enantiomer of [^{2}H]tyramine by decarboxylation in D_2O and the other by decarboxylation of [α-$^2H_2$]tyrosine in H_2O. The configuration of the deuterated amines was proved by asymmetric synthesis, showing that decarboxylation occurred with retention of

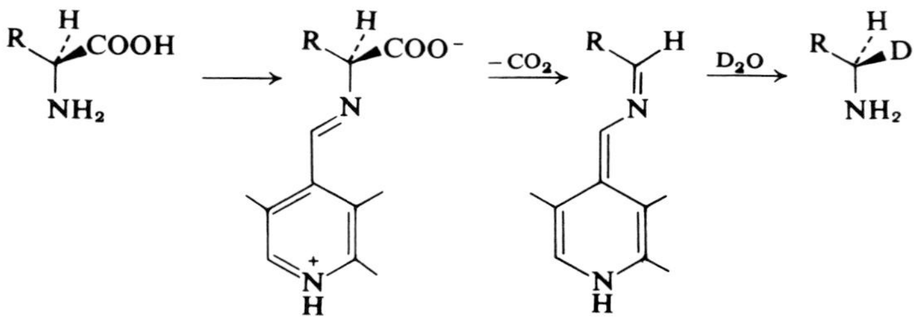

Scheme 18 Pyridoxal-catalyzed decarboxylation of amino acids.

configuration. Since the decarboxylase requires pyridoxal phosphate, decarboxylation is considered to lead to a pyridoxal-stabilized anion which is then protonated by solvent (Scheme 18); the retention observed is consistent with the stereochemical pattern of pyridoxal-catalyzed reactions discussed above.

A similar stereochemical result has been deduced by Chang and Snell for histidine decarboxylase [112], even though this enzyme does not require pyridoxal. The most recent of the decarboxylases whose stereochemistry has been elucidated is lysine. Leistner and Spenser [113] have shown, by oxidizing the [³H]cadaverine formed to (R)-[³H]glycine, that decarboxylation of L-lysine in T_2O takes place with at least 88% retention of configuration. The stereochemistry of this decarboxylation is crucial to the interpretation of biosynthesis of piperidine alkaloids from [α-³H]lysine [114].

In all cases of enzymatic decarboxylation of amino acids studied so far the prochiral center is created with retention of configuration, and this pattern seems likely to prove general.

PROCHIRAL CENTERS OF OTHER AMINO ACIDS

Proline

All three of the methylene groups of proline are prochiral, and the C-3 methylene is involved in enzymatic hydroxylation to hydroxyproline. Using (4R)- and (4S)-[4-³H]proline prepared by reduction of 4-tosylates with $LiAl^3H_4$, Witkop *et al.* and others [115] were able to show that hydroxylation in avian and microbial systems replaced the 4-pro-R hydrogen with retention of configuration.

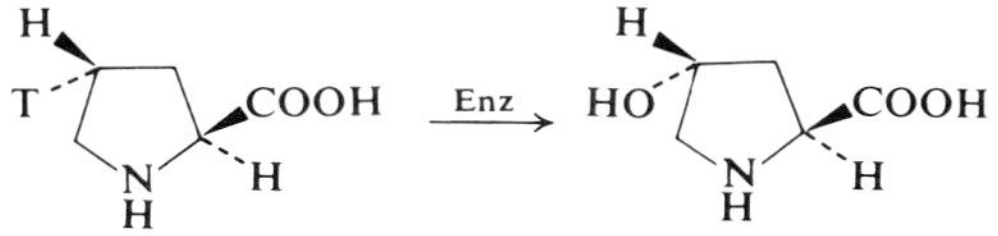

Threonine

Dehydration of threonine to α-ketobutyric acid is analogous to dehydration of serine to pyruvate (see the section on the stereochemistry of β-replacement reactions of serine) and indeed can be effected by serine dehydratase; a prochiral center is created at C-3 of the ketoacid. Snell and co-workers have shown that when the dehydration of D-threonine is carried out in D_2O, the α-[²H]ketobutyrate formed is the (S)-isomer of at least 80% optical purity [116]; the dehydration of L-threonine is also stereospecific [117]. Like serine

$$
\begin{array}{ccc}
\underset{\text{D-Threonine}}{\begin{array}{c} \text{COOH} \\ \text{H}\!\!-\!\!\text{NH}_2 \\ \text{HO}\!\!-\!\!\text{H} \\ \text{CH}_3 \end{array}}
& \xrightarrow[\text{D}_2\text{O}]{\text{D-serine dehydratase}} &
\underset{(S)}{\begin{array}{c} \text{COOH} \\ \text{C}\!\!=\!\!\text{O} \\ \text{D}\!\!-\!\!\text{H} \\ \text{CH}_3 \end{array}}
\quad
\underset{\text{Homoserine}}{\begin{array}{c} \text{COOH} \\ \text{H}_2\text{N}\!\!-\!\!\text{H} \\ \text{CH}_2 \\ \text{CH}_2\text{OH} \end{array}}
\end{array}
$$

dehydration, the replacement of hydroxyl by hydrogen occurs with retention of configuration, requiring protonation of an enzyme-bound enamine intermediate before its release from the chiral enzyme surface.

A key biosynthetic intermediate to threonine is homoserine, whose conversion to threonine involves both prochiral centers. The synthesis of stereospecifically labeled samples for a study of the stereochemistry of this transformation has been reported by Fuganti *et al.* [118].

Others

The stereospecific *trans* addition of ammonia to fumaric acid, which generates the prochiral center of aspartic acid, and the vitamin B_{12}-catalyzed interconversion of β-methylaspartic acid and glutamic acid, which is stereospecific at the prochiral center of the latter, have been thoroughly reviewed elsewhere [6–8].

CONCLUSION

This brief review has summarized recent uses of stereospecific labeling to explore biochemical reactions at prochiral centers of amino acids. Challenging synthetic problems of chiral labeling have been met with ingenuity, making stereospecifically labeled forms of most amino acids available. The results of these studies at prochiral centers have exposed an intricate world of enzymatic specificity at "nonasymmetric" sites, otherwise hidden to view, in which specificity at prochiral centers is routinely exercised, double bonds and planar intermediates are attacked from only one face, and even methyl groups are generated in a stereochemically discernible fashion. This growing body of evidence on the three-dimensional aspects of enzymatic catalysis is beginning to provide important details of mechanism; one example is the preponderance of overall substitutions cited above which take place with retention of configuration, implying that a single site on the enzyme is involved in removing the leaving group and delivering its replacement. Those interested in chemical reactions in living organisms will look forward to further applications of this approach, which are certain to follow.

ACKNOWLEDGMENTS

Our own work cited here has been supported by research grants from the National Institutes of Health, to whom I express my appreciation. My warmest thanks go to my collaborators: Dr. Shou-Jen Yan, Dr. Seiji Sawada, Dr. George R. Newkome, Professor S. M. Arfin, and Professor H. Yamada and his group at Kyoto. Dr. C. J. Sih, Dr. D. J. Aberhart, Dr. C. Walsh, Dr. Y. Cheung, and Dr. D. H. G. Crout kindly made results available before publication, and I thank them for helpful discussions. Finally, I wish to thank the North Atlantic Treaty Organization for a fellowship during which this review was completed and Professor A. R. Battersby, Cambridge University, for his hospitality during this period.

REFERENCES

1. A. G. Ogston, *Nature (London)* **162**, 963 (1948).
2. K. Mislow, "Introduction to Stereochemistry," p. 127. Benjamin, New York, 1965.
3. For a concise review, see R. Bentley, "Molecular Asymmetry in Biology," Vol. 2, p. 90. Academic Press, New York, 1970.
4. F. A. Loewus, F. H. Westheimer, and B. Vennesland, *J. Am. Chem. Soc.* **75**, 5018 (1953).
5. For a review, see R. Bentley, "Molecular Asymmetry in Biology," Vol. 2, p. 224. Academic Press, New York, 1970.
6. R. Bentley, "Molecular Asymmetry in Biology," Vols. 1 and 2. Academic Press, New York, 1969 and 1970.
7. W. L. Alworth, "Stereochemistry and Its Application in Biochemistry." Wiley (Interscience), New York, 1972.
8. G. Popjak, *in* "The Enzymes" (P. D. Boyer, ed.), 3rd ed., Vol. 2, p. 115. Academic Press, New York, 1970.
9. H. C. Dunathan, *Adv. Enzymol. Relat. Areas Mol. Biol.* **35**, 79 (1971).
10. I. A. Rose, *in* "The Enzymes" (P. D. Boyer, ed.), 3rd ed., Vol. 2, p. 281. Academic Press, New York, 1970.
11. D. Arigoni and E. L. Eliel, *Top. Stereochem.* **4**, 127 (1969).
12. A. R. Battersby and J. Staunton, *Tetrahedron* **30**, 1707 (1974); A. R. Battersby, *Acc. Chem. Res.* **5**, 148 (1972).
12a. D. H. G. Crout, *in Int. Sci. Rev., Ser.* **2** (D. H. Hey, Consultant ed.; H. N. Rydon, Volume ed.) **6**, 281 (1976).
13. E. L. Eliel and H. S. Mosher, *Science* **190**, 772 (1975); J. W. Cornforth, *Tetrahedron* **30**, 1515 (1974); *Chem. Soc. Rev.* **2**, 1 (1973); *Q. Rev., Chem. Soc.* **23**, 125 (1969); V. Prelog, *Pure Appl. Chem.* **9**, 119 (1964).
14. R. H. Abeles, *in* "The Enzymes" (P. D. Boyer, ed.), 3rd ed., Vol. 5, p. 481. Academic Press, New York, 1971; see also Bentley [6] and Popjak [8].
15. B. M. Babior, *Acc. Chem. Res.* **8**, 376 (1975); M. Sprecher, M. J. Clark, and D. B. Sprinson, *J. Biol. Chem.* **241**, 872 (1966).
16. K. R. Hanson and I. A. Rose, *Acc. Chem. Res.* **8**, 1 (1975); K. R. Hanson, *Annu. Rev. Plant Physiol.* **23**, 335 (1972); I. A. Rose, *Annu. Rev. Biochem.* **35**, 23 (1966).
17. K. Mislow and M. Raban, *Top. Stereochem.* **1**, 1 (1967).
18. K. R. Hanson, *J. Am. Chem. Soc.* **88**, 2731 (1966).
19. E. E. Snell and S. J. diMari, *in* "The Enzymes" (P. D. Boyer, ed.), 3rd ed., Vol. 2, p. 335. Academic Press, New York, 1970.

20. L. Schirch and W. T. Jenkins, *J. Biol. Chem.* **239**, 3801 (1964).
20a. D. Wellner, *Biochemistry* **9**, 2307 (1970).
21. P. M. Jordan and M. Akhtar, *Biochem. J.* **116**, 277 (1970); M. Akhtar and P. M., Jordan, *Chem. Commun.* p. 1691 (1968); *Tetrahedron Lett.* p. 875 (1969).
22. P. Besmer and D. Arigoni, *Chimia* **22**, 494 (1968); P. Besmer, Ph.D. Dissertation No. 4435. Eidgenössischen Technischen Hochschule, Zurich, 1970.
23. A. Streitweiser, Jr. and J. R. Wolfe, Jr., *J. Org. Chem.* **28**, 3263 (1963); H. Gerlach, *Helv. Chim. Acta* **49**, 2483 (1966).
23a. B. T. Golding, G. J. Gainsford, A. J. Herlt, and A. M. Sargeson, *Angew. Chem. Int. Ed.* **14**, 495 (1975); *Tetrahedron* **32**, 389 (1976).
23b. J. A. McClarin, L. A. Dressel, and J. I. Legg, *J. Am. Chem. Soc.* **98**, 4150 (1976).
23c. A. G. Palekar, S. S. Tate, and A. Meister, *J. Biol. Chem.* **248**, 1158 (1973).
24. E. M. Wilson and E. E. Snell, *J. Biol. Chem.* **237**, 3171 and 3180 (1962); L. G. Schirch and M. Mason, *J. Biol. Chem.* **238**, 1032 (1963).
25. N. Takamura, S. Terashima, K. Achiwa, and S. Yamada, *Chem. Pharm. Bull.* **15**, 1776 (1967).
26. R. K. Hill and R. Mothershed, in progress.
27. Z. Zaman, P. M. Jordan, and M. Akhtar, *Biochem. J.* **135**, 257 (1973); M. M. Abboud, P. M. Jordan, and M. Akhtar, *J. Chem. Soc., Chem. Commun.* p. 643 (1974).
28. H. C. Dunathan, L. Davis, P. G. Kury, and M. Kaplan, *Biochemistry* **7**, 4532 (1968); J. E. Ayling, H. C. Dunathan, and E. E. Snell, *ibid.* p. 4537; J. G. Voet, D. M. Hindenlang, T. J. J. Blanck, R. J. Ulevitch, R. G. Kallen, and H. C. Dunathan, *J. Biol. Chem.* **248**, 841 (1973).
29. D. Arigoni and P. Besmer, *Chimia* **23**, 190 (1969).
30. G. Barnard and M. Akhtar, *J. Chem. Soc., Chem. Commun.* p. 980 (1975).
31. J. W. Cornforth, J. W. Redmond, H. Eggerer, W. Buckel, and C. Gutschow, *Nature (London)* **221**, 1212 (1969); J. Lüthy, J. Rétey, and D. Arigoni, *ibid.* p. 1213.
31a. A. G. Palekar, S. S. Tate, and A. Meister, *Biochemistry* **9**, 2310 (1970); **10**, 2180 (1971).
32. H. Kochi and G. Kikuchi, *Arch. Biochem. Biophys.* **173**, 71 (1976).
33. J. F. Biellmann and F. J. Schuber, *Biochem. Biophys. Res. Commun.* **27**, 517 (1967); *Bull. Soc. Chim. Biol.* **52**, 211 (1970).
33a. For a later study, see C. M. Tatum, Jr., S. J. Benkovic, and H. G. Floss, *Fed. Proc., Fed. Am. Soc. Exp. Biol.* **34**, 496 (1975).
34. G. E. Skye, R. Potts, and H. G. Floss, *J. Am. Chem. Soc.* **96**, 1593 (1974).
35. H. G. Floss, D. K. Onderka, and M. Carroll, *J. Biol. Chem.* **247**, 736 (1972); D. K. Onderka and H. G. Floss, *J. Am. Chem. Soc.* **91**, 5894 (1969).
36. C. Fuganti, D. Ghiringhelli, D. Giangrasso, P. Grasselli, and A. Santopietro Amisano, *Chim. Ind. (Milan)* **56**, 424 (1974); *Chem. Abstr.* **81**, 116969 (1974).
37. M. Kainosho and K. Ajisaka, *J. Am. Chem. Soc.* **97**, 5630 (1975).
38. Y. Cheung and C. Walsh, *J. Am. Chem. Soc.* **98**, 3397 (1976).
39. C. Fuganti, D. Ghiringhelli, D. Giangrasso, and P. Grasselli, *J. Chem. Soc., Chem. Commun.* p. 726 (1974).
40. E. Schleicher, K. Mascaro, R. Potts, D. R. Mann, and H. G. Floss, *J. Am. Chem. Soc.* **98**, 1043 (1976).
41. L. Zee, U. Hornemann, and H. G. Floss, *Biochem. Physiol. Pflanz.* **168**, 19 (1975).
42. L. Davis and D. E. Metzler, *in* "The Enzymes" (P. D. Boyer, ed.), 3rd ed., Vol. 7, p. 33. Academic Press, New York, 1972; Y. Karube and Y. Matsushima, *J. Am. Chem. Soc.* **98**, 3725 (1976).

43. I. A. Rose, *J. Biol. Chem.* **245**, 6052 (1970).
44. A. Meister, "Biochemistry of the Amino Acids," 2nd ed., Vol. 1, pp. 78 and 402. Academic Press, New York, 1965.
45. D. J. Morecombe and D. W. Young, *J. Chem. Soc., Chem. Commun.* p. 198 (1975).
46. D. J. Aberhart, L. J. Lin, and J. Y. Chu, *J. Chem. Soc., Perkin Trans. 1* p. 2517 (1975).
47. R. K. Hill and G. R. Newkome, *J. Org. Chem.* **34**, 740 (1969).
48. The configuration is incorrectly given by the authors.
49. P. A. Lemke and D. R. Brannon, *in* "Cephalosporin and Penicillin Compounds: Their Chemistry and Biology" (E. H. Flynn, ed.), p. 370. Academic Press, New York, 1973.
50. J. A. Chan, F. Huang, and C. J. Sih, *Biochemistry* **15**, 177 (1976).
51. B. W. Bycroft, C. M. Wels, K. Corbett, and D. A. Lowe, *J. Chem. Soc., Chem. Commun.* p. 123 (1975).
52. See A. Meister, "Biochemistry of the Amino Acids," Vol. 2, pp. 793–797. Academic Press, New York, 1965.
53. J. E. Baldwin, J. Loliger, W. Rastetter, N. Neuss, L. L. Huckstep, and N. De La Higuera, *J. Am. Chem. Soc.* **95**, 3796 (1973).
54. H. Kluender, C. H. Bradley, C. J. Sih, P. Fawcett, and E. P. Abraham, *J. Am. Chem. Soc.* **95**, 6149 (1973).
55. H. Kluender, F. C. Huang, A. Fritzberg, H. Schnoes, C. J. Sih, P. Fawcett, and E. P. Abraham, *J. Am. Chem. Soc.* **96**, 4054 (1974).
56. R. K. Hill, S. Yan, and S. M. Arfin, *J. Am. Chem. Soc.* **95**, 7357 (1973).
57. K. Mislow, R. E. O'Brien, and H. Schaefer, *J. Am. Chem. Soc.* **84**, 1940 (1962).
58. (a) D. J. Aberhart and L. J. Lin, *J. Am. Chem. Soc.* **95**, 7859 (1973); (b) D. J. Aberhart, L. J. Lin, and J. Y. Chu, *J. Chem. Soc., Perkin Trans. 1* p. 2517 (1975).
59. In Aberhart *et al.* [58b] the data agree with this conclusion, but it is erroneously stated that the pro-*R* methyl is upfield.
60. R. A. Archer, R. D. G. Cooper, P. V. Demarco, and L. R. F. Johnson, *Chem. Commun.* p. 1291 (1970).
61. N. Neuss, C. H. Nash, J. E. Baldwin, P. A. Lemke, and J. B. Grutzner, *J. Am. Chem. Soc.* **95**, 3797 (1973).
62. See A. Meister, "Biochemistry of the Amino Acids," Vol. 2, p. 730. Academic Press, New York, 1965.
63. S. M. Arfin, *J. Biol. Chem.* **244**, 2250 (1969).
64. R. K. Hill and S. Yan, *Bioorg. Chem.* **1**, 446 (1971).
65. J. Sjolander, K. Folkers, E. A. Adelberg, and E. Tatum, *J. Am. Chem. Soc.* **76**, 1085 (1954); B. E. Nielsen, P. K. Larsen, and J. Lemmich, *Acta Chem. Scand.* **23**, 967 (1967).
66. R. K. Hill and S. Sawada, unpublished.
67. See A. Meister, "Biochemistry of the Amino Acids," Vol. 2, p. 747. Academic Press, New York, 1965.
68. D. J. Aberhart, *Tetrahedron Lett.* p. 4373 (1975).
69. T. W. Goodwin, "The Biosynthesis of Vitamins and Related Compounds," Chapter 5. Academic Press, New York, 1963.
70. N. M. Davies and D. H. G. Crout, *J. Chem. Soc., Perkin Trans. 1* p. 2079 (1974).
71. D. H. G. Crout, private communication.
72. J. P. Greenstein, S. M. Birnbaum, and L. Levintow, *Biochem. Prep.* **3**, 84 (1953).
73. R. K. Hill and S. Rhee, unpublished.

74. H. Weber, J. Seibl, and D. Arigoni, *Helv. Chim. Acta* **49**, 741 (1966); P. S. Skell, R. G. Allen, and G. K. Helmkamp, *J. Am. Chem. Soc.* **82**, 410 (1960).
75. R. H. Wightman, J. Staunton, A. R. Battersby, and K. R. Hanson, *J. Chem. Soc., Perkin Trans. 1* p. 2355 (1972); K. R. Hanson, R. H. Wightman, J. Staunton, and A. R. Battersby, *Chem. Commun.* p. 185 (1971).
76. G. W. Kirby and J. Michael, *J. Chem. Soc., Perkin Trans. 1* p. 115 (1973); *Chem. Commun.* pp. 187, 415 (1971).
77. K. Brocklehurst, R. P. Bywater, R. A. Palmer, and R. Patrick, *Chem. Commun.* p. 632 (1971).
78. S. Sawada, H. Kumagai, H. Yamada, R. K. Hill, Y. Mugibayashi, and K. Ogata, *Biochim. Biophys. Acta* **315**, 204 (1973).
79. R. Ife and E. Haslam, *J. Chem. Soc. C* p. 2818 (1971).
80. G. E. Clement and R. Potter, *J. Chem. Educ.* **48**, 695 (1971).
81. P. G. Strange, J. Staunton, H. R. Wiltshire, A. R. Battersby, K. R. Hanson, and E. A. Havir, *J. Chem. Soc., Perkin Trans. 1* p. 2364 (1972).
82. G. W. Kirby, S. Narayanaswami, and P. S. Rao, *J. Chem. Soc., Perkin Trans. 1* p. 645 (1975).
83. G. W. Kirby and M. J. Varley, *J. Chem. Soc., Chem. Commun.* p. 833 (1974).
84. G. W. Kirby, J. Michael, and S. Narayanaswami, *J. Chem. Soc., Perkin Trans. 1* p. 203 (1972).
85. V. E. Althouse, D. M. Feigl, W. A. Sanderson, and H. S. Mosher, *J. Am. Chem. Soc.* **88**, 3595 (1966).
85a. A. R. Battersby, J. Staunton, and H. R. Wiltshire, *J. Chem. Soc., Perkin Trans. 1* p. 1156 (1975).
86. A. Peterkofsky, *J. Biol. Chem.* **237**, 787 (1962).
87. I. L. Givot, T. A. Smith, and R. H. Abeles, *J. Biol. Chem.* **244**, 6341 (1969).
88. J. Retey, H. Fierz, and W. P. Zeylemaker, *FEBS Lett.* **6**, 203 (1970).
89. B. E. Ellis, M. H. Zenk, G. W. Kirby, J. Michael, and H. G. Floss, *Phytochemistry* **12**, 1057 (1973).
90. K. R. Hanson and E. A. Havir, *in* "The Enzymes" (P. D. Boyer, ed.), 3rd ed., Vol. 7, p. 75. Academic Press, New York, 1972.
91. R. J. Parry, *J. Chem. Soc., Chem. Commun.* p. 144 (1975).
92. G. W. Kirby and S. Narayanaswami, *J. Chem. Soc., Chem. Commun.* p. 322 (1973).
93. A. R. Battersby, J. E. Kelsey, J. Staunton, and K. E. Suckling, *J. Chem. Soc., Perkin Trans. 1* p. 1609 (1973).
94. L. Bachan, C. B. Storm, J. W. Wheeler, and S. Kaufman, *J. Am. Chem. Soc.* **96**, 6799 (1974).
95. K. B. Taylor, *J. Biol. Chem.* **249**, 454 (1974).
95a. A. R. Battersby, P. W. Sheldrake, J. Staunton, and D. C. Williams, *J. Chem. Soc. Perkin Trans. 1* p. 1056 (1976).
96. P. S. Steyn, R. Vleggaar, N. P. Ferreira, G. W. Kirby, and M. J. Varley, *J. Chem. Soc., Chem. Commun.* p. 465 (1975).
97. S. Sawada, H. Kumagai, H. Yamada, and R. K. Hill, *J. Am. Chem. Soc.* **97**, 4334 (1975).
98. N. Johns, G. W. Kirby, J. D. Bu'Lock, and A. P. Ryles, *J. Chem. Soc., Perkin Trans. 1* p. 383 (1975).
98a. J. C. Vederas and C. Tamm, *Helv. Chim. Acta* **59**, 558 (1976).
99. E. Haslam, "The Shikimate Pathway." Butterworth, London, 1974.
100. M. Cohn, J. E. Pearson, E. L. O'Connell, and I. A. Rose, *J. Am. Chem. Soc.* **92**, 4095 (1970).

101. M. J. Turner, B. W. Smith, and E. Haslam, *J. Chem. Soc., Perkin Trans. 1* p. 52 (1975).
102. B. W. Smith, M. J. Turner, and E. Haslam, *Chem. Commun.* p. 842 (1970).
103. K. R. Hanson and I. A. Rose, *Proc. Natl. Acad. Sci. U.S.A.* **50**, 981 (1963).
104. E. Haslam, M. J. Turner, D. Sargent, and R. S. Thompson, *J. Chem. Soc. C* p. 1489 (1971).
105. A. D. N. Vaz, J. R. Butler, and M. J. Nugent, *J. Am. Chem. Soc.* **97**, 5914 (1975).
106. R. K. Hill and G. R. Newkome, *J. Am. Chem. Soc.* **91**, 5893 (1969).
107. M. G. Bock, Ph.D. Dissertation, University of Georgia, Athens (1974).
108. R. K. Hill and A. G. Edwards, *Tetrahedron Lett.* p. 3239 (1964).
109. S. Mandeles, R. Koppelman, and M. E. Hanke, *J. Biol. Chem.* **209**, 327 (1954).
110. B. Belleau and J. Burba, *J. Am. Chem. Soc.* **82**, 5751 (1960).
111. B. Belleau, M. Fang, J. Burba, and J. Moran, *J. Am. Chem. Soc.* **82**, 5752 (1960).
112. G. W. Chang and E. E. Snell, *Biochemistry* **7**, 2005 (1968).
113. E. Leistner and I. D. Spenser, *J. Chem. Soc., Chem. Commun.* p. 378 (1975).
114. E. Leistner and I. D. Spenser, *J. Am. Chem. Soc.* **97**, 4715 (1973).
115. Y. Fujita, A. Gottlieb, B. Peterkofsky, S. Udenfriend, and B. Witkop, *J. Am. Chem. Soc.* **86**, 4709 (1964); L. A. Salzman, H. Weissbach, and E. Katz, *Proc. Natl. Acad. Sci. U.S.A.* **54**, 542 (1965).
116. I. Y. Yang, Y. Z. Huang, and E. E. Snell, *Fed. Proc., Fed. Am. Soc. Exp. Biol.* **34**, 496 (1975).
117. See Schleicher *et al.* [40, footnote 10].
118. D. Coggiola, C. Fuganti, D. Ghiringhelli, and P. Grasselli, *J. Chem. Soc., Chem. Commun.* p. 143 (1976).

CHAPTER

6

Bioformation and Biotransformation of Isoquinoline Alkaloids and Related Compounds

Tetsuji Kametani, Keiichiro Fukumoto, and Masataka Ihara

INTRODUCTION

The sole source of alkaloids has long been considered to be plants, but the evidence [1–7] that dopamine (1) with acetaldehyde (2a) or 3,4-dihydroxy-phenylacetaldehyde (2b) condenses in mammalian tissues to afford the 1-substituted tetrahydroisoquinolines (3) (Scheme 1) suggests that mammalian systems can synthesize alkaloids and substances that are recognized as intermediates in the biosynthesis of certain alkaloids. These *in vivo* transformations catalyzed by enzymes are of considerable importance from the pharmacological and biosynthetic points of view. In this review we describe

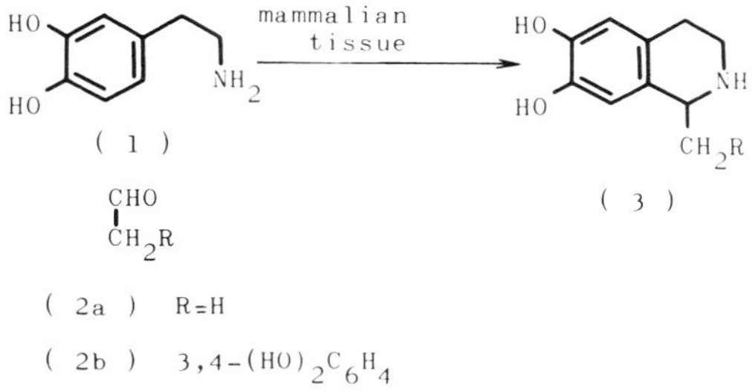

(1)

CHO
|
CH$_2$R

(3)

(2a)   R=H

(2b)   3,4-(HO)$_2$C$_6$H$_4$

Scheme 1

recent advances in the bioformation and biotransformation of some isoquinoline alkaloids.

ENZYMATIC OXIDATION

The enzymatic oxidation and coupling of phenols, which we call "phenol
oxidation" [8], is a subject of great importance in biochemistry and organic
chemistry. Biosynthetic pathways [9] to a wide range of isoquinoline alkaloids
[10] involve oxidation and coupling of simple phenolic isoquinolines as key
reactions [11].
The three main classes of enzymes known as catalysts for phenol oxidation
and coupling are the laccases, the tyrosinases, and the peroxidases, which
have a heavy metal as a coenzyme. For example, horseradish peroxidase
consists of protein together with an iron–porphyrin compound such as (**4**) as
a coenzyme. The enzyme also contains carbohydrate. The iron atom is
thought to be surrounded octahedrally by the porphyrin, the protein, and
another ligand, such as hydrogen peroxide (**5**). (See Scheme 2.)

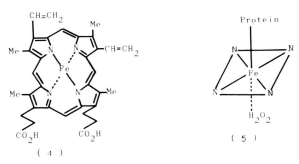

Scheme 2

One outcome of a deeper knowledge of the mechanism of enzymatic
oxidation and coupling of phenols will undoubtedly be the provision of
organic chemists with more efficient chemical oxidants. The chemical oxidant
that seems to come closest to the enzymatic oxidant is ferricyanide, because
the oxidation–reduction potential of ferricyanide (300 mV) is reasonably close
to that recorded for laccase (415 mV).
Potassium ferricyanide and a similar reagent, ferric chloride, have been
used as enzyme models for biogenetic-type synthesis of isoquinoline and
related alkaloids by phenol oxidation [11].
As shown in Schemes 3 and 3′, many types of isoquinoline alkaloids can
be synthesized by phenol oxidation with chemical oxidants along the

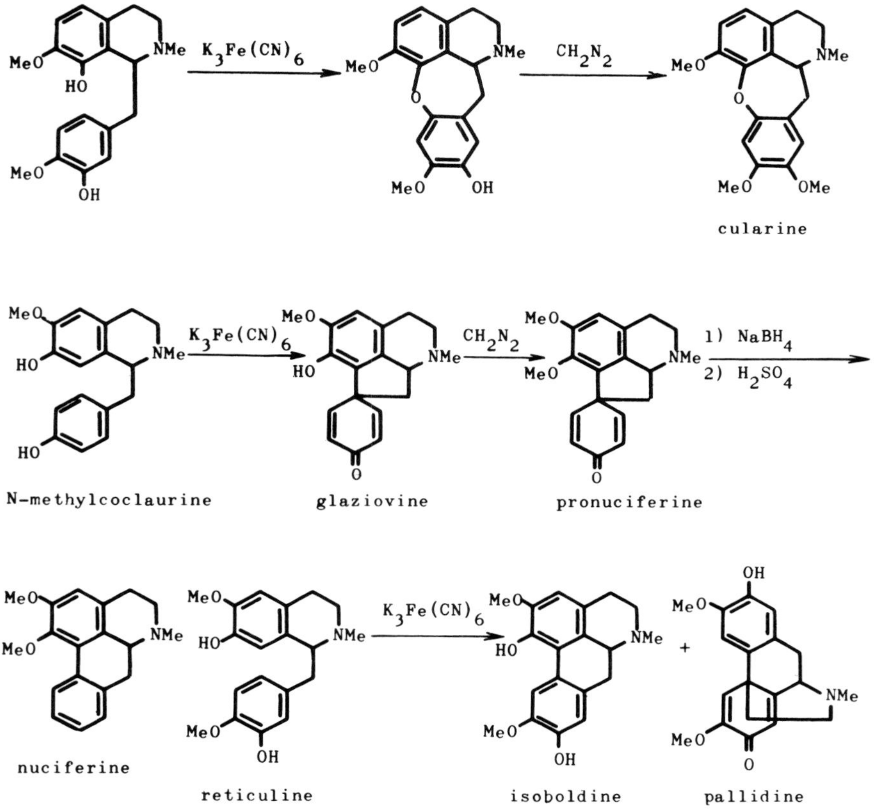
K3Fe(CN)6
CH2N2
cularine
N-methylcoclaurine
glaziovine
pronuciferine
1) NaBH4
2) H2SO4
nuciferine
reticuline
isoboldine
pallidine

Scheme 3

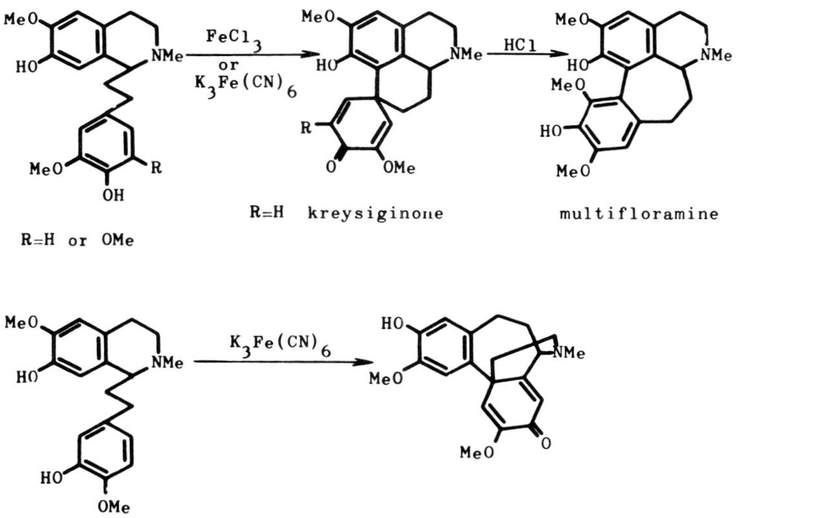
FeCl3
or
K3Fe(CN)6
HCl
R=H or OMe
R=H kreysiginone
multifloramine

Scheme 3′

biogenetic lines. In this oxidation, the coupling reaction proceeds in an intramolecular manner, but not intermolecularly. On the other hand, enzymatic
oxidation of phenolic isoquinolines proceeds in a manner that differs from the
case of oxidation with chemical reagents. Reticuline (6) was treated with
horseradish peroxidase in the presence of hydrogen peroxide at pH 7.5 and
20°C to give the isocarbostyril alkaloid thalifoline (7) [12]; oxidation of
reticuline (6) with homogenized *Papaver rhoeas* and hydrogen peroxide at
pH 5.0 and at 20°C afforded β-hydroxyreticuline (8), a key precursor of
phthalideisoquinoline (9) and rheadan alkaloids (10) [13]. On the other hand,
chemical oxidation of reticuline gave isoboldine and pallidine, as shown in
Scheme 4.

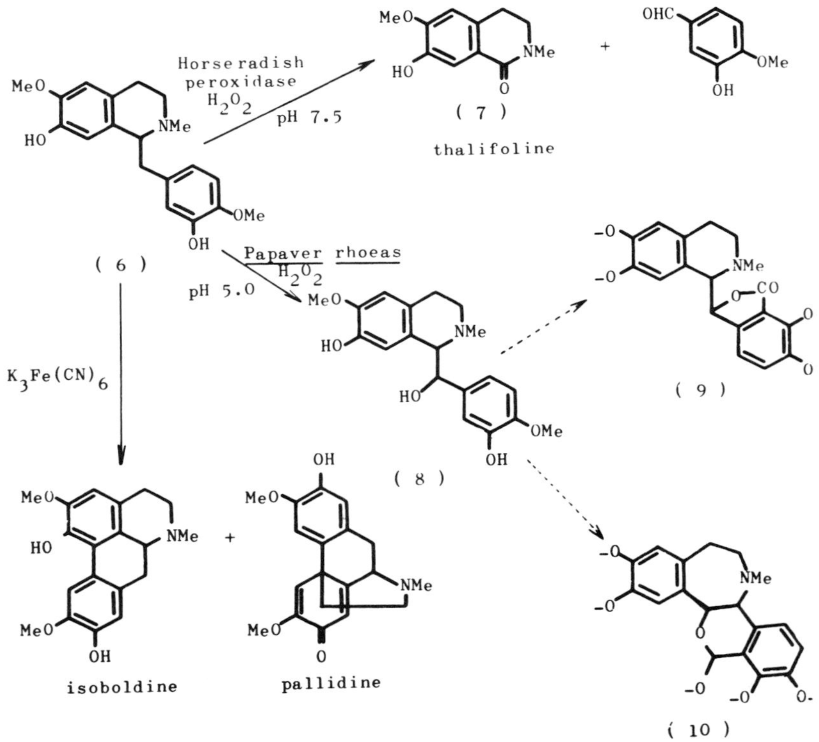

Scheme 4

Similarly, oxidation of O-benzyl-N-methylcoclaurine (11) with horseradish
peroxidase and hydrogen peroxide at pH 6.6 for 30 hr at 25–28°C and then
reduction of this crude product with sodium borohydride afforded O-benzyl-
corypalline (13) [12]. Similar oxidation of (1R)-(−)-N-methylcoclaurine (12)

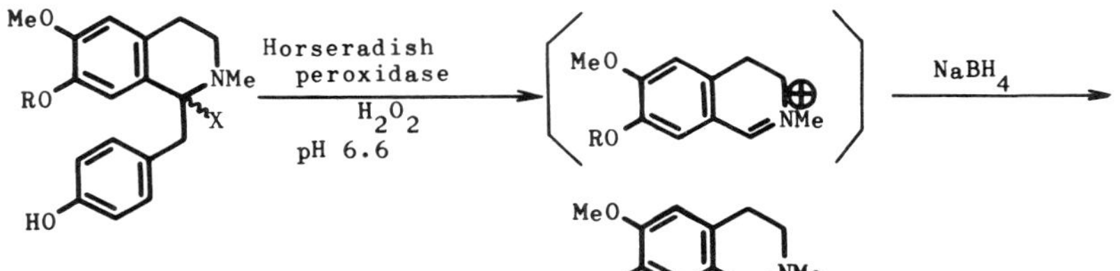

(11)   R=CH₂Ph, X=∼H

(12)   R=H, X=—H

(13)   R=CH₂Ph

(14)   R=H

Scheme 5

followed by sodium borohydride reduction gave corypalline (**14**) [12]. (See Scheme 5.)

Inubushi [14] also reported *in vitro* oxidation of some phenolic isoquinolines with horseradish peroxidase and hydrogen peroxide as follows. N-Methylisosalsoline (**15**) was oxidized in 3% aqueous ammonium acetate solution with peroxidase and hydrogen peroxide at pH 7.0 for 14 hr at 20–25°C, and the

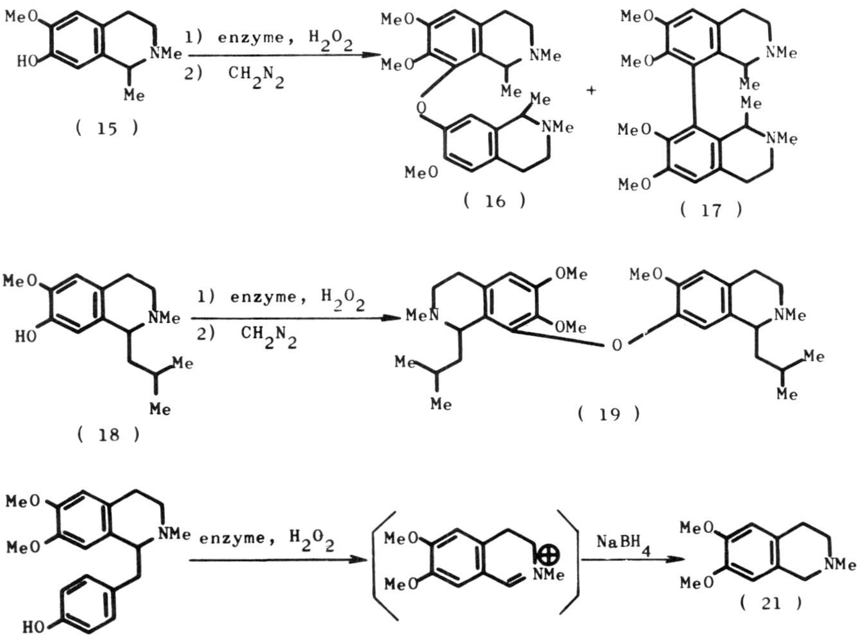

(15)

(16)

(17)

(18)

(19)

(20)

(21)

Scheme 6

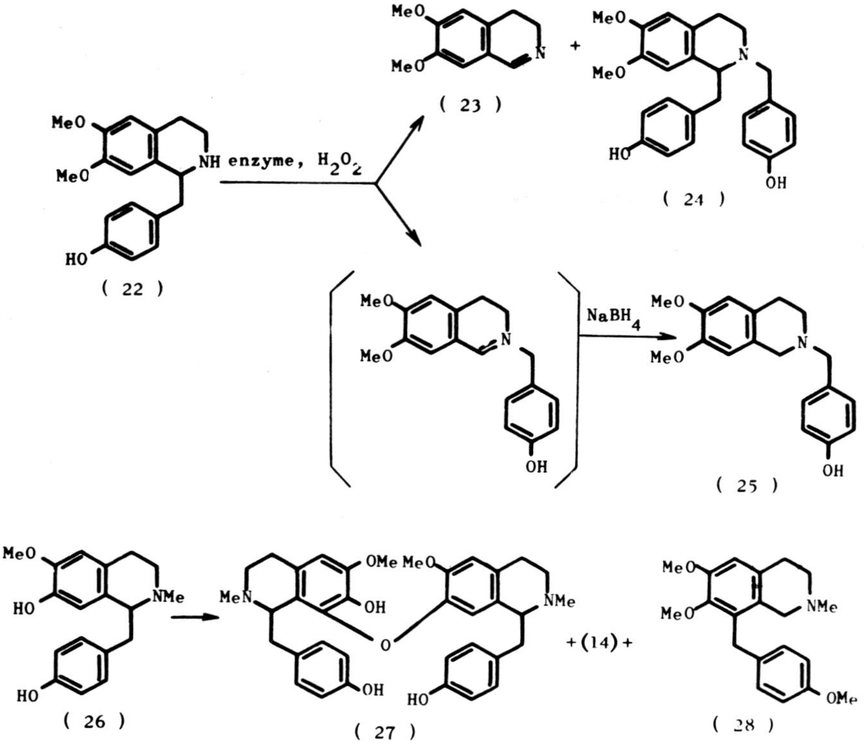

Scheme 6'

crude product was treated with diazomethane to afford the biphenyl ether (**16**) as a diastereoisomeric mixture, along with biphenyl (**17**). The oxidation of lophocerine (**18**) by the same method followed by *O*-methylation yielded *O*-methylisopilocereine (**19**) and its diastereoisomer. In the oxidation of armepavine (**20**) at pH 7.0, *O*-methylcorypalline (**21**) was isolated from the quaternary base fraction after sodium borohydride reduction. (See Scheme 6.) On the other hand, the oxidation of *N*-norarmepavine (**22**) at pH 6.5 gave a somewhat different result; thus, from the tertiary base fraction the non-phenolic base (**23**) and the phenolic base (**24**) were isolated, and the quaternary base fraction gave *O*-methylsendaverine (**25**) after sodium borohydride reduction of the crude product. The same oxidation of *N*-methylcoclaurine (**26**) afforded the biphenyl ether (**27**), corypalline (**14**), and the rearranged benzylisoquinoline (**28**). (See Scheme 6'.)

Inubushi suggested that the elimination of the *p*-hydroxybenzyl group in an oxidation might involve participation of the nitrogen cation radical as well as the phenolic hydroxyl group. Furthermore, *N*-benzylisoquinoline (**25**) and

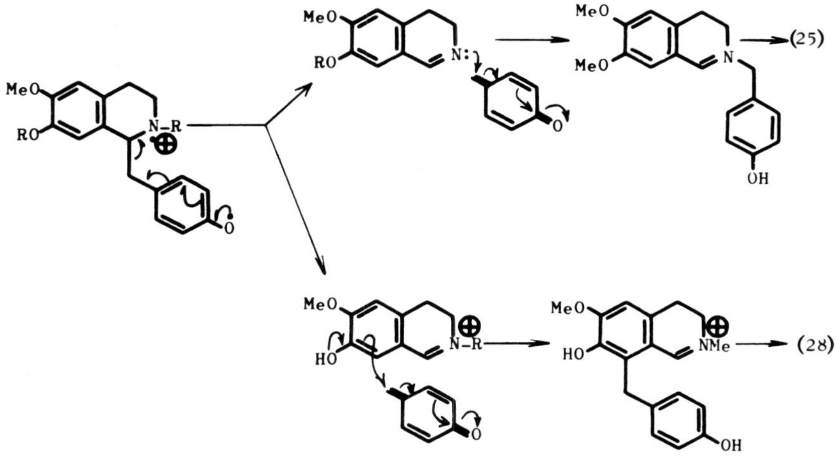

Scheme 7

8-benzylisoquinoline (**28**) were assumed to result from the attack of the nitrogen lone pair or phenolic nucleus on a quinonoid form in the elimination reaction. (See Scheme 7.)

In 1967, Frömming [15] effected the conversion of laudanosoline metho-bromide (**29**) to 1,2,9,10-tetrahydroxyaporphine methobromide (**30**) by oxidation with the enzymes tyrosinase, laccase, and peroxidase (Scheme 8), but the yield was very poor, and the identification of the product was carried out only by ultraviolet spectral and thin-layer chromatography (tlc) comparison. In contrast, Brossi [16] reported that by the use of the purified

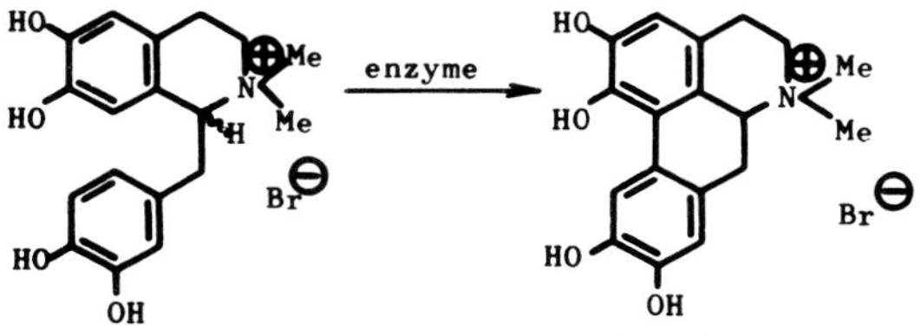

Scheme 8

enzyme horseradish peroxidase under controlled reaction conditions, oxidative coupling of (1S)-(+)-laudanosoline hydrobromide (**31**) and (1R)-(−)-laudanosoline methiodide (**33**) could be effected with great facility and in a preparative manner at pH 5 to afford the quaternary dibenzopyrrocoline (**32**) in 81% yield and the quaternary aporphine (**34**) in 50% yield, with retention of configuration (Scheme 9).

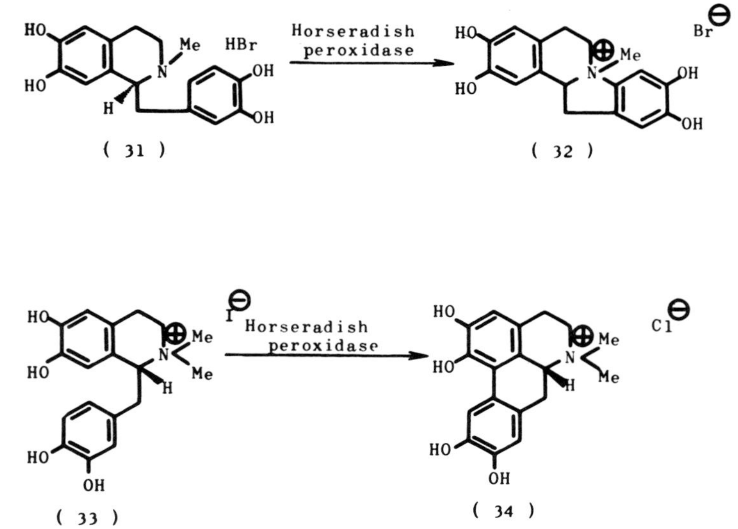

Scheme 9

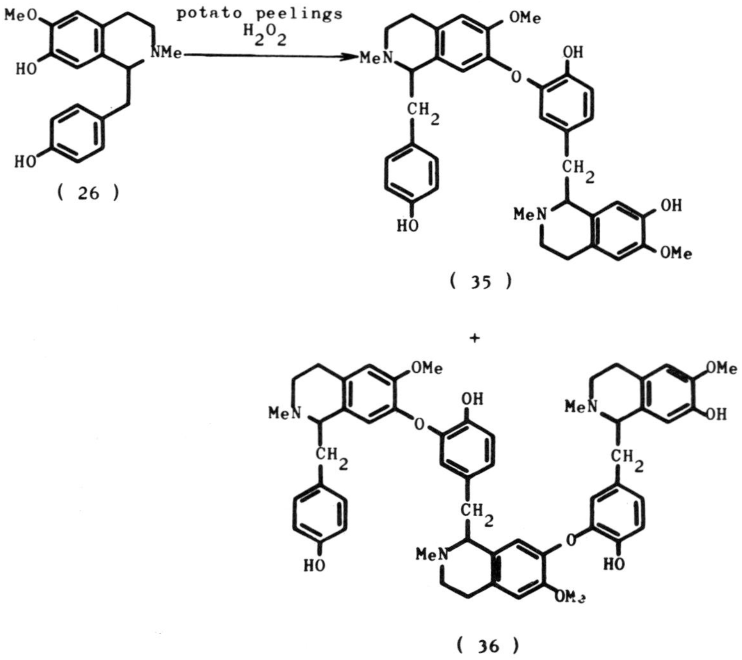

Scheme 10

N-Methylcoclaurine (**26**), which furnished glaziovine by potassium ferricyanide oxidation as shown in Scheme 3, on treatment with homogenized
potato peelings and hydrogen peroxide at pH 4.8 for 8 days at room temperature afforded the liensinine-type compound (**35**) in addition to a small amount
of the trimer (**36**) (Scheme 10). Both compounds were isolated as the corresponding acetates and showed a C—O—C head-to-tail coupling pattern
[17]. This reaction constitutes a biogenetic-type synthesis of liensinine-type
bisbenzylisoquinoline alkaloids. Similarly, treatment of *N*-methylhomoco-
claurine (**37**), which gives the homoproaporphine (**39**) by potassium ferricyanide or ferric chloride oxidation [11], with homogenized potato peelings
and hydrogen peroxide at pH 4.8 for 4 days at room temperature, gave

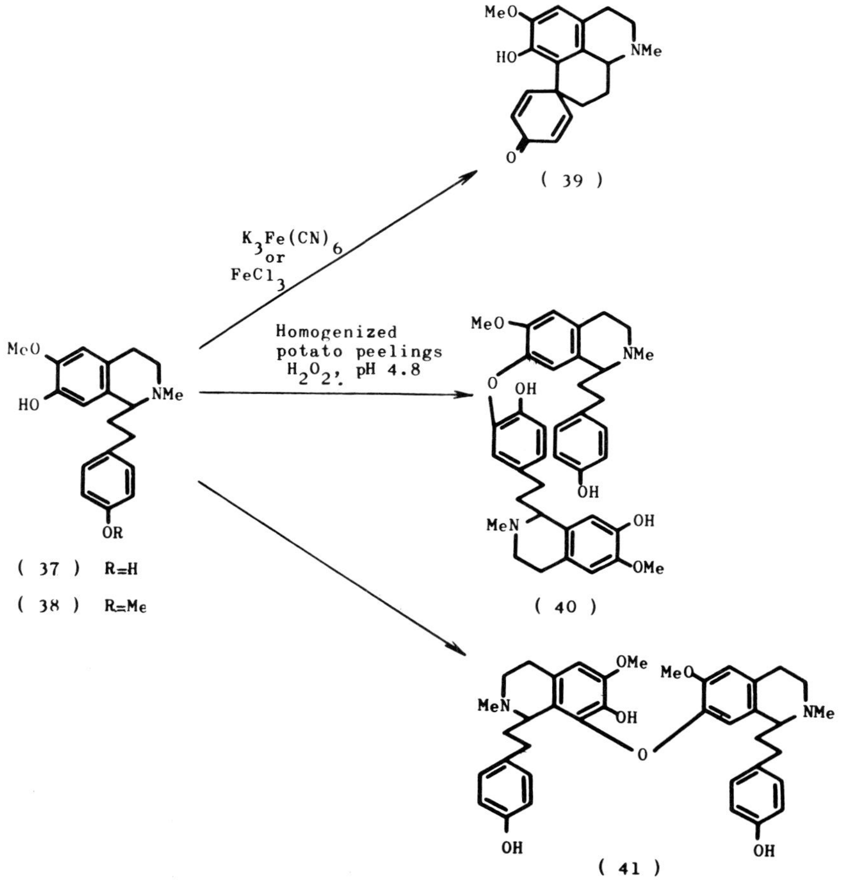

Scheme 11

promelanthioidine (**40**) [18]. Oxidation of (**37**) with homogenized horseradish and hydrogen peroxide at pH 4.8, under the same conditions as those described above, afforded, interestingly, the isomeric bisphenethylisoquinoline (**41**) (Scheme 11) [19]. On the other hand, monophenolic isoquinoline (**38**) gave no oxidized products under the same reaction conditions [19].

In the above reactions, C—C coupling products were not observed. Potato peel homogenates appear to contain the enzyme needed for head-to-tail coupling, while horseradish and *Wasabia japonica* seem to contain the enzyme for head-to-head coupling. Bisbenzylisoquinolines resulting from both head-to-head and tail-to-tail C—O—C coupling (oxyacanthine–berbamine-type alkaloids) and only tail-to-tail C—O—C coupling (dauricine-type alkaloids) have been found in the same plants. Natural alkaloids arising from only head-to-head C—O—C coupling, however, have not yet been observed. In contrast, bisbenzylisoquinolines resulting from head-to-tail C—O—C coupling (liensinine-type alkaloids) and alkaloids of the dauricine and oxyacanthine types have not been observed in the same plants. In the light of this evidence, the protein part of the enzyme which accepts the substrate seems to be different in the enzymes of potato and *Wasabia japonica*. It is interesting that no phenethylisoquinolines with two ether linkages have been observed in this enzymatic oxidation. It is possible that one pair of hydroxy groups in the 1-phenethyl groups participates in the approach to two isoquinoline molecules through the enzyme active site, whereas another two hydroxy groups in the isoquinoline nucleus seem to oxidize each other by phenol oxidation.

In order to investigate these points further, 1,2,3,4-tetrahydro-7-hydroxy-6-methoxy-1-(4-methoxyphenethyl)-2-methylisoquinoline (**38**) was synthesized

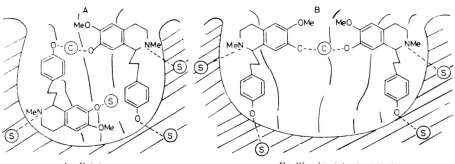

A: Potato enzyme   B: *Wasabia japonica* enzyme
S: Specific site
C: Catalytic site

Scheme 12

for use as a modified substrate. Although conditions identical to those described above were used, (38) was recovered unchanged in attempted oxidation with *Wasabia japonica* and potato homogenates. On the basis of the above evidence one of the present authors suggested that the probable active center in the enzyme is that shown in Scheme 12 [19].

Homoorientaline (42) was oxidized with homogenized potato peelings at pH 4.8 without hydrogen peroxide at room temperature for 3 days and then acetylated in the usual way to give 1-hydroxyhomoorientaline triacetate (43) [20], but chemical oxidation of (42) afforded kreysiginone (44) in good yield (Scheme 13) [11]. It is interesting that enzymatic oxidation in the absence of hydrogen peroxide gave the hydroxylated product.

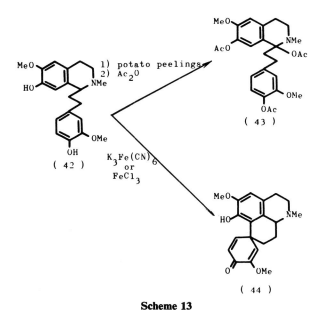

Scheme 13

In summary, many differences were found in the behavior of enzymatic models and enzymes; chemical reagents were more suitable for intramolecular C—C coupling than for intermolecular reaction, but enzymes led to the intermolecular C—O coupling and intramolecular C—N coupling products. In order to understand these differences, it will be necessary to engage in thorough physicochemical and biochemical studies of the relationship between the metal ion, a type of enzymatic model, and the enzyme.

MICROBIAL TRANSFORMATION OF ISOQUINOLINE ALKALOIDS

One of the most striking and significant developments in synthetic organic chemistry is the use of biological systems to effect chemical reactions [21], because their high specificity in the above systems have distinct advantages over the chemical reactions. Therefore, microbial transformation is one of the most important synthetic methods that would be expected to be developed in the future.

The important discovery of the microbial 11α-hydroxylation of progesterone by Peterson [22] resulted in surprising progress in steroid chemistry. However, systematic studies cannot be pursued in the field of microbial transformation of alkaloids as in the case of steroids because of the wide variation in skeletal structures. Some examples have been reported by Tsuda and co-workers; namely, *Trametes sanguinea* transformed thebaine (45) and codeinone (46) to the corresponding 14-hydroxycodeinone (47) (Scheme 14). Tsuda proved that this transformation is an enzymatic reaction by showing that the oxygen of the 14-hydroxyl group originated from air, and not from water [23].

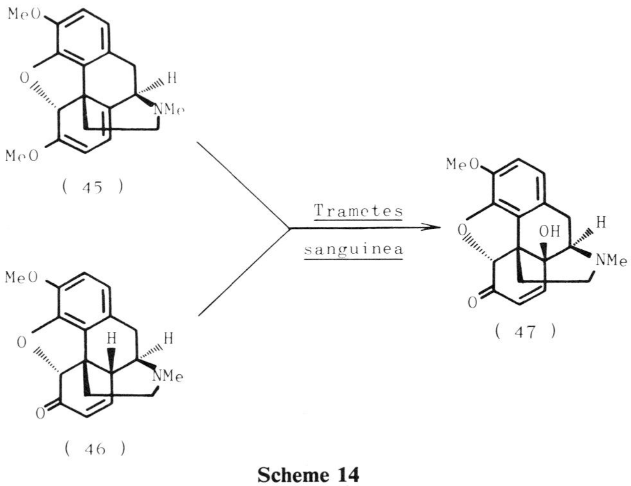

Scheme 14

Microbial transformation of *N*-methylcoclaurine (26) has also been examined by the Tsuda group through use of *Poria inermis* IAM 9050. The clear supernatant from the culture broth of the logarithmic phase was fractionated with ammonium sulfate (0–50, 50–80, and 0–80% saturated fraction), and *N*-methylcoclaurine (26) was treated with these enzyme fractions in the presence of pyridine nucleotide. The oxidation with a mixture containing 50% saturation fraction (0.27 mg/ml, protein concentration) and NAD as the electron acceptor converted (26) to glaziovine (48) in poor yield (Scheme 15) [24]. Recently, Smith and Rosazza [25] described microbial models of mammalian metabolism and discussed hydroxylation, N-dealkylation, O-dealkylation, and N-oxidation as microbial transformations.

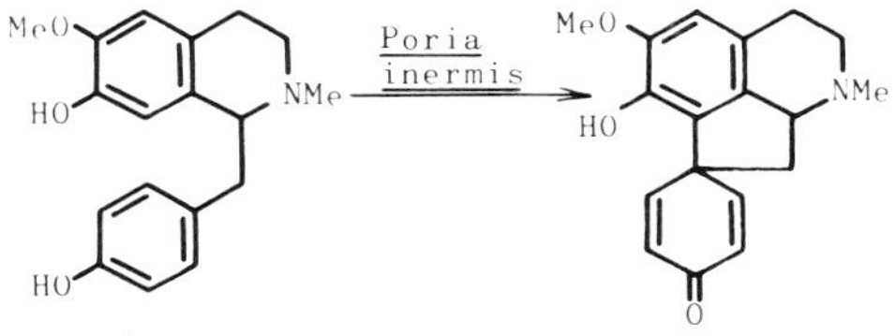

Scheme 15

BIOFORMATION AND BIOTRANSFORMATION OF ISOQUINOLINE ALKALOIDS IN MAMMALIAN SYSTEMS

As mentioned earlier in this review, plants have long been considered to be the sole source of isoquinoline alkaloids. In 1964, Holtz and co-workers [1,26] suggested that precursors of some isoquinoline alkaloids might be formed in mammalian tissues. Thus, this first communication reported that dopamine (1) was converted to norlaudanosoline (3b) in experimental animals in the presence of a monoamine oxidase concentrate. Later, Sandler isolated salsolinol (3a) and norlaudanosoline (3b) as urinary excretion products from patients treated with L-dopa (49) [27], and Collins reported formation of salsolinol in the brain of rats after oral administration of L-dopa and ethanol [28]. He also reported that cow adrenal glands containing epinephrine (50) and norepinephrine (51) were perfused with acetaldehyde to form the corresponding tetrahydroisoquinolines (52) and (53) [5]. Furthermore, using rat brain stem homogenates, it was found that an incubation mixture of dopamine (1) and acetaldehyde or ethanol, together with the tissue homogenate, produced salsolinol (3a) (see Scheme 16) [4].

Interestingly, using dopamine (1) as a substrate with rat tissue homogenates, Davis found that addition of ethanol or acetaldehyde to the substrate, NAD, and aldehyde dehydrogenase gave N-norlaudanosoline (36) and that, furthermore, N-norlaudanosoline was intravenously administered to the rats to form morphinelike alkaloids (54). Thus, alcohol inhibits the oxidation of 3,4-dihydroxyphenylacetaldehyde to phenylacetic acid, the normal

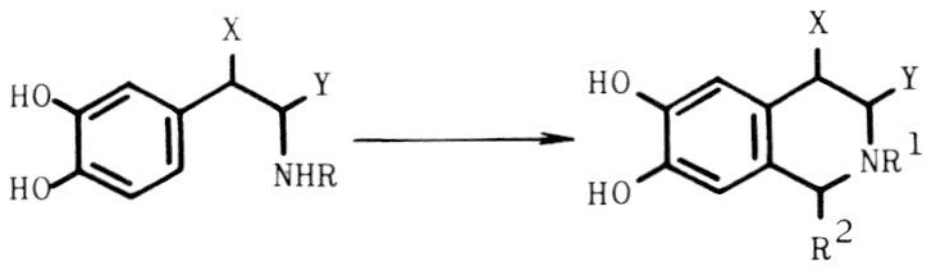

(1)   X=Y=R=H   (3a)   X=Y=R^1=R^2=H

(49)   X=R=H, Y=CO$_2$H   (3b)   X=Y=R^1=H, R^2=CH$_2$C$_6$H$_3$(OH)-3,4

(50)   X=OH, Y=H, R=Me   (52)   X=OH, Y=H, R^1=R^2=Me

(51)   X=OH, Y=R=Me   (53)   X=OH, Y=H, R^1=H, R^2=Me

Scheme 16

metabolic pathway, and promotes the formation of *N*-norlaudanosoline. On the basis of the foregoing information, Scheme 17 was presented by Davis [3,29]. However, the Davis report has been criticized by Seevers [30].

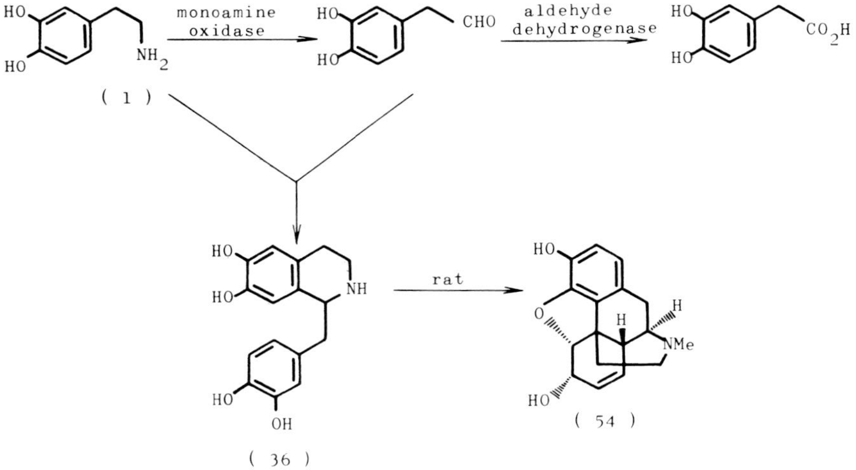

Scheme 17

As mentioned earlier, reticuline (**6**) plays an important role in the biogenesis of isoquinoline alkaloids. Direct oxidative intramolecular coupling of reticuline gives aporphine alkaloids (isoboldine and corytuberine), morphinandienone alkaloids (pallidine and salutaridine), and protoberberine alkaloids (coreximine and scoulerine), which may be precursors of benzophenanthridine alkaloids (chelidonine), phthalideisoquinoline alkaloids (narcotine), protopine alkaloids (protopine), spirobenzylisoquinoline alkaloids (ochotensimine), and rheadan alkaloids (rhoeadine). (See Scheme 18.) In order to investigate the biotransformation of reticuline with mammalian tissue, a solution of (+)-reticuline in propylene glycol was injected intraperitoneally into rats, and for

isobolud ine   pallidine   salutaridine   chelidonine

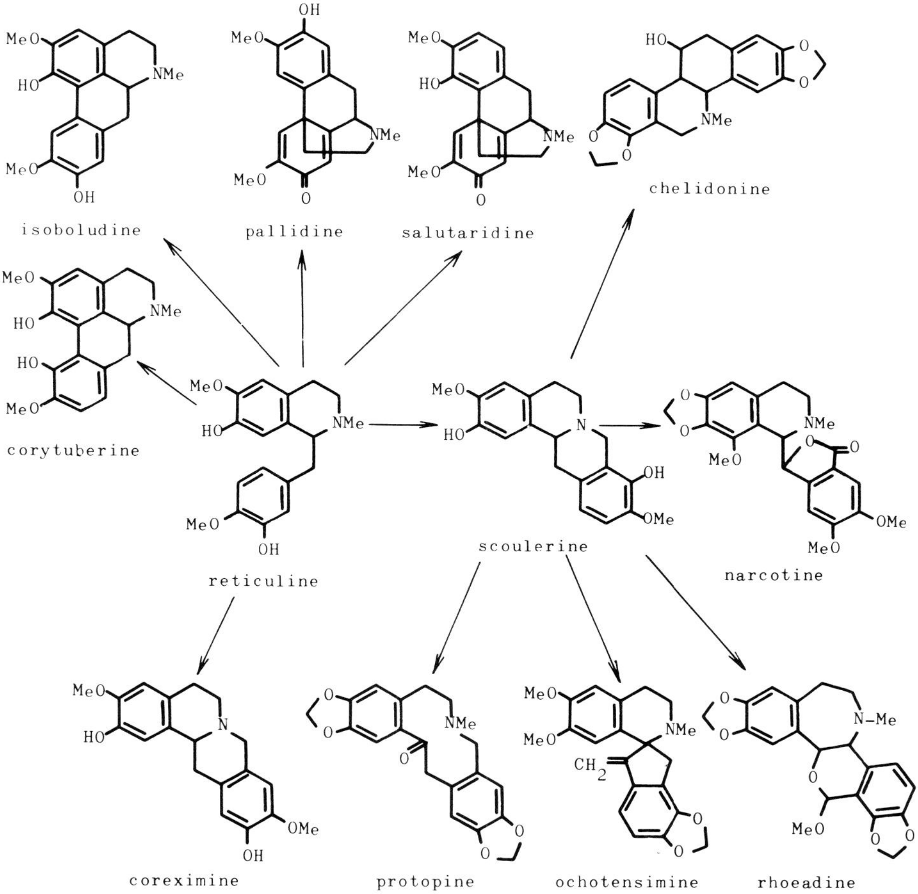

corytuberine   reticuline   scoulerine   narcotine

coreximine   protopine   ochotensimine   rhoeadine

Scheme 18

4 days after the injection the urine was collected in a bottle containing a few drops of toluene. The pooled urine was adjusted to pH 5 with dilute sulfuric acid and then to pH 4.5 with 0.1 mole acetate buffer and incubated with β-glucuronidase. The crude basic extract was separated by preparative tlc on silica gel to yield the starting reticuline and a protoberberine, the structure of which was shown to be coreximine by thin layer chromatography (tlc) and gas chromatographic comparison and mass spectrometry [31].

The biotransformation of 1-benzyl-1,2,3,4-tetrahydro-2-methylisoquinolines to tetrahydroprotoberberines was substantiated by tracer experiments. The radioactive substrates were synthesized as follows. (±)-Reticuline was prepared by methylation of 7-benzyloxy-1-(3-benzyloxy-4-methoxybenzyl)-3,-4-dihydro-6-methoxyisoquinoline with methyl iodide, followed by the reduction of the methiodide with sodium borohydride and then debenzylation of the resulting (±)-*O,O*-dibenzylreticuline with a mixture of concentrated hydrochloric acid and benzene. Tritiation of (±)-reticuline with tritiated water in the presence of potassium *tert*-butoxide according to Kirby's method [32] gave (±)-[2′,6′,8-³H]reticuline (**55**). A parallel experiment using deuterated water instead of tritiated water gave (**56**), the nuclear magnetic resonance spectrum of which showed that three hydrogens at the ortho and para positions of the phenolic hydroxyl groups were equally substituted with deuterium. Methylation of (±)-[2′,6′,8-³H]reticuline (**55**) with diazomethane afforded (±)-[2′,6′,8-³H]laudanosine (**57**), which was free from radioactive reticuline (**55**) or laudanine (**58**) on the basis of thin-layer radiochromatographic analysis. (See Scheme 19.)

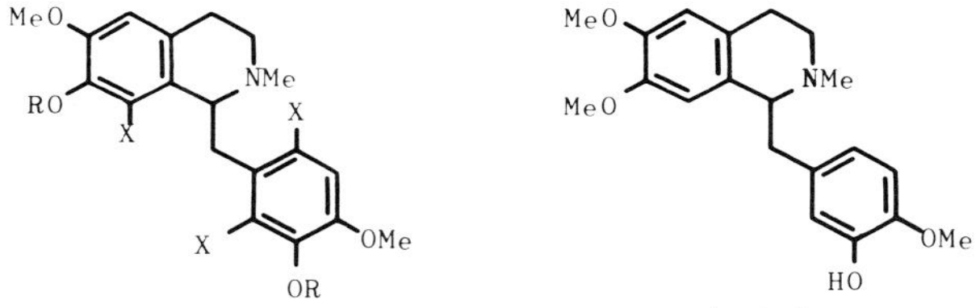

(55)   X = ³H, R = H

(56)   X = ²H, R = H

(57)   X = ³H, R = Me

Scheme 19

(±)-[*N*-¹⁴CH₃]Reticuline (**61**) was prepared by the reductive methylation of (±)-*O,O*-dibenzylnorreticuline (**59**) with [¹⁴C]formalin and sodium borohydride in methanol, followed by debenzylation of (±)-*O,O*-[*N*-¹⁴CH₃]-dibenzylreticuline (**60**) (Scheme 20).

A solution of (±)-[*N*-¹⁴CH₃]reticuline (**61**) was administered to a female rat of Wistar strain. After treatment of the collected urine and dilution with nonradioactive (±)-coreximine (**64**), radioactive (±)-coreximine was isolated in a pure state in 0.03% yield.

The biotransformation was also demonstrated by incubating a 20% homogenate of rat liver, prepared in phosphate buffer at pH 7.4, with (±)-[2′,6′,8-³H]- (**55**) and (±)-[*N*-¹⁴CH₃]reticuline (**61**). After incubation at 37°C for 2 hr, carrier alkaloid was added to each reaction mixture and then purified by preparative tlc, followed by recrystallization to constant activity. The results of the above tracer experiments are shown in Table 1. In experiment 4, yields were calculated as 1.5 times greater than that of the total

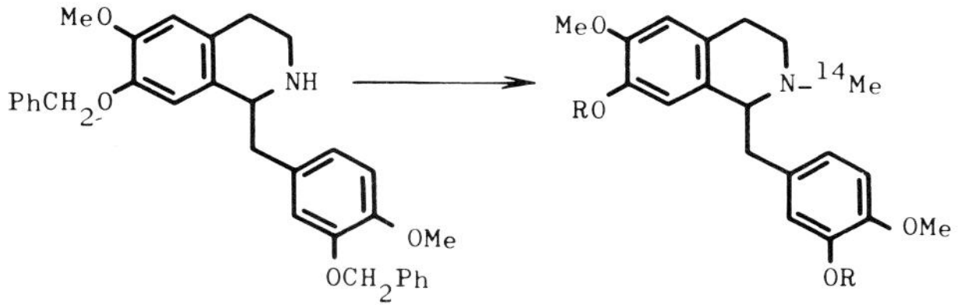

(59)

(60) R=CH$_2$Ph

(61) R=H

Scheme 20

activities determined for pure product, since one tritium was lost in the cyclization. Formation of corexime (**64**) and scoulerine (**66**) from (±)-reticuline with homogenized rat liver was proved by experiments with (±)-reticuline labeled with carbon-14 or tritium. In the experiments, in which ring-tritiated precursors were utilized, percent incorporations observed may possibly be lower than actual values because some tritium may be expected to exchange with hydrogen. No radioactivity was found in pure sinoacutine (**70**).

The amounts of alkaloids formed by the above enzymatic reaction were too small for measurement of optical rotation. Since the optically active corexime was separable from the racemate by repeated recrystallizations, it was expected that, if a small amount of the radioactive product, which is optically

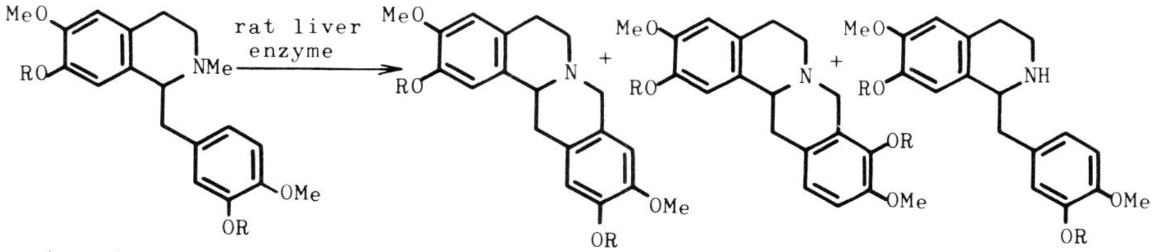

(62) R=H

(63) R=Me

(64) R=H

(65) R=Me

(66) R=H

(67) R=Me

(68) R=H

(69) R=Me

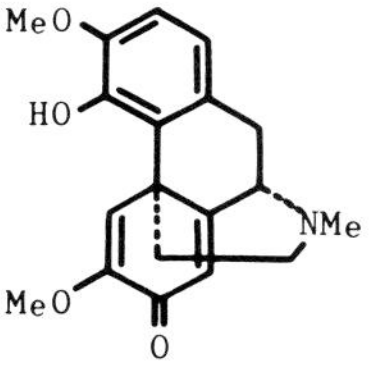

(70)

Scheme 21

TABLE 1

Biotransformation of Radioactive Reticuline Using a Whole Rat or Homogenized Rat Liver

Experiment No.	Substrate	Total activity fed (dpm)	Carrier alkaloid	Yield (%)
1[a]	(±)-[N-$^{14}CH_3$]Reticuline	3.27×10^6	(±)-Coreximine (64)	0.030
2[b]	(±)-[N-$^{14}CH_3$]Reticuline	3.79×10^6	(±)-Coreximine (64)	0.083
3[b]	(±)-[N-$^{14}CH_3$]Reticuline	5.52×10^6	(−)-Scoulerine (66)	0.042
			Sinoacutine (70)	0
4[b]	(±)-[2′,6′,8-^{3}H]Reticuline	1.57×10^7	(±)-Coreximine (64)	0.110
			(±)-Scoulerine (66)	0.034
5[b]	(±)-[N-$^{14}CH_3$]Reticuline	1.06×10^7	(±)-Coreximine	0.090
			(+)-Coreximine	0.044
			(−)-Coreximine	0.050

[a] A whole-rat experiment.
[b] A homogenized rat liver experiment.

TABLE 2

Biotransformation of Tritium-Labeled Reticuline and Laudanosine Using a Rat Liver 9000 g Supernatant

Experiment No.	Radioactive substrate	Cofactors	Total activity fed (dpm)	Carrier alkaloid	Yield (%)
6	($\pm$)-Reticuline (55), 15.3 μmoles	—	1.16×10^7	($\pm$)-Coreximine (64)	0.160
				($\pm$)-Norreticuline (68)	0.410
7	($\pm$)-Reticuline (55), 16.5 μmoles	NADPH, 16.5 μmoles	1.26×10^7	($\pm$)-Coreximine (64)	3.31
				($\pm$)-Norreticuline (68)	1.35
8	($\pm$)-Reticuline (55), 8.7 μmoles	NADPH, 8.7 μmoles; MgCl$_2$, 1 mmole	6.65×10^6	($\pm$)-Coreximine (64)	11.68
9	($\pm$)-Reticuline (55), 8.0 μmoles	NADP, 8.0 μmoles; MgCl$_2$, 1 mmole; nicotinamide, 1 mmole; G-6-P, 30 μmoles	6.05×10^6	($\pm$)-Coreximine (64)	5.99
10	($\pm$)-Laudanosine (57), 46.5 μmoles	NADPH, 55.2 μmoles	3.54×10^7	($\pm$)-Xylopinine (65)	0.019
				(−)-Tetrahydropalmatine (67)	0.007
				($\pm$)-Norlaudanosine (69)	14.13

active, was diluted with a large amount of the enantiomeric coreximine, the repeated recrystallizations would eventually show no radioactivity. Thus, after incubation, the resulting homogenate was equally separated into three fractions, to which (−)-, (+)-, and (±)-coreximine were added. The radiochemically pure bases were obtained by rigorous preparative tlc, followed by repeated recrystallizations. The yields, which were calculated on the basis of the amount of "cold" carrier added, are shown in the last experiment of Table 1. The radiochemical yield of the product from dilution with (±)-coreximine was nearly twice that obtained from dilution with (−)- or (+)-coreximine, respectively, a result which suggests that the coreximine formed was the racemate. It is interesting that the above enzymatic transformation of reticuline to coreximine was not stereospecific, while in a similar case Battersby and co-workers have found that reticuline undergoes rapid racemization in plants [33].

The transformation of (±)-reticuline was further studied using a 9000 *g* supernatant of rat liver homogenates (Table 2). Without any cofactor, the radioactivity of (±)-[2′,6′,8-³H]reticuline (55) was incorporated into (±)-coreximine (64) in 0.160% and (±)-norreticuline (68) in 0.410% yield (experiment 6). Addition of NADPH to the supernatant increased the formation of (±)-coreximine and (±)-norreticuline to 3.31 and 1.35%, respectively (experiment 7). Furthermore, the yield of coreximine was enhanced by the addition of magnesium chloride. Thus, in the presence of NADPH and magnesium chloride, the incorporation of (±)-reticuline into (±)-coreximine was 11.68% (experiment 8). When the supernatant together with glucose 6-phosphate, NADP, nicotinamide, and magnesium chloride was used, (±)-reticuline was converted to (±)-coreximine in 5.99% yield (experiment 9).

In the case of nonphenolic tetrahydroisoquinolines, the radioactivity of (±)-[2′,6′,8-³H]laudanosine (57) was incorporated into (±)-norlaudanosine (69) in 14.13% yield in the presence of NADPH. The incorporations into (±)-xylopinine (65) and (−)-tetrahydropalmatine (67) were very small, but not zero (experiment 10). (See Scheme 21.)

It is reasonable on the basis of the above results that 1-benzyl-1,2,3,4-tetrahydro-2-methylisoquinolines (71) would give rise to immonium cations (72), which afford tetrahydroprotoberberines by cyclization or an *N*-demethylated product by hydrolysis. All possible mechanisms for the formation of the immonium cation (72) are outlined in Scheme 22. The responsible rat liver enzyme would require NADPH and molecular oxygen as in the known enzymatic *N*-demethylation, suggesting the participation of the cytochrome *P*-450 enzyme system [25]. The formation of xylopinine (65) and tetrahydropalmatine (67) from laudanosine (63) would exclude the possibility that the radical (73) is enzymatically oxidized to the diradical (74), which affords a

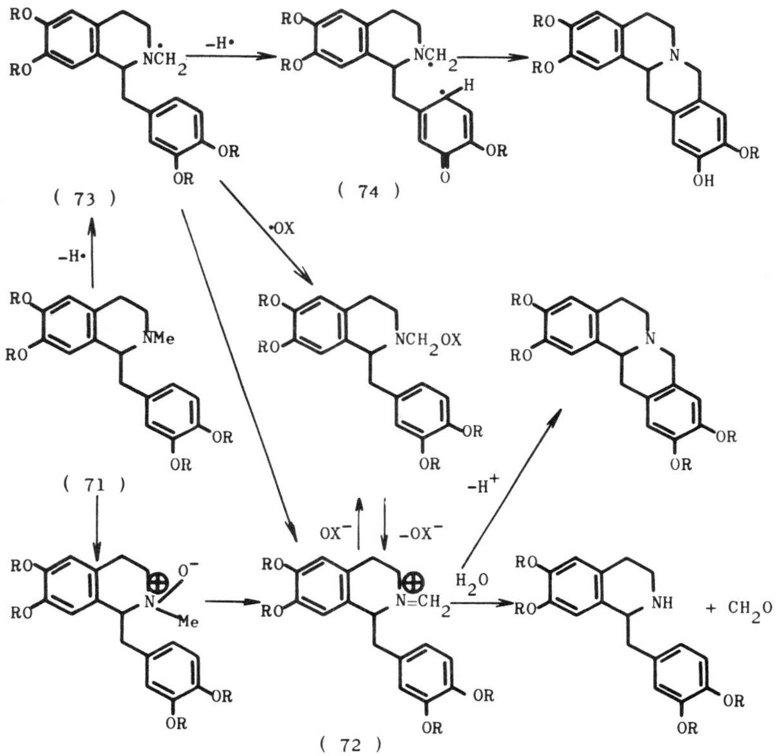

Scheme 22

protoberberine by a radical coupling. It is difficult to determine whether the cyclization step of immonium cation (**72**) is enzymatic or nonenzymatic [34].

Recently, Davis and co-workers observed the *in vivo* and *in vitro* conversion of tetrahydropapaveroline to coreximine and related berberines by mammalian systems in the presence of *S*-adenosylmethionine [35,36]. We believe the cyclization occurred after *N*-methylation.

REFERENCES

1. P. Holtz, K. Stock, and E. Westermann, *Arch. Exp. Pathol. Pharmakol.* **248**, 387 (1964).
2. P. V. Halushka and P. C. Hoffmann, *Biochem. Pharmacol.* **17**, 1873 (1968).
3. V. E. Davis and M. S. Walsh, *Science* **167**, 1005 (1970).
4. H. Yamanaka, M. S. Walsh, and V. E. Davis, *Nature* (*London*) **227**, 1143 (1970).
5. G. Cohen and M. Collins, *Science* **167**, 1749 (1970).

6. T. L. Sourkes, *Nature (London)* **229**, 413 (1970).
7. R. Heikkila, G. Cohen, and D. Dembiec, *J. Pharmacol. Exp. Ther.* **179**, 250 (1971).
8. W. I. Taylor and A. R. Battersby, "Oxidative Coupling of Phenols." Dekker, New York, 1967; T. Kametani and K. Fukumoto, "Phenolic Oxidation," Gihodo, Tokyo, 1970.
9. K. Mothes and H. R. Schütte, "Biosynthese der Alkaloide." BVEB Deutsche Verlag Wissenschaften, Berlin, 1969.
10. T. Kametani, "The Chemistry of the Isoquinoline Alkaloids." Elsevier Publ., Amsterdam, 1968; T. Kametani, "The Chemistry of the Isoquinoline Alkaloids," Vol. 2. Sendai Inst., Heterocycl. Chem., Sendai, Japan, 1974.
11. T. Kametani and K. Fukumoto, *Synthesis* p. 657 (1972); T. Kametani, K. Fukumoto, and F. Satoh, *Bioorg. Chem.* **3**, 430 (1974).
12. T. Kametani, K. Fukumoto, K. Kigasawa, and K. Wakisaka, *Chem. Pharm. Bull.* **19**, 714 (1971).
13. T. Kametani, S. Takano, and T. Kobari, *J. Chem. Soc. C* p. 1030 (1971).
14. Y. Inubushi, Y. Aoyagi, and M. Matsuo, *Tetrahedron Lett.* p. 2363 (1969).
15. K.-H. Frömming, *Arch. Pharm. (Weinheim, Ger.)* **300**, 977 (1967).
16. A. Brossi, A. Ramel, J. O'Brien, and S. Teitel, *Chem. Pharm. Bull.* **21**, 1839 (1973).
17. T. Kametani, H. Nemoto, T. Kobari, and S. Takano, *J. Heterocycl. Chem.* **7**, 181 (1970).
18. T. Kametani, S. Takano, and T. Kobari, *Tetrahedron Lett.* p. 4565 (1968); *J. Chem. Soc. C* p. 131 (1969).
19. T. Kametani, S. Takano, and T. Kobari, *J. Chem. Soc. C* p. 2770 (1969).
20. T. Kametani, M. Mizushima, S. Takano, and K. Fukumoto, *Tetrahedron* **29**, 2031 (1973).
21. K. Tsuda, "List of Papers Published by Professor Kyosuke Tsuda and His Co-workers." Hirokawa Publ. Co. Inc., Tokyo, 1967.
22. D. H. Peterson and H. C. Murray, *J. Am. Chem. Soc.* **74**, 1871 (1952).
23. K. Aida, K. Uchida, K. Iizuka, S. Okuda, K. Tsuda, and T. Uemura, *Biochem. Biophys. Res. Commun.* **22**, 13 (1966).
24. K. Tsuda, H. Iizuka, S. Okuda, M. Iida, H. Isaka, and Y. Minemura, *Z. Allg. Mikrobiol.* **7**, 239 (1967).
25. R. V. Smith and J. P. Rosazza, *J. Pharm. Sci.* **64**, 1737 (1975).
26. A. Brossi, *Heterocycles* **3**, 343 (1975).
27. M. Sandler and S. B. Carter, *Nature (London)* **241**, 439 (1973).
28. M. A. Collins and M. G. Bigdeli, *Abstr. 5th Annu. Meet. Am. Soc. Neurochem.* p. 160 (1974).
29. V. E. Davis and M. J. Walsh, *Science* **170**, 1114 (1970).
30. M. H. Seevers, *Science* **170**, 1113 (1970).
31. T. Kametani, M. Ihara, and K. Takahashi, *Chem. Pharm. Bull.* **20**, 1587 (1972).
32. G. W. Kirby and L. Ogunkoya, *J. Chem. Soc.* p. 6914 (1965).
33. A. R. Battersby, D. M. Foulkers, and R. Binks, *J. Chem. Soc. C* p. 3323 (1965).
34. T. Kametani, M. Takemura, M. Ihara, K. Takahashi, and K. Fukumoto, *J. Am. Chem. Soc.* **98**, 1956 (1976).
35. J. L. Cashaw, K. D. McMurtrey, H. Brown, and V. E. Davis, *J. Chromatogr.* **99**, 567 (1974).
36. L. R. Meyerson and V. E. Davis, *Fed. Proc., Fed. Am. Soc. Exp. Biol.* **34**, 508 (1975).

Carbanions as Substrates in Biological Oxidation Reactions

Daniel J. Kosman

INTRODUCTION

The catalytic mechanisms employed by enzymes must be compatible with an aqueous environment at a statistically "cold" temperature which is confined to a narrow and rather (chemically) unremarkable pH range. Although mechanistic models for hydrolytic reactions, for example, are, in this context, conceptually easy to formulate, such schemes for many other reaction types are generally less obviously compatible with the medium in which the reactions presumably occur. Certainly one such reaction type is biological oxidation at the substrate level (i.e., other than terminal oxidation of electron transport factors). Depending on the nature of the redox pairs in any reaction of this type, certain conceptual problems arise in the formulation of hypothetical reaction pathways. One such problem is the spin-forbidden nature of direct two-electron transfer between singlet-state substrate and triplet-state molecular oxygen; two one-electron transfers are required along with the intermediacy of, in many cases, thermodynamically unstable and in general kinetically reactive free radicals. Another general concern is the mode of electron transfer in these (minimally) two-electron transfer reactions. Hamilton [1,2] has described the various ways in which the $2e^-$ and $2H^+$ can be transferred between redox pairs; he has persuasively made the case for the importance of the proton transfer in the oxidation reactions, an emphasis first made by Westheimer [3]. Of course, if an initial step in substrate oxida-

TABLE 1

TABLE 1

Some Enzymes That Generate Putative Carbanion Intermediates

EC number	Name	Reference
2.1.1.21	N-Methylglutamate synthase	4
2.2.1.1(2)	Transaldolase	5
2.6.1.n	Aminotransferases	6
4.1.1.1	Pyruvate decarboxylase	7
4.1.1.12	Aspartate β-decarboxylase	8
4.1.1.39	Ribulosebisphosphate carboxylase	9
4.1.2.13	Aldolase	10
4.1.3.2	Malate synthase	11
4.2.1.11	Enolase	12
4.2.1.24	δ-Aminolevulinic-acid dehydratase	13
5.1.1.n	Amino-acid racemases	14
5.3.1.n	Aldose-ketose isomerases (intramolecular oxidoreductases)	15
6.4.1.n	Acyl (CoA) carboxylases	16

tion is H^+ transfer from substrate, then the actual species oxidized is the conjugate base, a carbanion. Proton transfer and carbanion oxidation are both compatible with the reaction conditions, the former more obviously than the latter. The purpose of this review is to present the evidence for the possible and, in some cases, probable importance of carbanions as substrates in biological oxidation reactions involving organic oxidants as well as O_2.

With the exception of the hydrolases, examples of putative carbanion formation can be found among the reactions catalyzed by enzymes of all other classes. That is, for the several reaction types involving saturated carbon, a carbanion is demonstrably a viable intermediate. Some of these examples are presented in Table 1 [4–16]. In some, the carbanion per se is not fully formed; however, the tautomerization of the substrate catalyzed by the enzyme in these cases must involve ionization of a C—H bond with carbanion formation truncated by conjugation [Eq. (1)]. This important consideration is relevant to the discussion of certain of the oxidoreductases.

$$\underset{\substack{|\\HA}}{\overset{\substack{B:\quad H\\|}}{-C}}-C=X \;\longleftrightarrow\; \underset{\substack{|\quad|\\HA}}{\overset{\substack{B:\text{---}H\\|\quad|}}{-C\text{---}C\text{---}X}} \;\longleftrightarrow\; \underset{\substack{|\quad\;\;^-A}}{\overset{\substack{BH^+\\|}}{-C=C-XH}} \qquad (1)$$

Prosthetic-group-mediated oxidation of carbanions is, in many cases, simply one of many alternative reaction pathways available to the substrate intermediate. For example, the hydroxyethylthiamine pyrophosphate (HETP) intermediate can break down directly to CH_3CHO, or the corresponding carbanion can condense with a suitable carbonyl group to yield an acetoin-type product in a typical lyase reaction. However, this same nucleophilic

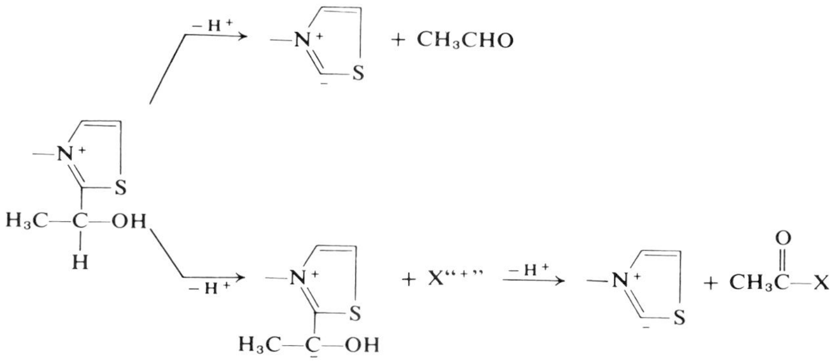

$$
\begin{array}{c}
\xrightarrow{-H^+} \quad -N^+ \diagdown_S \quad + \; CH_3CHO \\[2em]
\end{array}
\tag{2}
$$

(Equation 2: thiamine carbanion structures)

$$-N^+\diagdown_S \quad H_3C-\underset{H}{\overset{}{C}}-OH$$

$$\xrightarrow{-H^+} \quad -N^+\diagdown_S,\; H_3C-\underset{}{C}-OH \quad + \; X^{''+''} \quad \xrightarrow{-H^+} \quad -N^+\diagdown_S \quad + \; CH_3\overset{O}{\overset{\|}{C}}-X$$

carbon center can also react with suitable electron acceptors, "oxidants," to yield covalent intermediates in which the carbon center has been effectively oxidized. These reactions can be illustrated as shown in Eq. (2) [7].

This is typical of oxidation mechanisms described by Hamilton as PPC types—transfer of two protons concurrent with the formation of a covalent bond between nucleophile (carbanion) and electrophile ($X^{''+''}$), the redox pair [1,2]. The oxidant, $X^{''+''}$, can be lipoic acid (mammalian pyruvate dehydrogenase) or Fl_{ox} (pyruvate oxidase and pyruvate: cyt b_1 oxidoreductase). The fine line between oxidase and lyase, isomerase, transferase, and ligase reactions persists among examples of all of these reaction types. For example, pyridoxal phosphate-mediated aminotransferases are, in fact, oxidoreductases (transaminating) [17], catalyzing reactions in which amino acid and keto acid represent the redox pair.

The subtlety of this distinction is illustrated by Christen's investigation of the oxidation of known carbanion intermediates by common one- and two-electron acceptors [18–21]. A summary of his data is given in Table 2, and a mechanistic scheme might be represented by Eq. (3).

TABLE 2

Trapping by Oxidation of Enzyme-Generated Carbanions[a]

Enzyme (EC number)	Oxidants (E_0', V)	
	One-electron	Two-electron
Aldolase (4.1.2.13)	Porphyrexide (0.725)	Porphyrindin (0.565)
Aspartate aminotransferase (2.6.1.1)	$Fe(CN)_6^{3-}$ (0.360)	2,6-Dichlorophenol-
Phosphogluconate dehydrogenase	Cytochrome c (0.255)	indophenol (0.217)
(1.1.1.44)		Tetranitromethane[b]
Pyruvate decarboxylase (4.1.1.1)		

[a] Taken from Healey and Christen [21].
[b] See Riordan and Christen [18].

$$R-\underset{\underset{Y-Enz}{\overset{\displaystyle\|}{}}}{\overset{\displaystyle OH}{\underset{\displaystyle H}{C}}}-R' \;\underset{H^+}{\rightleftharpoons}\; R-\underset{\underset{Y-Enz}{\overset{\displaystyle\|}{}}}{\overset{\displaystyle OH}{C}}-R' \;\xrightarrow{\;Ox\;\;Ox^{2-}\;}\; R-\underset{\underset{Y-Enz}{\overset{\displaystyle\|}{}}}{\overset{\displaystyle OH}{\overset{\displaystyle |}{C}}}-\overset{+}{C}-R'$$

$$R-\underset{\overset{\displaystyle |}{^-Y-Enz}}{\overset{\displaystyle OH}{C}}=CHR'$$

$$R-\underset{\underset{Y-Enz}{\overset{\displaystyle\|}{}}}{C}-\overset{\overset{\displaystyle O}{\displaystyle\|}}{C}-R' \qquad H^+$$

(3)

The results are remarkable. For example, the aldolase-catalyzed oxidation of dihydroxyacetone by porphyrindin ($E'_0 = 0.565$ V) is ca. 5% of the rate at which the enzyme cleaves fructose 1,6-diphosphate, the normal lyase reaction [21]. That is, these oxidation reactions can be effectively catalyzed. The stoichiometry of the reaction is as shown in Eq. (4). No linear relationship

$$CH_2OHCOCH_2OPO_3{}^{2-} + 2Fe(CN)_6{}^{3-} \longrightarrow$$
$$CHOCOCH_2OPO_3{}^{2-} + 2Fe(CN)_6{}^{4-} + 2H^+ \tag{4}$$

exists between substrate activity of oxidant and oxidant E'_0. Two factors can account for this. First, the relative Marcus theory [22] when applied to biological electron transfer systems indicates the importance of specificity in interaction between redox pairs [23]. The size and charge of the oxidant, as well as its potential, are factors that determine the efficiency of electron transfer. Second, the rate of a two-electron oxidation by a two-electron oxidant is likely to be inherently faster than the same oxidation effected by a one-electron acceptor in as much as the latter reaction is of necessity a two-step process. Thus, the smaller and more electronegative one-electron oxidant porphyrexide ($E'_0 = 0.725$ V) is less than one-fifth as effective as porphyrindin in the aldolase-catalyzed oxidation of dihydroxyacetone [21]. Another and very important consequence of the utilization of one-electron acceptors as substrate is that an enzyme-bound radical species is an obligatory intermediate. Surprisingly, perhaps, this radical does not "leak out" of the reaction pathway. The stoichiometry of the redox reaction indicates this [Eq. (4)]. On the other hand, the same factors in these enzymes that serve to stabilize the catalytically normal carbanion intermediates will certainly stabilize the corresponding radical. Thus, the redox activity of one-electron oxidants can be viewed simply as another manifestation of enzymatic stabilization of nucleophilic carbon.

The following question remains, however. Are carbanions viable substrates for biological oxidation reactions? In particular, are they involved in reactions in which no C—H bond activation is provided, e.g., by thiamine, pyridoxal phosphate, or other electrophilic centers?

SUBSTRATE-LEVEL OXIDATION BY FLAVINS

The original suggestion by Hamilton that flavin-mediated oxidation could be effected via adduct formation was illustrated by a mechanistic proposal whose essential features are shown in Eq. (5) [1,2]. That is, attack by the

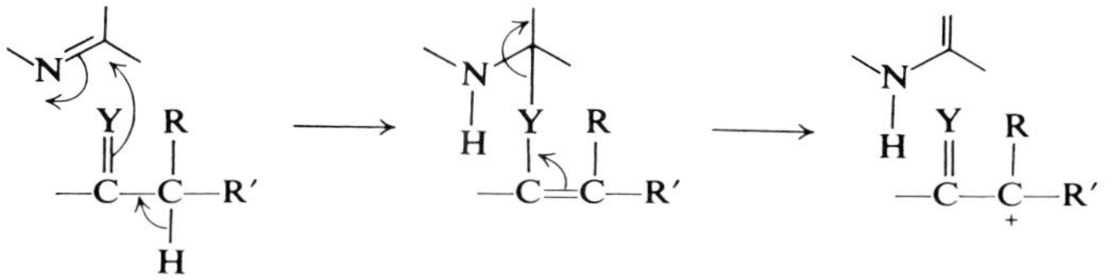

$$(5)$$

nucleophile Y at the electron-deficient C-4 is followed by what is basically a tautomerization of the adduct, resulting, however, in C—Y bond breaking. As a result of the electron flow, the carbon vicinal to the C=Y function is oxidized from a putative carbanion to a carbonium ion. The fate of the carbonium ion is determined by the groups R and R′. A similar reaction mechanism was proposed for geminal oxidation, that is, the oxidation of the same carbon to which the nucleophile Y is bonded, e.g., as in the reactions catalyzed by amine and alcohol flavoprotein oxidases. This is represented by Eq. (6) [1,2].

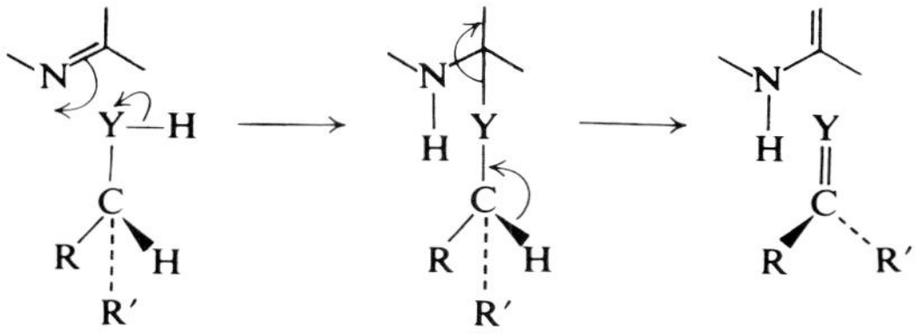

$$(6)$$

Although the concept of a covalent redox reaction has been experimentally verified, the details of the reaction itself are apparently somewhat different from those outlined above. Of particular importance to this discussion is that the nucleophilic species apparently is not the heteroatom Y but the carbon itself, as pictured in Eq. (7) [24,25]. As indicated the electrophile is probably the N-5 nitrogen of the isoalloxazine ring [24]. The significant relationship of this mechanistic postulate to the nonphysiological oxidations discussed above is that the species directly involved in the electron transfer in both is a carbanion or a nacsent one; it is the carbon center itself that is the nucleophile.

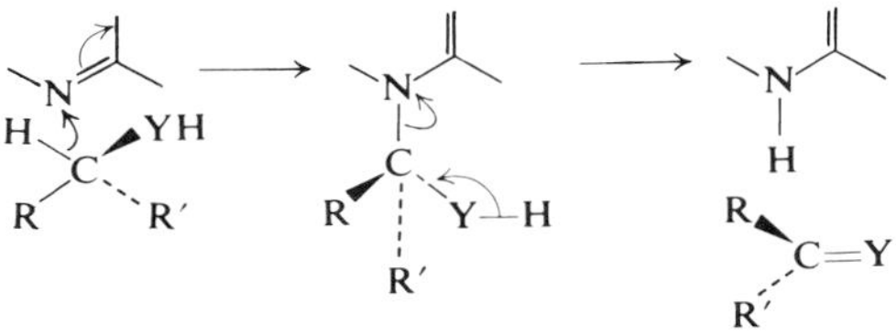

$$\text{(7)}$$

There is adequate precedent for carbanion oxidation by flavins. For example, α-hydroxy- and aminoketone enolates are readily oxidized by 3-benzylumiflavin [Eq. (8)] [26]. Dihydrophthalates are dehydrogenated by

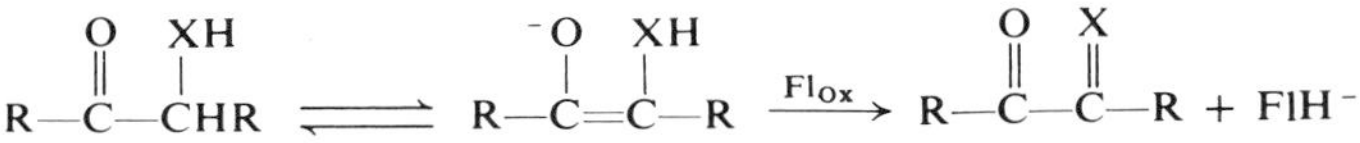

$$
\underset{\substack{O \quad XH \\ \| \quad | \\ R-C-CHR}}{} \rightleftharpoons \underset{\substack{^-O \quad XH \\ | \quad | \\ R-C=C-R}}{} \xrightarrow{F_{lox}} \underset{\substack{O \quad X \\ \| \quad \| \\ R-C-C-R}}{} + FlH^- \qquad \text{(8)}
$$

$$X = O, NH$$
$$R = C_6H_5, CH_3, (CH_2)_2$$

riboflavin to the corresponding pthalate derivatives [27]. The pH dependence of this reaction suggests that the initiating event is the nucleophilic attack by a dihydropthalate carbanion on the isoalloxazine. Subsequent proton transfer and adduct dissociation as in Eq. (7) would complete the reaction. The overall transformation is reminiscent of the succinic dehydrogenase reaction.

However, the work that really has provided insights into flavin-mediated substrate-level oxidation are the recent contributions by Bright and Abeles and their respective co-workers. In brief summary, Bright [24,28,29] has established that (1) a carbanion can form an adduct with an isoalloxazine; (2) the electrophile is the N-5 nitrogen; (3) this adduct is kinetically competent in the enzymatic turnover of both carbanion and flavin cofactor. Specifically, the reaction, as catalyzed by D-amino-acid oxidase, is as depicted in Eq. (9). Both glucose oxidase and L-amino-acid oxidase are also reactive with nitroalkanes [24].

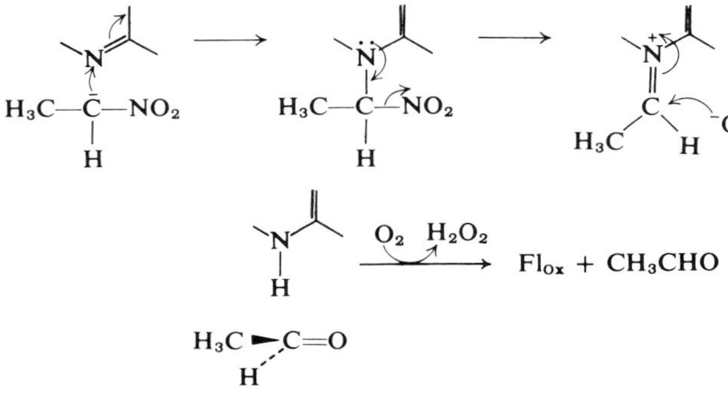

$$\text{(9)}$$

$$\xrightarrow{O_2 \quad H_2O_2} Fl_{ox} + CH_3CHO$$

Abeles, Massey, and their co-workers have exploited an entirely different approach to delineate the steps in flavin-mediated oxidations [30–33]. They reason that, if the normal reaction [see Eq. (10)] proceeds via path 1 (X = H), path 2 is a viable alternate when X represents a good leaving group, e.g., Cl or OAc. The enzyme thus catalyzes a nonoxidative elimination reaction which

$$
\begin{array}{c}
\underset{\overset{|}{\underset{X-C-}{|}}}{\overset{HY}{\underset{|}{\overset{|}{C}-H}}}
\ \underset{\overset{-H^+}{\rightleftharpoons}}{}\
\underset{\overset{|}{\underset{X-C-}{|}}}{\overset{HY}{\underset{|}{\overset{|}{C}{}^-}}}
\ \xrightarrow[-2e^-]{\overset{\textcircled{1}}{-H^+}}\
\underset{\overset{|}{\underset{X-C-}{|}}}{\overset{Y}{\underset{}{\overset{\|}{C}}}}
\end{array}
\qquad (10)
$$

$$
\textcircled{2}\ \xrightarrow{-X^-}\
\underset{\overset{\|}{\underset{C-}{}}}{\overset{HY}{\underset{}{\overset{|}{C}}}}
\ \longrightarrow\
\underset{\overset{|}{\underset{H-C-}{}}}{\overset{Y}{\underset{}{\overset{\|}{C}}}}
$$

does, however, result in Y group oxidation. In effect, X^- represents the *reduced* oxidant. A summary of the results is given in Table 3 [30,32–35]. Stopped-flow kinetic studies employing an α-deuterated substrate revealed that the formation of the first flavin–substrate complex required the loss of this proton [32,36]. The relevance of this proton abstraction as a first step is also suggested by the covalent inhibition of lactate oxidase by 2-hydroxy-3-butynoic acid [see Eq. (11)] [31].

$$
\left[\begin{array}{c} Fl_{ox} \\[4pt] CH\equiv CH-\underset{\overset{|}{H}}{\overset{\overset{OH}{|}}{C}}-CO_2^- \\[6pt] B: \end{array}\right]
\longrightarrow
\left[\begin{array}{c} Fl_{ox} \\[4pt] CH\equiv CH-\underset{}{\overset{\overset{OH}{|}}{C}}-CO_2^- \\[6pt] BH \end{array}\right]
\longrightarrow
$$

$$
\left[\begin{array}{c} Fl_{ox} \\[4pt] {}^-CH=C=\underset{}{\overset{\overset{OH}{|}}{C}}-CO_2 \\[6pt] BH \end{array}\right]
\longrightarrow
\left[\begin{array}{c} Fl_{ox} \\[4pt] H_2C=C-\overset{\overset{O}{\|}}{C}-CO_2^- \\[6pt] B: \end{array}\right]
\qquad (11)
$$

The effectiveness of acetylinic substrate analogs as allenic precursors in enzymatic reactions known or postulated to proceed via carbanion intermediates has been well established [37]. Thus, although no direct experimental evidence is currently available which demonstrates unequivocally the carbanion nature of the "simple" flavoprotein oxidase mechanism, it remains the most attractive model. The only conceptual difficulty of this model is its requirement for the ionization of an extremely weak acid at neutral pH. Unlike the electrophilically activated C—H bonds illustrated previously, the αC—H bonds in the substrates for these enzymes appear to be relatively inert.

TABLE 3

Elimination Reactions Catalyzed by Oxidases

Enzyme (EC number)	Substrate	Product		Reference
		Aerobic	Anaerobic	
D-Amino-acid oxidase (1.4.3.3)	$ClCH_2CHNH_2CO_2H$	$ClCH_2\overset{O}{\overset{\|}{C}}CO_2H$ + $CH_3\overset{O}{\overset{\|}{C}}CO_2H$	$CH_3\overset{O}{\overset{\|}{C}}CO_2H$	30
	$CH_3CHClCHNH_2CO_2H$	$CH_3CH_2\overset{O}{\overset{\|}{C}}CO_2H$ (no O_2 uptake)	$CH_3CH_2\overset{O}{\overset{\|}{C}}CO_2H$	32
	$AcOCHCHNH_2CO_2H$	—	$CH_3\overset{O}{\overset{\|}{C}}CO_2H$	32

Enzyme	Substrate		Product	Ref.
L-Amino-acid oxidase (1.4.3.2)	ClCH$_2$CHNH$_2$CO$_2$H	$\overset{O}{\overset{\|}{ClCH_2CCO_2H}}$ + $\overset{O}{\overset{\|}{CH_3CCO_2H}}$	$\overset{O}{\overset{\|}{CH_3CCO_2H}}$	30
Lactate oxidase (1.1.3.2)	ClCH$_2$CHOHCO$_2$H	ClCH$_2$CO$_2$H + $\overset{O}{\overset{\|}{CH_3CCO_2H}}$	$\overset{O}{\overset{\|}{CH_3CCO_2H}}$	33
Amine oxidase (1.4.3.6) (porcine plasma)	φCHClCH$_2$NH$_2$	φCH$_2$CHO (no O$_2$ uptake)	φCH$_2$CHO	34
	CH$_2$NH$_2$CO$_2$-p-X-φ (X = H, NO$_2$)	Enzyme inactivated by reaction with presumed intermediate, CHNH$_2$=C=O		35

If, however, the ionization of such protons is biologically feasible, then it becomes important to consider more completely the reactivity of the resultant carbanions.

ARE SUCH CARBANIONS FEASIBLE?

In the absence of any resonance stabilization of the conjugate base, the pK_a of an sp^3 carbon acid is > 40. In fact, the pK_a for CH_4, for example, is likely to be closer to 60 [38]. Thus, an intermediate with even a small degree of carbanionic character seems unreasonable for the substrates utilized by some simple oxidases, as listed in Table 4 [33,39–44]. However, the comparison to hydrocarbon pK_a as above is misleading, in as much as each *substrate* C—H bond is geminal to a heteroatom which is more electronegative than carbon and is a reasonably strong Brönsted base. Thus, consider the possible effect of protonation of the amino group in alanine on the pK_a of the αC—H bond. A model for this might be the relative pK_a values of CH_4 and $H-CH_3{}^+N(CH_3)_3$; the latter has a pK_a of ca. 33, similar to triphenylmethane [45–47]. This is due, of course, to the electrostatic stabilization of the carbanion; the conjugate base is a nitrogen ylid, $(CH_3)_3N^+-{}^-CH_2$. Another factor that could enhance the acidity of the C—H group is geometric distortion of the reaction center causing an increase in the s character of this bond. The acidity of carbon acids is known to increase in this way even as the degree of unsaturation remains constant [45]. For example, note the values given below [48].

The distortion energy necessary to achieve any degree of catalysis in this way would be provided by the binding enthalpy. The fact that the Michaelis

TABLE 4
Substrate Michaelis Constants for Selected Oxidases

Enzyme	Substrate (K_m, mM)	K_{mO_2} (mM)	Reference
Amine oxidase (porcine plasma)	Benzylamine (0.5)	0.2	39
D-Amino-acid oxidase	D-Alanine (4.1)	0.33	40
L-Amino-acid oxidase	L-Leucine (~ 2)	~ 0.2	41
Lactate oxidase	β-Chlorolactate (~ 2)	~ 0.6	33
Glucose oxidase	2-Deoxyglucose (29)	ns[a]	42
Urate oxidase	Urate (3)	1	43
Galactose oxidase	β-Methyl-D-galactopyranoside (175)	3.1	44

[a] Not stated.

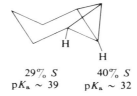

$$29\% \; S \qquad\qquad 40\% \; S$$
$$pK_a \sim 39 \qquad\qquad pK_a \sim 32$$

constants for the substrates in Table 4 indicate a rather small observed free energy of binding is certainly consistent with this suggestion. The difference between the intrinsic and observed binding energies is an indication of how far the substrate has been brought along the reaction coordinate in the Michaelis complex [49]. One component of the activation energy that is accommodated by the binding process could be a geometric distortion of the reaction center. This factor, plus protonation of the heteroatom by an enzymatic group, could provide the activation of the C—H bond required.

A third catalytic factor could be specific solvation or electrostatic stabilization of an incipient carbanion [49]. The rate of formation of highly localized carbanions is, in part, attenuated by the extensive solvent reorganization required in the transition state and correspondingly large, negative activation entropies [45]. For example, at least half of the difference in ionization rates of a proton in $(CH_3)_4N^+$ compared to either $(CH_3)_4P^+$ or $(CH_3)_3S^+$ is associated with a $\Delta\Delta S^{\ddagger}$ of ~ 15 eu [47]. The less negative $\Delta S^{\ddagger}$ values for the ionization of the latter two species are associated with the greater effective charge delocalization in the carbanion. This appears to be associated with the polarizability of these second-row elements rather than valence shell expansion [50,51]. The important point is, however, that kinetic acidity can be enhanced by appropriate transition-state solvation (or electrostatic stabilization by appropriate enzymatic groups).

In this context, the fact that the fast step in the reaction catalyzed by D-amino-acid oxidase does involve the breaking of the αC—H bond [32] gains added meaning. One would expect this step to be fast. Highly localized carbanions are reprotonated at rates near the diffusion-controlled limit [52]. The covalent mechanisms putatively utilized by the flavoproteins are based on the Lewis concept of acid–base behavior, electron donation. That is, the degree to which the nascent nucleophilic carbanion is engaged in electron donation to the electrophilic isoalloxazine will affect the kinetic acidity of the C—H bond. In effect, the formation of the adduct "pulls" the ionization step to the right, although this implies a thermodynamic equilibrium which has not been verified nor even suggested.

Taken together, these considerations do not in any way prove the intermediacy of carbanions in these or any other similar biological oxidation reaction; they do, however, illustrate the catalytic mechanisms that could be employed to effect, at least, the concerted ionization of an apparently inert C—H bond.

OXIDATION BY ELECTRON ACCEPTORS OTHER THAN RIBOFLAVIN

Of some interest would be the demonstration that a deflavooxidase was catalytically active with another electron acceptor such as $Fe(CN)_6^{3-}$, porphyrexide, or $IrCl_6^{2-}$. This could be interpreted to indicate the presence of an essential concentration of an enzyme-generated carbanion intermediate.

$$ROH + Ox^n(2Ox^n) \longrightarrow R{=}O + Ox^{n-2}(2Ox^{n-1}) + 2H^+$$

$$R = R'CH_2{-},\ HO{-}$$

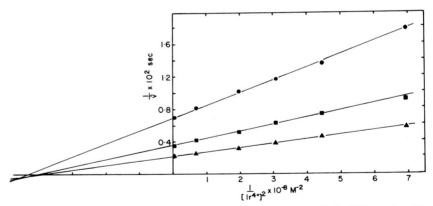

(12)

$$Ox = Fe(CN)_6^{3-},$$

In our work with a copper oxidase [44,53–59], galactose oxidase, we have investigated the activity of such oxidants relative to the assumed biological substrate, molecular oxygen. The reaction catalyzed is that shown in Eq. (12)

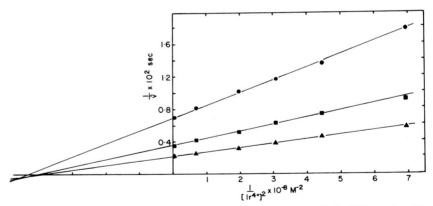

Fig. 1   Initial molecular velocity of the oxidation of H_2O_2 by $IrCl_6^{2-}$ (IV) catalyzed by galactose oxidase. The millimolar concentrations of H_2O_2 are 1.49 (●—●); 2.98 (■—■); and 4.47 (▲—▲) in 0.1 M sodium phosphate, pH 7.0. The reaction was monitored by the decrease in $IrCl_6^{2-}$ absorbance at 490 nm using a Durram stopped-flow instrument with data collection by a PDP 8. The second-order rate constant of the H_2O_2 reaction is $9.7 \times 10^4\ sec^{-1}\ M^{-1}$; $K_A'(IrCl_6^{2-}) = 5.0 \times 10^{-5}\ M^2$. Enzyme concentration is $4.11 \times 10^{-7}\ M$.

[60–62]. All three oxidants do replace O_2; $IrCl_6{}^{2-}$, in fact, is a better substrate (compare data in Table 4 with those in Fig. 1). The enzyme also catalyzes the oxidation by $IrCl_6{}^{2-}$ of H_2O_2 to yield O_2, in effect the reverse of the normal reaction. The initial velocity data for this reaction (Fig. 1) as well as for the galactose/$IrCl_6{}^{2-}$ reaction indicate that these reactions are kinetically sequential, with the redox reaction occurring in a complex containing all the reactants. The galactose/oxygen reaction behaves similarly [44].

In view of the fact that the copper does not change valence during turnover [53,63], these data taken together are consistent with the idea that, like the flavoprotein oxidases, a copper oxidase can also catalyze a reaction involving a carbanion susceptible to an oxidative attack by a suitable electrophile. The pH–rate data and the chemical modification of an essential histidine residue provide the type of evidence needed to suggest the presence of a catalytically functional general base at the active site [55,56]. The type of carbanion stabilization provided by the enzyme is as before. In particular, competitive fluorescence binding and $^{19}F^-$ nuclear magnetic resonance relaxation studies indicate that galactose binds to the Cu(II) through a water molecule [57–59],

thus providing the type of electrostatic stabilization discussed above. The pK_a of this water molecule is ca. 7.8 in the free enzyme; it decreases to 7.3 in the enzyme–sugar complex [55]. Furthermore, the Michaelis constant for galactose is 175 mM [44]; the pH-independent value for k_{cat}, however, is 2500 sec^{-1} [44]. Thus, the catalytic efficiency of the enzyme may well derive from a large substrate distortion upon binding which occurs at the expense of the intrinsic binding energy [49]. Nonetheless, a carbanion mechanism for galactose oxidase remains at present only a hypothesis.

The argument that is commonly made against such a hypothesis concerning a reaction involving triplet oxygen is that the redox reaction is spin forbidden [64]. The generally held viewpoint is that the lifetime of collisional complexes is too short to allow for the necessary spin inversion to occur; thus, the complex either reverts to reactants or dissociates to radical species (R· and $O_2{}^{\bar{\cdot}}$). This is a problem in leucoflavin oxidation, as well; before the second electron can be transferred to the $O_2{}^{\bar{\cdot}}$ intermediate (or the FlH_2O_2 covalent complex can form [65–67]) the triplet complex, $FlH \cdot O_2{}^{\bar{\cdot}}$, must relax to a singlet; the relaxation time for this process would appear to be less than a millisecond [68]. The oxidation of leucoflavin by O_2 "works" since the flavosemiquinone radical intermediate is stable with an electronic structure that allows for mixing in of vibrational excited states, which presumably decreases the inversion lifetime [66–68].

For an enzymatic reaction, however, the problem of this spin inversion is not necessarily so severe. After all, enzymes are designed to hold substrate complexes together in some fashion. Furthermore, an inspection of the literature uncovers some work which suggests that single electron transfers (including those to O_2) do not need to be from "carbanions" which are highly delocalized, e.g., those species with relatively low pK_a values. For example, Russell has demonstrated rapid electron transfer from a number of carbanions to nitrobenzene [69]. Guthrie has shown that even triphenylmethide ion is efficiently trapped in this way, the electron transfer competing effectively with proton transfer ($k_e/k_H \approx 10^3$) [70]. This is significant since, as argued earlier, H^+ transfer (Brönsted) to such localized carbanions is fast, often near the diffusion limit. The fact that electron transfer (Lewis) is even faster suggests the possible efficiency of carbanionlike redox reactions.

Russell has also investigated the oxidation of triphenylmethide ion (and other benzylic anions [71]) by molecular oxygen. The reaction definitely proceeds via a one-electron transfer from carbanion to O_2; in fact, the slow step is the ionization of the hydrocarbon. The work leaves unresolved the events that occur subsequent to the first electron transfer. The results are, however, consistent with a cage-type reaction such as that shown in Eq. (13).

$$R_3C:^- + O_2\uparrow \longrightarrow \left[R_3C\cdot\uparrow + O_2^{\cdot\,-}\uparrow \longleftarrow \longrightarrow R_3C\cdot\downarrow O_2^{\cdot\,-}\uparrow \right] \longrightarrow R_3CO_2 \quad (13)$$

This scheme is not entirely speculative, but is based on work by Nelson and Bartlett [72]. They had previously compared the thermal and photochemical decomposition of azocumene. These reactions proceed from excited singlet and triplet states, respectively. Thus, the comparison affords an evaluation of the triplet $\rightarrow$ singlet conversion required for the excited state of the second process to collapse back to azocumene. The comparison actually showed that the spin inversion occurred faster than the separation of the parallel spins. Nelson and Bartlett concluded that spin inversion is rapid and pointed out, quite correctly, that this suggestion had never been proved to be unreasonable. On the contrary, more recent experimental and theoretical evidence has helped delineate the molecular factors which actually favor spin inversion [73–77] or allow for mechanisms which "avoid the spin barrier" [78]. Salem [76,77], in particular, has clearly demonstrated that heavy atom effects, one center interactions, (near) orthogonality of those orbitals involved in the spin inversion, and a high degree of ionic character in the state undergoing inversion favor spin inversion. As pointed out by Turro and Lechten, singlet–triplet conversion can occur during a concerted reaction, in apparent violation of Wigner's "spin conservation rule" [74]. In summary, experimental evidence (and this report does not purport to be complete) is consistent with the hypothesis that molecular $^3\Sigma$ oxygen can be utilized directly in the oxidation of carbanionlike nucleophilic carbon.

Other enzymatic examples are also consistent with this proposal. Uricase, for example, catalyzes the random, sequential oxidation of uric acid by O_2 [43]. The mechanism can be pictured as in Eq. (14). Compare this with the obviously similar oxidation of leucoflavin [Eq. (15)].

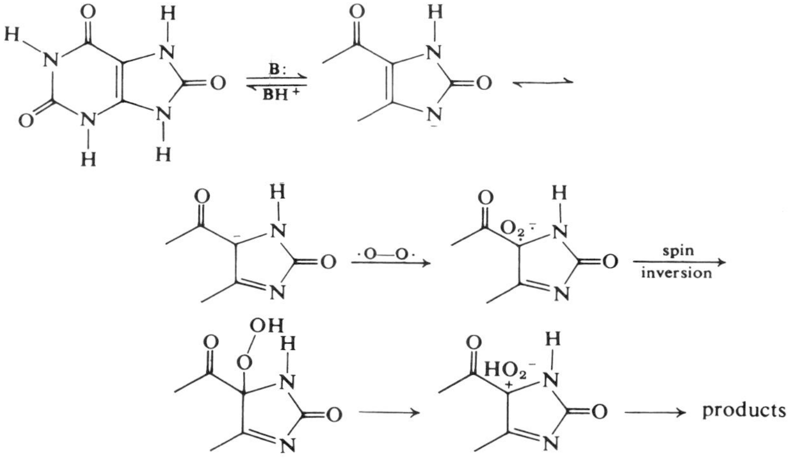

$$\text{products} \qquad (14)$$

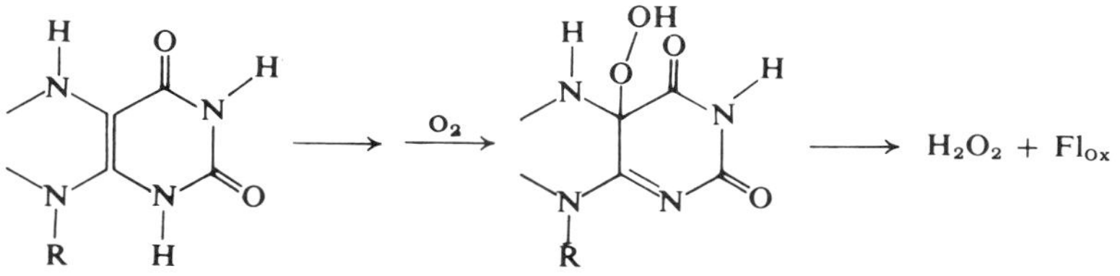

$$\xrightarrow{O_2} \qquad \longrightarrow H_2O_2 + Fl_{ox} \qquad (15)$$

The nature of any metal prosthetic group in uricase is unclear; by flameless atomic absorption spectrophotometry, we have eliminated all transition elements except zinc as intrinsic to this enzyme [79]. It is definitely not a copper oxidase. However, inasmuch as leucoflavin oxidation does not depend on a metal catalyst, uricase activity does not necessarily depend on one either.

The basic oxidation mechanism of plasma monoamine oxidase may be strikingly similar to that of galactose oxidase. The oxidation step per se is studied by following the oxidation of aromatic alcohols, e.g., benzyl alcohol [39]. This avoids the mechanistic uncertainties [34,80] associated with the deamination of the corresponding amines. Also, galactose oxidase catalyzes the oxidation of the same alcohols [44,81], further justifying this comparison.

Carbanion formation in the amine substrate, at least, has been strongly suggested by the work of Abeles and co-workers, as indicated in Eq. (10). Although the base that would catalyze carbanion formation has not been identified in this protein, affinity labeling does indicate its presence [35]. Since changes in the copper oxidation state have not been detected in this enzyme reaction [82], an oxidation mechanism like that suggested for galactose oxidase [56] is consistent with the available data on this amine oxidase.

CONCLUSION: BIOLOGICAL SIGNIFICANCE OF CARBANION OXIDATION

The view that carbanions are possible substrates for enzymes that utilize O_2 as oxidant would appear to have some biological significance. A possible inference to be drawn from this idea is that this type of substrate oxidation (by O_2) evolved essentially accidentally from nonoxidative reactions involving intermediate carbanions, reactions that were biologically important literally billions of years before aerobiosis occurred. Thus, this oxidative mechanism would be of the most primitive type and would exhibit many similarities with its precursor.

In fact, one enzyme does appear to catalyze both a typical lyase reaction and an oxidation reaction utilizing O_2, apparently with the same substrate intermediate. This enzyme is ribulosebisphosphate carboxylase (oxygenase), which catalyzes the two reactions shown in Eq. (16) [9,83]. These reactions are presumed to possess a common intermediate, the enol [Eq. (17)].

The reaction intermediate for oxidation is not tenable; a more likely

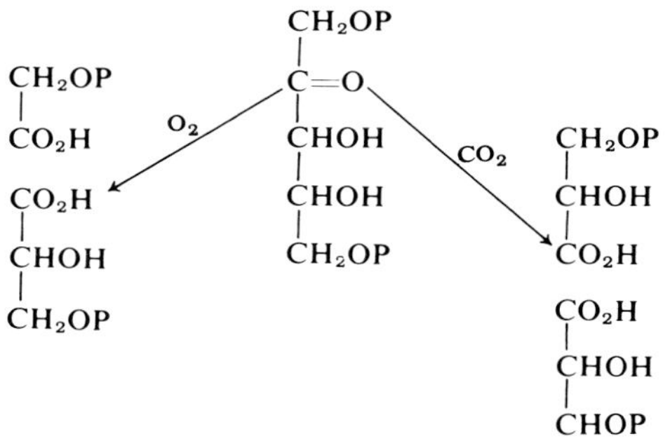

$$\tag{16}$$

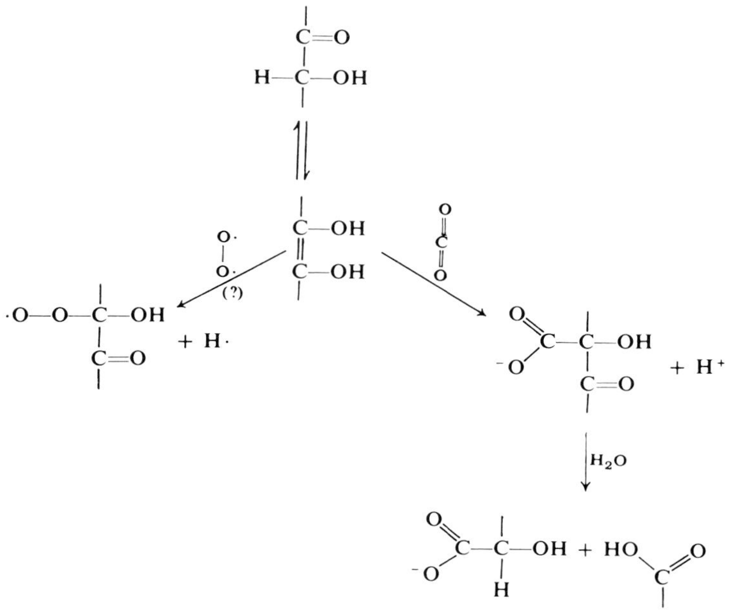

$$
\begin{array}{c}
\overset{|}{C}=O \\
H-\overset{|}{C}-OH \\
|
\end{array}
$$

(17)

reaction scheme is one that recalls the base-catalyzed air oxidation of any enediol [Eq. (18)] (hydroquinone, ascorbic acid).

As such, the stability of the radical intermediate allows for the requisite spin inversion followed by adduct formation. Although not demonstrated for the RuDP (ribulose bisphosphate carboxylase) reaction, the intermediacy of $HO_2\cdot$ has been indicated in other oxygenase reactions [84]. The point is,

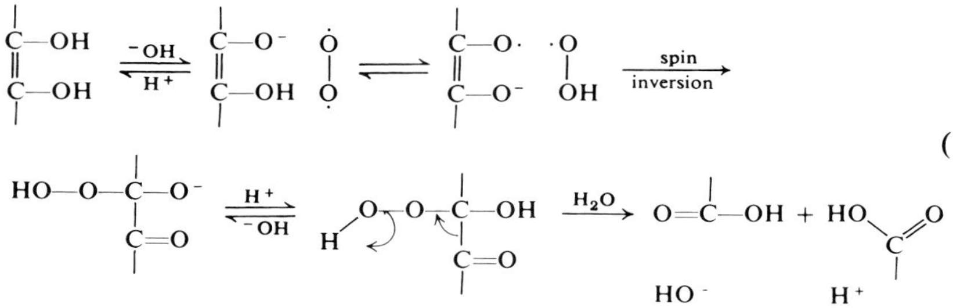

(18)

however, that a nucleophilic carbon can be both air oxidized and carbonylated by the same enzyme, an enzyme that evolved at least a billion years before aerobiosis [85]. The implication is that the oxidase activity of RuDP carboxylase is serendipitous and that any enzymatic reaction involving a nucleophilic carbon could, potentially, be diverted in a similar fashion.

The conservative nature of the evolutionary process is expressed in many ways: the similarity of the primary sequence of the same enzyme among diverse organisms, the retention of secondary and tertiary structural features in a particular class of enzyme, the use of a protein or a closely conserved copy of a protein for a purpose distinct from the template protein's function, and the ubiquity of mechanism type throughout the reactions catalyzed by a specific enzymatic class. This conclusion merely submits that an oxidation mechanism as utilized by certain oxidases can be viewed as evolved from chemical mechanisms extant before aerobiosis. This evolutionary process required only that molecular oxygen could be made a kinetically competent oxidant of biological substrates. If this proposal is experimentally viable, what perhaps is most remarkable is not that such an evolutionary event has occurred, but that it has not occurred more frequently. Either the structural changes required to generate this catalytic behavior are extremely difficult to effect, or there is a great selective pressure against such evolutionary developments. Indeed, as an event associated with aerobiosis, the appearance of enzymes catalyzing many *noncoupled* substrate (and cofactor) oxidations by molecular oxygen would be lethal to a living organism. What is significant, then, is that the reducing potential of the cell is conserved so well despite the presence of molecular oxygen and the availability of simple mechanisms for its direct utilization.

ACKNOWLEDGMENTS

The research of this group is supported by the National Science Foundation (B040662) and the graduate school of the State University of New York at Buffalo.

REFERENCES

1. G. A. Hamilton, *Prog. Bioorg. Chem.* **1**, 83–157 (1971).
2. G. A. Hamilton, *Adv. Enzymol.* **32**, 55 (1969).
3. F. H. Westheimer, *in* "The Mechanism of Enzyme Action" (W. D. McElroy and B. Glass, eds.), p. 321. Johns Hopkins Press, Baltimore, Maryland, 1954.
4. R. J. Pollock and L. B. Hersh, *J. Biol. Chem.* **248**, 6724 (1973).
5. O. Tsolas and B. L. Horecker, *in* "The Enzymes" (P. D. Boyer, ed.), 3rd ed., Vol. 7, pp. 259–280. Academic Press, New York, 1974.
6. A. E. Braunstein, *in* "The Enzymes" (P. D. Boyer, ed.), Vol. 10, pp. 379–481. Academic Press, New York, 1974.
7. R. Breslow, *J. Am. Chem. Soc.* **79**, 1762 (1957).
8. S. S. Tate, N. M. Relyea, and A. Meister, *Biochemistry* **8**, 5016 (1969).
9. M. I. Siegel, M. Wishnick, and M. D. Lane, *in* "The Enzymes" (P. D. Boyer, ed.), 3rd ed., Vol. 6, pp. 169–192. Academic Press, New York, 1972.
10. B. L. Horecker, O. Tsolas, and C. Y. Lai, *in* "The Enzymes" (P. D. Boyer, ed.), 3rd ed., Vol. 7, pp. 213–258. Academic Press, New York, 1974.

11. H. Eggerer and A. Kettle, *Eur. J. Biochem.* **1**, 447 (1967).
12. F. Wold, *in* "The Enzymes" (P. D. Boyer, ed.), 3rd ed., Vol. 5, pp. 499–538. Academic Press, New York, 1971.
13. D. Shervin, *in* "The Enzymes" (P. D. Boyer, ed.), 3rd ed., Vol. 7, pp. 323–337. Academic Press, New York, 1972.
14. E. Adams, *in* "The Enzymes" (P. D. Boyer, ed.), Vol. 6, pp. 479–507. Academic Press, New York, 1972; E. Snell, *ibid.* Vol. 2, pp. 335–370 (1970).
15. I. A. Rose, *Adv. Enzymol.* **43**, 491 (1975).
16. J. Retey and F. Lynen, *Biochem. Z.* **342**, 256 (1965); A. S. Mildvan and M. C. Scrutton, *Biochemistry* **6**, 2978 (1967).
17. "Enzyme Nomenclature," p. 10. Am. Elsevier, New York, 1972.
18. J. F. Riordan and P. Christen, *Biochemistry* **7**, 1525 (1968); **8**, 2381 (1969).
19. P. Christen, *Experientia* **26**, 337 (1970).
20. M. J. Healy and P. Christen, *Experientia* **28**, 736 (1972); *J. Am. Chem. Soc.* **94**, 7911 (1972).
21. M. J. Healy and P. Christen, *Biochemistry* **12**, 35 (1973).
22. R. A. Marcus, *J. Phys. Chem.* **72**, 891 (1968).
23. See, for example, R. A. Holwerda and H. E. Gray, *J. Am. Chem. Soc.* **97**, 6036 (1975).
24. D. J. T. Porter, J. G. Voet, and H. J. Bright, *J. Biol. Chem.* **248**, 4400 (1973).
25. M. Akhtar and D. C. Wilson, *Annu. Rep. Chem. Soc.* (*B*) **71**, 98 (1974).
26. J. A. Rynd and M. J. Gibian, *Biochem. Biophys. Res. Commun.* **41**, 1097 (1970).
27. G. D. Weatherby and D. O. Carr, *Biochemistry* **9**, 344 and 351 (1970).
28. J. G. Voet, D. J. T. Porter, and H. J. Bright, *Z. Naturforsch., Teil B* **27**, 1054 (1972).
29. D. J. T. Porter, J. G. Voet, and H. J. Bright, *J. Biol. Chem.* **247**, 1951 (1972).
30. C. T. Walsh, A. Schonbrunn, and R. H. Abeles, *J. Biol. Chem.* **246**, 6855 (1971).
31. C. T. Walsh, A. Schonbrunn, O. Lockridge, V. Massey, and R. H. Abeles, *J. Biol. Chem.* **247**, 6004 (1972).
32. C. T. Walsh, E. Krodel, V. Massey, and R. H. Abeles, *J. Biol. Chem.* **248**, 1946 (1973).
33. C. Walsh, O. Lockridge, V. Massey, and R. H. Abeles, *J. Biol. Chem.* **248**, 7049 (1973).
34. R. Neumann, R. Hevey, and R. H. Abeles, *J. Biol. Chem.* **250**, 6362 (1975).
35. A. L. Maycock, R. H. Suva, and R. H. Abeles, *J. Am. Chem. Soc.* **97**, 5613 (1975).
36. K. Yagi, M. Nishikimi, N. Ohishi, and A. Takai, *FEBS Lett.* **6**, 2209 (1970).
37. K. Bloch, *Acc. Chem. Res.* **2**, 193 (1969); R. H. Abeles and C. T. Walsh, *J. Am. Chem. Soc.* **95**, 6124 (1973).
38. F. G. Bordwell and W. S. Matthews, *J. Am. Chem. Soc.* **96**, 1216 (1974).
39. C. E. Taylor, R. S. Taylor, C. Rasmussen, and P. F. Knowles, *Biochem. J.* **130**, 713 (1972).
40. M. Dixon and K. Kleppe, *Biochim. Biophys. Acta* **96**, 368 (1965).
41. V. Massey and B. Curti, *J. Biol. Chem.* **242**, 1259 (1967).
42. H. J. Bright and Q. H. Gibson, *J. Biol. Chem.* **242**, 994 (1967); H. J. Bright and M. Appleby, *ibid.* **244**, 3625 (1969).
43. O. M. Pitts and D. G. Priest, *Arch. Biochem. Biophys.* **163**, 359 (1974).
44. L. D. Kwiatkowski and D. J. Kosman, *Arch. Biochem. Biophys.* submitted for publication; a preliminary report of these data has been made, *Fed. Proc., Fed. Am. Soc. Exp. Biol.* **32**, 550 (abstr.) (1973).
45. D. J. Cram, "Fundamentals of Carbanion Chemistry." Academic Press, New York. 1965.
46. D. J. Cram and R. T. Uyeda, *J. Am. Chem. Soc.* **80**, 5466 (1964).

47. W. von Doering and A. K. Hoffman, *J. Am. Chem. Soc.* **77**, 521 (1955).
48. G. L. Closs and L. E. Closs, *J. Am. Chem. Soc.* **85**, 2022 (1963).
49. See W. P. Jencks, *Adv. Enzymol.* **43**, 219 (1975), for a recent discussion.
50. A. Streitweiser, Jr. and J. E. Williams, Jr., *J. Am. Chem. Soc.* **97**, 191 (1975).
51. F. Bernardi, I. G. Csizmadia, A. Mangini, H. B. Schlegal, M.-H. Whangbo, and S. Wolfe, *J. Am. Chem. Soc.* **97**, 2209 (1975).
52. A. J. Kresge, *Acc. Chem. Res.* **8**, 354 (1975).
53. D. J. Kosman, R. D. Bereman, M. J. Ettinger, and R. S. Giordano, *Biochem. Biophys. Res. Commun.* **54**, 856 (1973).
54. M. J. Ettinger and D. J. Kosman, *Biochemistry* **13**, 1247 (1974).
55. L. D. Kwiatkowski, L. Siconolfi, and D. J. Kosman, *Arch. Biochem. Biophys.* (1977).
56. D. J. Kosman, M. J. Ettinger, R. E. Weiner, R. S. Giordano, and R. D. Bereman, *Arch. Biochem. Biophys.* (1977).
57. R. E. Weiner, M. J. Ettinger, and D. J. Kosman, *Biochemistry* (1977).
58. D. J. Kosman, M. J. Ettinger, R. S. Giordano, and R. D. Bereman, *Biochemistry* (1977).
59. B. J. Marwedel, R. J. Kurland, D. J. Kosman, and M. J. Ettinger, *Biochem. Biophys. Res. Commun.* **63**, 773 (1975).
60. J. A. D. Cooper, W. Smith, M. Bacila, and H. Medina, *J. Biol. Chem.* **234**, 445 (1959).
61. G. Avigad, D. Amaral, C. Asensio, and B. L. Horecker, *J. Biol. Chem.* **237**, 2736 (1962).
62. D. Amaral, F. Kelly-Falcoz, and B. L. Horecker, *in* "Methods in Enzymology" (W. A. Wood, ed.), Vol. 9, p. 87. Academic Press, New York, 1966.
63. W. Blumberg, B. L. Horecker, F. Kelly-Falcoz, and J. Peisach, *Biochim. Biophys. Acta* **99**, 187 (1965).
64. H. Taube, *in* "Oxygen," p. 29. New York Heart Association, New York, 1965.
65. V. Massey, F. Müller, R. Feldberg, M. Schuman, D. A. Sullivan, L. G. Howell, S. G. Mayhew, R. G. Matthews, and G. P. Foust, *J. Biol. Chem.* **244**, 3999 (1969).
66. P. Hemmerich, A. P. Bhaduri, G. Blankenhorn, M. Brüstlein, W. Haas, and W.-R. Knappe, *Oxidases Relat. Redox Syst., Proc. Int. Symp., 2nd*, pp. 3–24 (1973).
67. V. Massey, G. Palmer, and D. Ballou, *Oxidases Relat. Redox Syst., Proc. Int. Symp., 2nd, 1971* pp. 25–43 (1973).
68. See the discussion following Massey *et al.* [67, pp. 47–48].
69. G. A. Russell, E. G. Janzen, and E. T. Strom, *J. Am. Chem. Soc.* **86**, 1807 (1964).
70. R. D. Guthrie, *J. Am. Chem. Soc.* **91**, 6201 (1969).
71. G. A. Russell and A. G. Bemis, *J. Am. Chem. Soc.* **88**, 5491 (1966).
72. S. F. Nelson and P. D. Barlett, *J. Am. Chem. Soc.* **88**, 143 (1966).
73. J. R. Thomas, *J. Am. Chem. Soc.* **88**, 2064 (1966).
74. N. J. Turro and P. Lechtken, *J. Am. Chem. Soc.* **95**, 264 (1973).
75. N. J. Turro, V. Ramamurthy, K.-C. Liu, A. Krebs, and R. Kemper, *J. Am. Chem. Soc.* **98**, 6758 (1976).
76. L. Salem and C. Rowland, *Angew. Chem. Int. Ed. Eng.* **11**, 92 (1972).
77. L. Salem, *Pure Appl. Chem.* **33**, 317 (1973).
78. D. H. R. Barton, R. K. Haynes, G. Leclerc, P. D. Magnus, and I. D. Menzies, *J. Chem. Soc., Perkin Trans. 1* 2055 (1975).
79. M. J. Ettinger and L. Schallinger, unpublished results.
80. M. Inamasu, K. T. Yasunobu, and W. A. Konig, *J. Biol. Chem.* **249**, 5265 (1974).
81. D. J. Kosman, unpublished data.

82. B. Mondovi, G. Rotillio, M. T. Costa, A. Finazzi-Agro, E. Chiacone, R. E. Hanson, and H. Beinert, *J. Biol. Chem.* **242**, 1160 (1967).
83. N. E. Tolbert, *Annu. Rev. Curr. Top. Cell. Reg.* **7**, 21 (1973).
84. See I. Fridovich, *Adv. Enzymol.* **41**, 35 (1974).
85. L. Margulis, "Origin of Eukaryotic Cells." Yale Univ. Press, New Haven, Connecticut, 1970.

Synthetic Studies in Indole Alkaloids: Biogenetic Considerations

James P. Kutney

INTRODUCTION

The indole alkaloids constitute a very large family of natural products. The impressive array of structures in this group has stimulated much thought about their mode of biosynthesis and provided a considerable challenge to the synthetic chemist to devise synthetic approaches to these systems. Our own interest in biosynthesis (for recent reviews and collection of references see Kutney [1]) and synthesis in this area stimulated us to consider the development of synthetic strategies that encompass reactions popularly accepted as being of some biogenetic interest, at least in certain late stages of the biosynthetic pathways in the *Aspidosperma*, *Iboga*, and *Vinca* alkaloids. The purpose of this chapter is to summarize some of our synthetic studies with particular emphasis on the phases of the pathway in which such biogenetic considerations are invoked.

The diversity of structures that are inherent in the *Aspidosperma*, *Iboga*, and *Vinca* families and that were of primary consideration in our synthetic objectives are shown in Fig. 1. This figure also summarizes the possible biogenetic interrelationships among the various alkaloid systems, and such interrelationships were seriously considered for the possible development of versatile synthetic routes to the various members of these three alkaloid families.

It was of interest to recognize at the outset that there was a fundamental

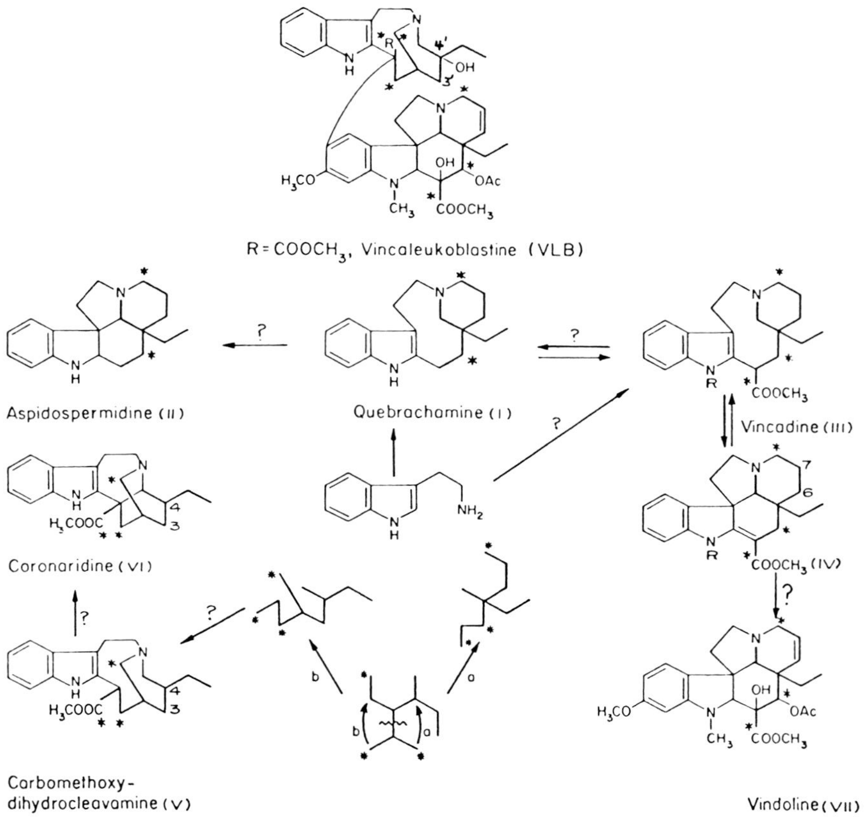

R = COOCH₃, Vincaleukoblastine (VLB)

Aspidospermidine (II)

Quebrachamine (I)

Vincadine (III)

Coronaridine (VI)

(IV)

Carbomethoxy-
dihydrocleavamine (V)

Vindoline (VII)

Fig. 1 A summary of some *Aspidosperma*, *Vinca*, and *Iboga* alkaloid structures and possible biogenetic interrelationships that may exist among them.

nine-membered ring system, exemplified by quebrachamine (**I**) and vincadine (**III**, R = H) in the *Aspidosperma* and *Vinca* series and by carbomethoxy-dihydrocleavamine (**V**) in the *Iboga* series, which could be considered the building unit for the more rigid cyclic members exemplified by aspidospermidine (**II**), vincadifformine (**IV**, R = H), and coronaridine (**VI**). Such considerations were also stimulated by our biosynthetic interests, since Wenkert [2] (Fig. 2) had already invoked such nine-membered ring intermediates in the late stages of the biosynthetic pathways leading to such alkaloid systems.

The nine-membered ring system of carbomethoxydihydrocleavamine (**V**) was also of interest in the synthesis of the highly complex bisindole alkaloids of the vincaleukoblastine series shown in Fig. 1. It should be noted that the

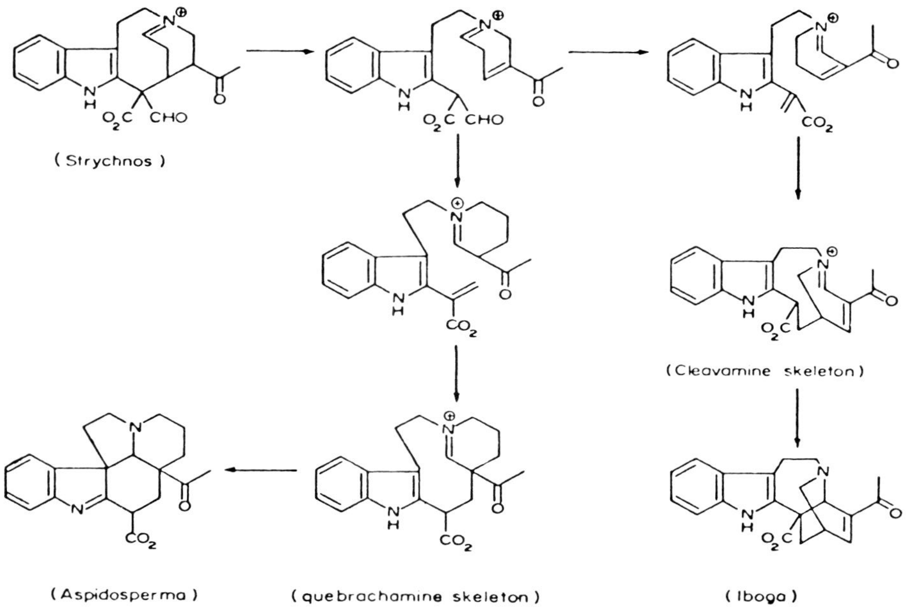

Fig. 2 Wenkert's postulates as they relate to later stages of *Aspidosperma* and *Iboga* alkaloid biosynthesis.

latter series is constituted from a cleavamine-like indole unit and a dihydro-indole unit, vindoline (**VII**), the latter being one of the most highly oxygenated members of the *Vinca* alkaloids. The eventual synthesis of these bisindole systems would obviously involve the appropriate linking of these two units at the requisite sites. For the sake of clarity and chronological development of the synthetic program, the following discussion is divided into several sections.

THE TRANSANNULAR CYCLIZATION APPROACH

Our initial considerations [3–12] directed toward the laboratory synthesis of the pentacyclic *Aspidosperma* and *Iboga* bases involved a series of reactions that we have generally termed the "transannular cyclization approach." During these studies, attempts to interrelate appropriate alkaloids bearing the medium-sized nine-membered ring system with the more rigid cyclic bases were considered in terms of the biogenetic relationships implicated in Figs. 1 and 2.

The success of this approach depends on the generation of an appropriate electrophilic site, the iminium system

$$\diagdown \underset{\diagdown}{C} = \overset{\oplus}{N} \diagup$$

in one portion of the molecule and its subsequent reaction, via a transannular cyclization process, with the electron-rich indole ring to form the requisite carbon–carbon bond. Ideally such a process should occur in a general and stereospecific manner so that a variety of functional groups could be tolerated in the starting substrates while the chiral centers generated in the products would be identical with those in the natural alkaloids. Fortunately these two requirements could be amply satisfied, and the approach could be utilized in the syntheses of a considerable variety of alkaloid systems.

The reagent of choice for generating the required iminium system from the alkaloid substrates was mercuric acetate, although in some instances oxygen in the presence of a catalyst, a reagent employed by Schmid [13] in related studies, could be employed.

In the initial studies [3], the dihydrocleavamine system **VIII** was converted to the intermediate **IX**, which underwent transannular cyclization to the indolenine **X**, the stereochemistry as depicted being subsequently established by X-ray analysis [7,14,14a].*

VIII   IX   X

The stereospecificity of the cyclization reaction was quickly established when it was shown [7] that quebrachamine (**I**), with antipodal stereochemistry at C-5 (see **VIII**), provides aspidospermidine (**II**) with chirality antipodal to that portrayed in **X**. Clearly the stereochemistry at C-5 in the starting nine-membered ring intermediates was the determining factor in the final stereochemistry of the resultant product.

The extension of the transannular cyclization approach to other alkaloids was then undertaken [4–6], with the carbomethoxydihydrocleavamine system **XI** being the substrate employed. Figure 3 summarizes the results obtained in this study.

---

* For the sake of clarity and facile comparisons with various publications, the numbering system and nomenclature employed in this chapter are those classically employed in these alkaloid families. The more recent proposals have not been adopted here (Le Men and Taylor [14] and Trojanek and Blaha [14a]).

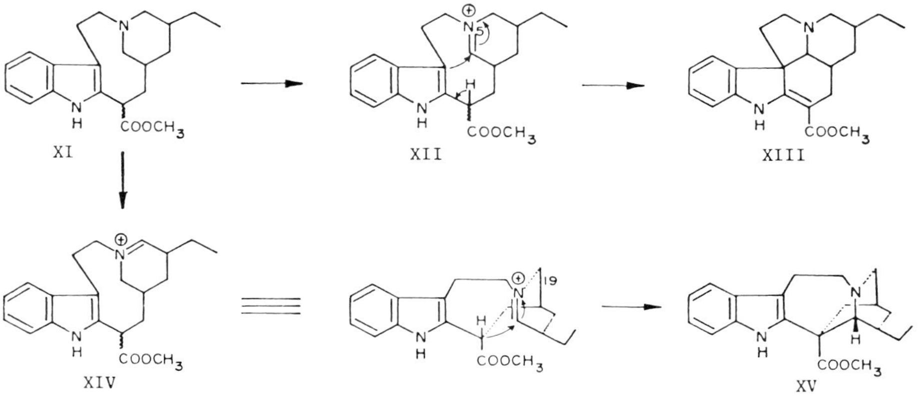

Fig. 3 Utilization of a carbomethoxydihydrocleavamine system (**XI**) in the transannular cyclization approach.

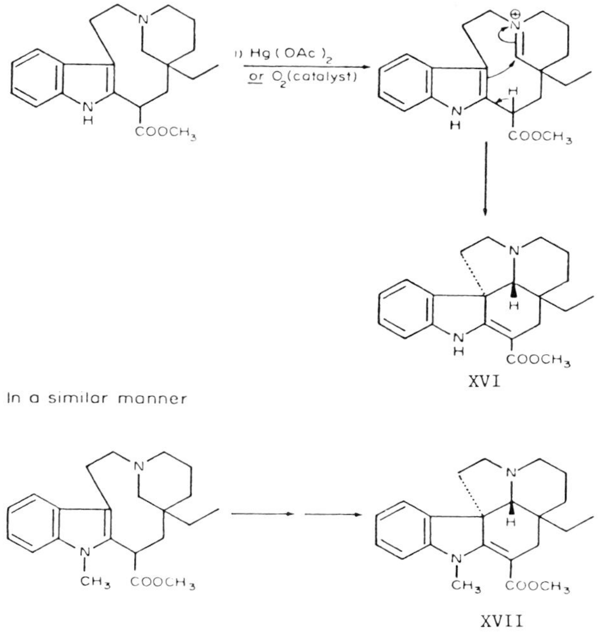

Fig. 4 The synthesis of vincadifformine (**XVI**) and minovine (**XVII**) via the transannular cyclization approach.

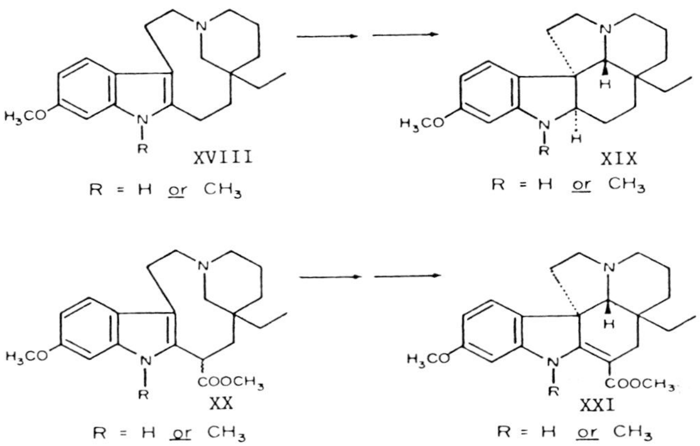

Fig. 5 The synthesis of alkaloids containing methoxyl substituents in the aromatic ring via the transannular cyclization approach.

The oxidation of **XI** with mercuric acetate proceeds in several directions, thereby generating intermediates **XII** and **XIV**, which in turn cyclize to pseudovincadifformine (**XIII**), a skeletal system present in many *Vinca* alkaloids, and dihydrocatharanthine (**XV**), a member of the *Iboga* family.

The application of this approach to the natural nine-membered ring alkaloids provided a laboratory synthesis of vincadifformine (**XVI**) and minovine (**XVII**) (Fig. 4) [10].

There are a variety of alkaloids that possess oxygen functionality in the aromatic ring of the indole or dihydroindole system, and it was necessary to extend the transannular cyclization approach to this series as well. Figure 5 illustrates that such substrates (**XVIII** and **XX**) are also capable of cyclization to provide **XIX** and **XXI**, respectively, the latter substance (R = CH₃) being an important intermediate in our laboratory synthesis of vindoline (**VII**).

The above-mentioned studies had established an important role of the nine-membered ring indole alkaloids in the laboratory synthesis of a variety of *Aspidosperma, Vinca,* and *Iboga* bases. Clearly the laboratory synthesis of these substances, when coupled with the above studies, would provide totally synthetic routes to these various natural products. Such investigations were undertaken, and a brief discussion of these studies follows. Figure 6 illustrates a typical synthetic route to the required nine-membered ring substrates.

As is seen in Fig. 6, the early stages of the synthesis involve the condensation of tryptamine or tryptamine analogs (for example, **XXII**) with an appropriate aldehydo ester (**XXIII**) in a straightforward manner to provide the tetracyclic

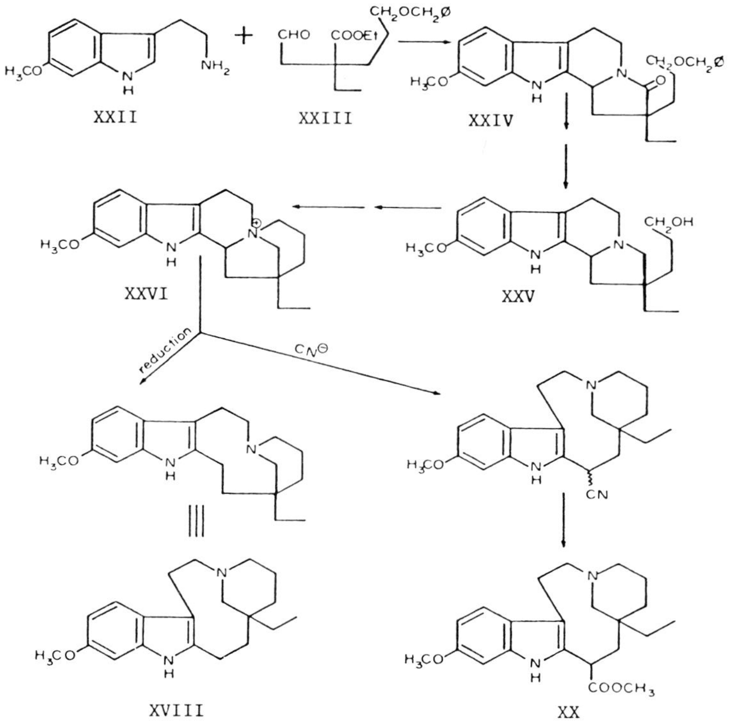

Fig. 6 The total synthesis of nine-membered ring substrates **XVIII** and **XX** for subsequent utilization in transannular cyclization studies.

lactam **XXIV** in high yield. Subsequent hydride reduction of the latter to **XXV**, conversion of the primary alcohol function in **XXV** to the mesylate (or tosylate) derivative, and spontaneous cyclization leads to the quaternary ammonium salt **XXVI**. This type of salt has a crucial role in all of our synthetic studies in this area since this intermediate can be converted in good overall yield to the desired nine-membered ring system. Thus, reductive cleavage of **XXVI** (lithium aluminum hydride in N-methylmorpholine or alkali metal in anhydrous ammonia) leads directly to **XVIII**, while reaction with cyanide provides a route to the ester derivatives (**XX**). The application of this strategy has led to the synthesis of quebrachamine (**I**) [8,15,16], a series of monomeric *Vinca* alkaloids [10,16], dihydrocleavamine (**VIII**), and carbomethoxydihydrocleavamine (**V**) [9,17]. In summary, the above investigations have provided the total syntheses of a variety of *Aspidosperma* and *Vinca* bases (Fig. 7) as well as various members of the *Iboga* family (Fig. 8).

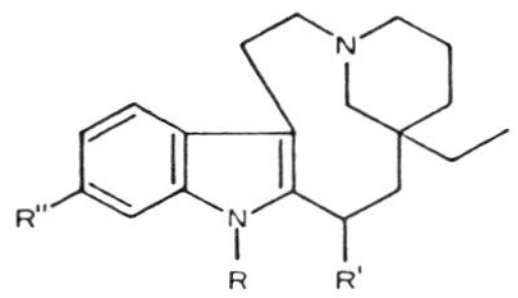

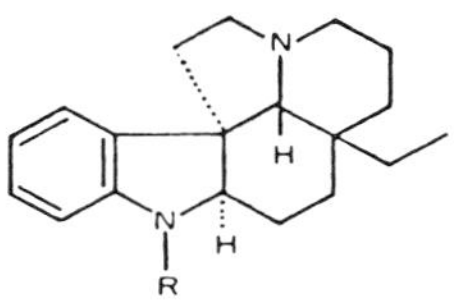

R = R' = R'' = H ; quebrachamine   R = H , (+) - aspidospermidine

R = R''= H , R' = COOCH₃, vincadine

R = CH₃,R' = COOCH₃, R'' = H , vincaminoreine

R = CH₃,R' = COOCH₃ R'' = OCH₃ ; vincaminoridine

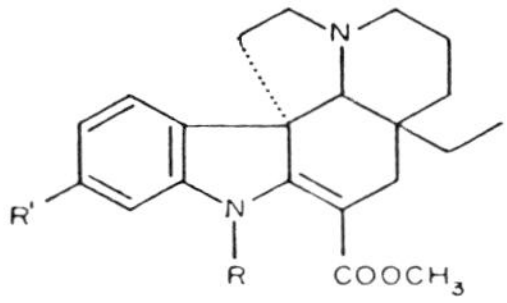

R = R' = H , vincadifformine

R = CH₃,R' = H , minovine

R = CH₃,R' = OCH₃, important intermediate for <u>vindoline</u>
                                              synthesis

Fig. 7 A summary of structures in the *Aspidosperma* and *Vinca* families for which total syntheses have been completed.

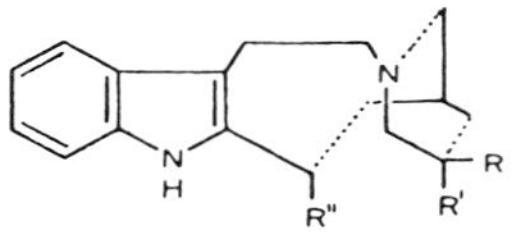

R = R'' = H ; R' = CH₂CH₃ ;   4 β- dihydrocleavamine

R' = R'' = H ; R = CH₂CH₃ ,   4 a - dihydrocleavamine

R''= COOCH₃; R <u>or</u> R'= CH₂CH₃, 18a-(and β)- carbomethoxy-

                                     4 a-(and β)- dihydrocleava-
                                                          mine

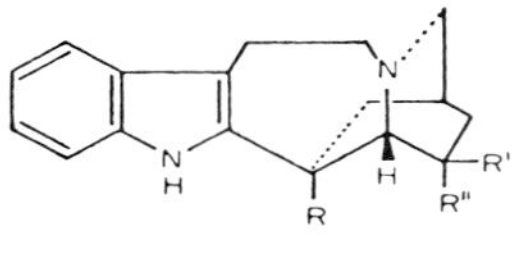

R = H , R'= CH₂CH₃; R''= H   ibogamine

R = COOCH₃; R' = CH₂CH₃; R''= H   coronaridine

R = COOCH₃; R' = H ; R''= CH₂CH₃ dihydrocatharanthine

Fig. 8 A summary of structures in the *Iboga* family for which total syntheses have been completed.

THE FRAGMENTATION APPROACH

While the above studies involving the transannular cyclization approach were underway, we also undertook some investigations of reactions that we felt might have both biosynthetic and synthetic significance, particularly in the bisindole alkaloids of the vincaleukoblastine (VLB) series (Fig. 1). As is obvious from the structural analysis of the VLB family, all of these bisindole or "dimeric" alkaloids contain an indole unit which is of the cleavamine type but which possesses oxygen functionality at the C-3′ and/or C-4′ positions. The synthetic strategy outlined in Fig. 6 did not provide direct introduction of such functionality into these positions of the cleavamine system so that other approaches appeared to be necessary. Furthermore, it was of interest that catharanthine (**VI**, 3,4 double bond), one of the major components in *Catharanthus roseus* G. Don (*Vinca rosea* L.), the plant from which these various alkaloids are isolated, belongs to the rigid pentacyclic *Iboga* family (other examples are coronaridine and ibogamine; see Fig. 8) and does not contain the nonrigid nine-membered ring system of the cleavamine series. The latter series cannot be isolated from this plant in its parent state but is always found coupled with the dihydroindole unit as seen in VLB. On this basis it was of interest to consider that catharanthine, or a closely related analog, is a biosynthetic precursor of the cleavamine-like unit present in VLB. Consequently, studies involving fragmentation of the C-5—C-18 bond in the catharanthine skeleton (see **XXVII**) would not only provide a possible synthetic entry into the required C-3′ and/or C-4′ functionalized cleavamine units, but perhaps shed some light on the "biosynthetic" fragmentation

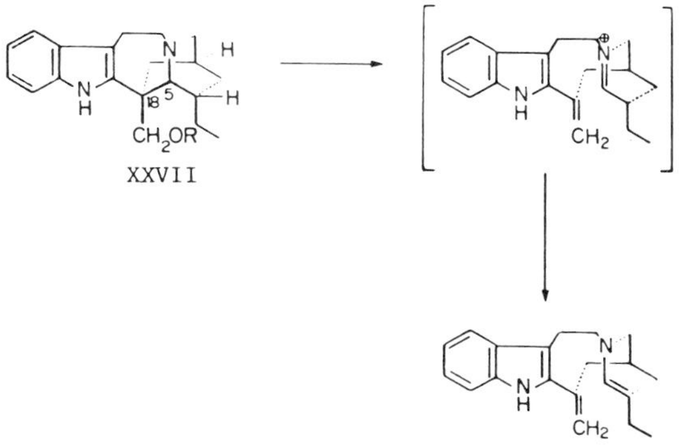

Fig. 9 Grob type fragmentation of dihydrocatharanthinol tosylate (**XXVII**, R = Tos) to the *seco*-diene (**XXVIII**).

process as well. Several "fragmentation approaches" were developed in the subsequent investigations, and these are discussed below.

In one of these investigations the dihydrocatharanthine system (Fig. 8) was utilized as the starting material. It was clear that the rigid quinuclidine unit present in this molecule possessed the essential steric requirements for a Grob type of fragmentation reaction [18–20], and appropriate derivatization of dihydrocatharanthine was considered (Fig. 9). Hydride reduction to the primary alcohol **XXVII** (R = H) and reaction of the latter with tosyl chloride provided dihydrocatharanthinol tosylate (**XXVII**, R = Tos), which was in turn exposed to the fragmentation reaction conditions (triethylamine, benzene, 70°C). The isolated 5,18-*seco*-diene (**XXVIII**) possesses the necessary activation via the enamine system for introduction of functionality at the C-3 and/or C-4 positions. Figure 10 summarizes the successful pathway [21,22] that led to the completion of the synthesis of isovelbanamine (**XXXIII**), cleavamine (**XXXIV**), and velbanamine (**XXXV**).

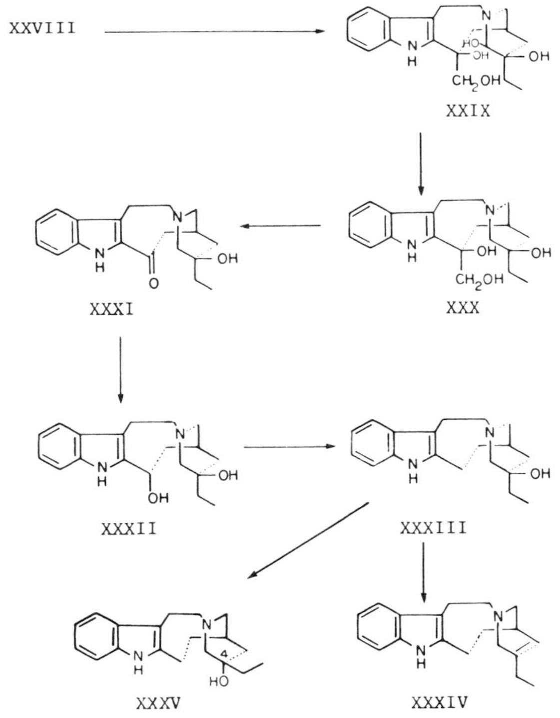

Fig. 10 The synthesis of isovelbanamine (**XXXIII**), cleavamine (**XXXIV**), and velbana-mine (**XXXV**) via osmium tetroxide oxidation of diene **XXVIII**.

Carefully controlled osmylation of **XXVIII** afforded the tetrol **XXIX**, which in turn is readily converted to the triol **XXX** by virtue of facile and selective removal (sodium borohydride) of the carbinolamine hydroxyl group in **XXIX**. Glycol cleavage of **XXX** provides **XXXI**, and the latter, via mild reduction (sodium borohydride) to **XXXII** and then more drastic reduction (lithium aluminum hydride) of **XXXII**, affords isovelbanamine (**XXXIII**), the C-4 hydroxy epimer of velbanamine (**XXXV**), which is the indole unit obtained from the cleavage of VLB and related bisindole alkaloids. An interesting difference in the reaction of **XXXIII** with acid affords either cleavamine (**XXXIV**) or velbanamine (**XXXV**). Thus, treatment of **XXXIII** with concentrated sulfuric acid affords dehydration to **XXXIV**, while aqueous sulfuric acid causes C-4 epimerization to **XXXV**.

The introduction of ester functionality at C-18 of the cleavamine system was accomplished via chloroindolenine intermediates generated in the reaction of the indole ring of these substances with positive halogen (*tert*-butyl hypochlorite or 1-chlorobenzotriazole being most frequently employed). Figure 11 illustrates the sequence that led to the synthesis of 18β-carbomethoxycleavamine (**XXXIX**) from cleavamine (**XXXIV**) [22]. The *in situ* generation of the chloroindolenine intermediate **XXXVI** was performed with *tert*-butyl hypochlorite at −15°C, and **XXXV** was immediately subjected to reaction with fused sodium acetate in acetic acid to provide the quaternary salt

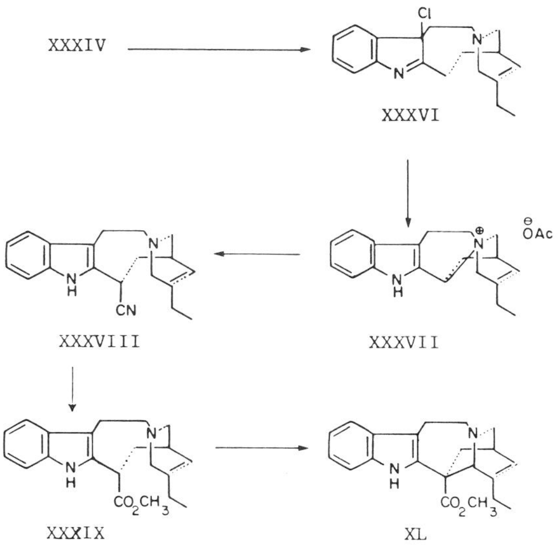

Fig. 11 The synthesis of 18β-carbomethoxycleavamine (**XXXIX**) and catharanthine (**XL**) via the chloroindolenine method.

XXXVII. Reaction of **XXXVII** with potassium cyanide in refluxing dimethyl-formamide provides 18β-cyanocleavamine (**XXXVIII**), which after alkaline hydrolysis and diazomethane treatment provides the desired 18β-carbo-methoxycleavamine. Transannular cyclization of **XXXIX** employing mercuric acetate, as in the previous study, completed the synthesis of catharanthine (**XL**).

Now that the total synthesis of catharanthine had been completed it was appropriate to pursue the chemistry of this readily available alkaloid, particularly in terms of the fragmentation approach. Earlier investigations in the Lilly laboratories [23] and in our own studies [24] had demonstrated that acid-catalyzed fragmentation of catharanthine to the cleavamine system could be accomplished, although the yields were too low (generally 10–15%) for this approach to be of synthetic utility. However, the obvious importance of the cleavamine series in any synthetic objectives directed toward the bisindole family demanded a detailed study of this fragmentation approach [25]. Refinements in reaction conditions, as shown in Fig. 12, allowed the genera-tion of 18β-carbomethoxycleavamine in 90% yield. This important break-through provided, for the first time, the ready availability of various cleavamine analogs for further synthetic utility. Thus, catalytic reduction of the double bond in **XXXIX** (R = CO$_2$CH$_3$) allowed a facile entry into the dihydro-cleavamine series, while acidic hydrolysis and/or decarboxylation provides a high-yielding synthesis of the parent cleavamine (**XXXIX**, R = H) family of compounds. The utilization of such intermediates in the laboratory synthesis of the bisindole alkaloids is now discussed.

The X-ray analysis of leurocristine (vincristine) methiodide by Lipscomb [26] provided the conformational structure for this and the related bisindole

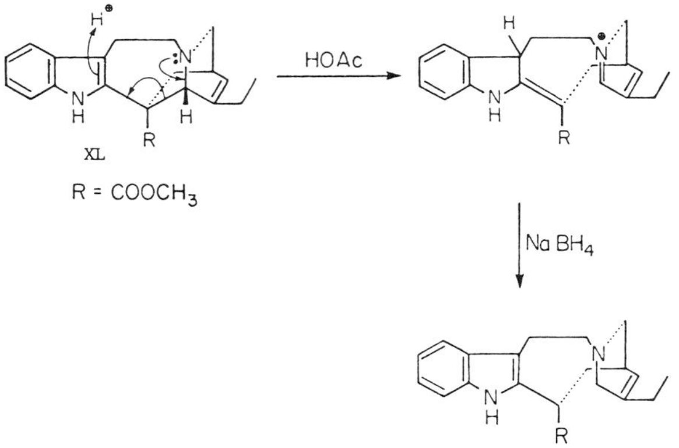

Fig. 12   The acid-catalyzed fragmentation of catharanthine (**XL**) to 18β-carbomethoxy-cleavamine (**XXXIX**).

alkaloids, for example, vinblastine, as shown in **XLI**. Since two subsequent X-ray analyses on our synthetic compounds [27,28] (see later) have also revealed similar conformational structures it is best to discuss our results in the bisindole area in terms of such conformational representations.

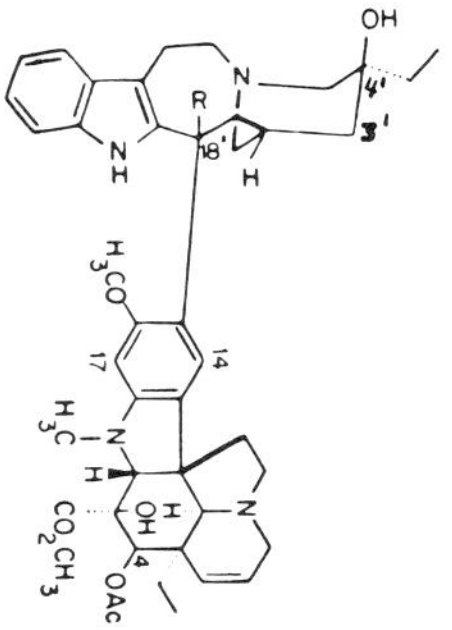

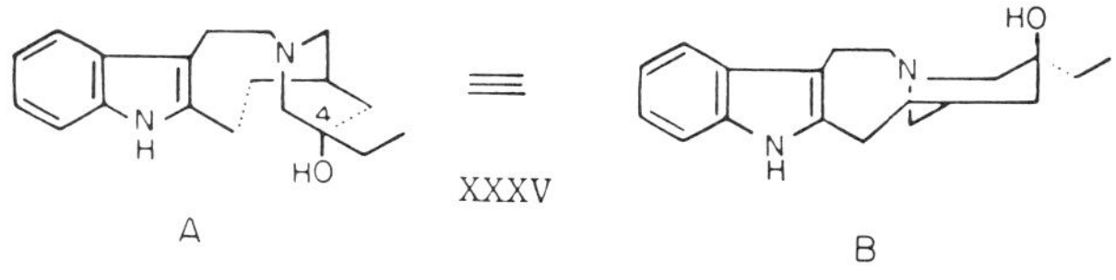

It should be emphasized at this point that, although catharanthine (**XL**) and the related *Iboga* bases exist in a rigid conformation, their corresponding fragmentation products, the "cleavamines," may exist in various conformational forms. Two of the most important conformational expressions that allow a rationalization of all of our results in the bisindole area are shown for velbanamine, namely, the "*Iboga*" conformation **XXXVA**, which was employed in the above discussion and which reveals the piperidine unit in a boat conformation, and **XXXVB**, in which this unit exists in a chair conformation similar to that in the bisindole alkaloids.

As mentioned earlier, biogenetic speculations as they relate to the bisindole area include the possibility that the catharanthine system undergoes fragmentation to the cleavamine-type skeleton and the latter, in some activated form, couples with the dihydroindole unit (vindoline) to provide the natural dimers. Clearly the above-mentioned studies involving fragmentation of catharanthine and the conversion of the resulting cleavamine analogs to chloroindolenine intermediates had some relevance in such biogenetic considerations. It seemed reasonable to assume that oxidative enzymes would be capable of converting the indole ring of the cleavamine family to appropriate hydroxyindolenines, and these intermediates would be expected to behave in a manner similar to that of the chloroindolenine system. On this basis a study involving the "chloroindolenine approach" in the synthesis of bisindole alkaloids was undertaken [27,29,30].

In previous studies [16,17,22] (Fig. 11) we investigated in some detail the reaction of appropriate chloroindolenine intermediates with a variety of nucleophilic reagents (Fig. 13, where $N^\ominus = {}^\ominus CN, {}^\ominus OH, {}^\ominus OAc, CH_3OH, H^\ominus$)

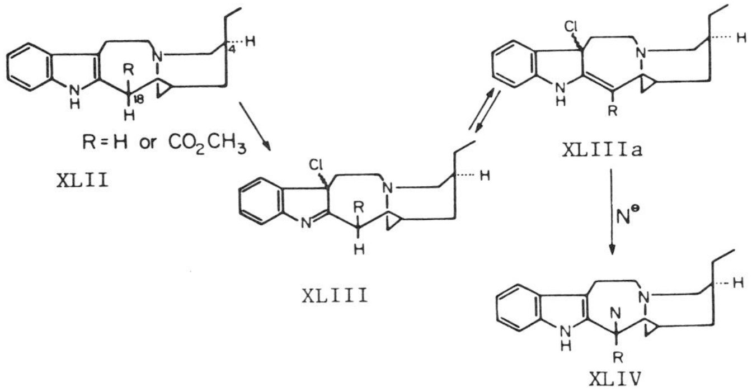

Fig. 13 The synthesis of various C-18-substituted dihydrocleavamine derivatives (XLIV) via the chloroindolenine approach.

and succeeded in introducing such substituents at the C-18 position of the cleavamine system. It appeared appropriate to consider the application of such an approach in the synthesis of bisindole alkaloids. In this instance it was hoped that the dihydroindole ring system in vindoline would possess sufficient nucleophilic character to achieve the coupling required for the dimeric series (Fig. 14). Indeed, subsequent experiments illustrated that this approach can provide a general and versatile synthetic entry to a series of novel synthetic analogs in the dimeric series.

It had been shown by Biemann [31,32] in his elegant mass spectrometric studies of the vinblastine series that complications in the fragmentation patterns are minimized when hydrazide derivatives of the dimeric alkaloids are employed and, for this reason, our earliest studies in this area involved the coupling of deacetylvindoline hydrazide (XLVI) with the chloroindolenine

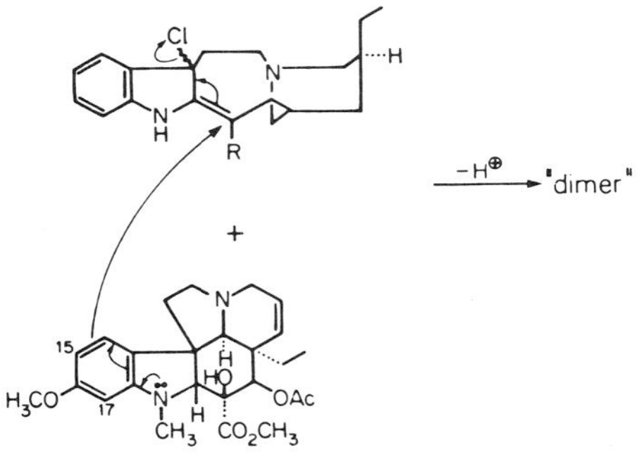

Fig. 14 A proposal outlining the possible coupling of a chloroindolenine intermediate with vindoline to provide a dimeric product.

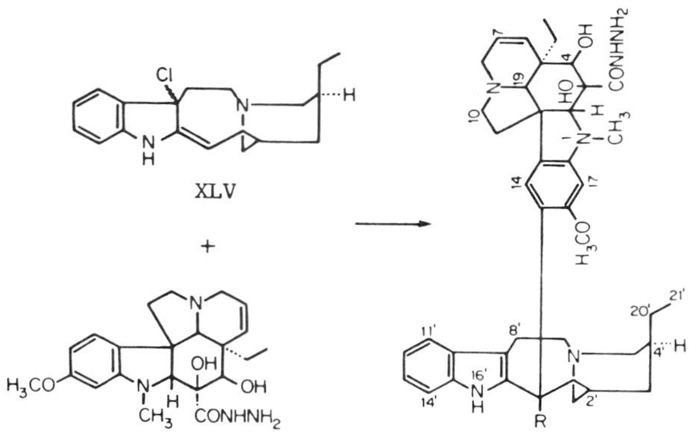

Fig. 15 Coupling of deacetylvindoline hydrazide (**XLIV**) with the chloroindolenine derivative of 4β-dihydrocleavamine (**XLV**) to provide dimer **XLVII** (R = H).

derivative of 4β-dihydrocleavamine (**XLV**) (Fig. 15). The resultant dimer (**XLVII**, R = H) was obtained in 77% yield when refluxing methanolic hydrogen chloride was employed as the coupling reagent. It should be noted that at this point the stereochemistry shown in **XLVII** (R = H) was not known with certainty, but subsequent X-ray analyses (see later) on several synthetic products [28] established the absolute configuration of this and the other synthetic dimers discussed below.

The extension of the chloroindolenine method to the 18β-carbomethoxy-4β-dihydrocleavamine series (**XLIII**, R = CO₂CH₃) was then considered since a successful coupling with this system would provide a dimer possessing a carbomethoxygroup at C-18′ (**XLVII**, R = CO₂CH₃), a feature of vinblastine. This study indeed provided the desired dimer, but the yield was significantly reduced (36.5%).

Similar studies with vindoline (**VII**), the natural dihydroindole unit in vinblastine, and the chloroindolenine derivatives **XLV** and **XLIII** (R = CO₂CH₃) provided, respectively, the dimers **XLVIII** (R = R₁ = H) and **XLVIII** (R = CO₂CH₃; R₁ = H). In general, optimum conditions of dimerization [30] provide yields in excess of 50%. The latter substance, being a close analog of vinblastine, was converted to a crystalline methiodide derivative and subjected to X-ray analysis [28] in order to establish, beyond doubt, its complete structure and absolute stereochemistry. This study revealed that this product is 18′-*epi*-4′-deoxo-4′-*epi*-vinblastine, as shown in **XLVIII** (R = CO₂CH₃; R₁ = H), thus establishing that it possesses the incorrect stereochemistry at C-18′, the crucial chiral center involved in linking the two "halves" of the dimeric system. A similar X-ray study [28]

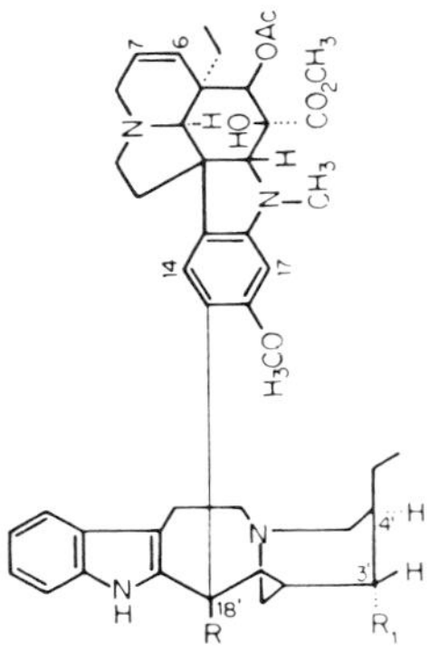

XLVIII

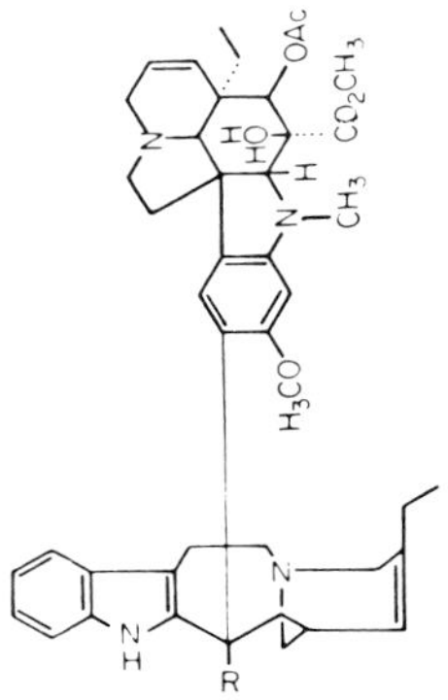

XLIX

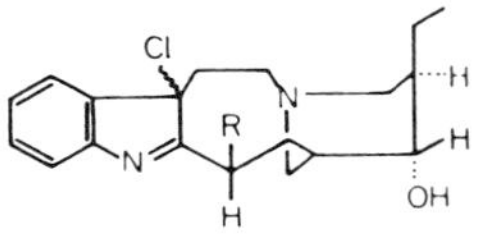

L

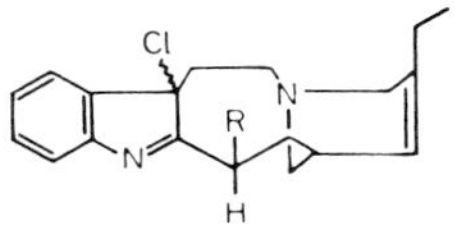

LI

on the dihydrobromide salt of **XLVIII** ($R = R_1 = H$) established that it also possessed the incorrect stereochemistry at C-18' and could be given the name 18'-decarbomethoxy-18'-*epi*-4'-deoxo-4'-*epi*-vinblastine.

It was now clear that the chloroindolenine method was providing, in reasonable yields, dimeric substances but with incorrect stereochemistry at the important C-18' center.

Apart from our synthetic objectives, we were also interested in obtaining some information about structure–activity relationships in this clinically important area and therefore proceeded to prepare other dimeric substances via the chloroindolenine method. Thus, the chloroindolenines **L** ($R = CO_2CH_3$) and **LI** ($R = CO_2CH_3$) provided, respectively, the dimers **XLVIII** ($R = CO_2CH_3$; $R_1 = OH$) and **XLIX** ($R = CO_2CH_3$). This entire series of novel synthetic dimers was then submitted for biological evaluation in several tumor systems (P388 lymphocytic leukemia and L1210 lymphoid leukemia) in animals. None of these compounds showed significant activity when compared with the clinical drugs vinblastine and vincristine (**XLI**, *N*-formyl instead of *N*-methyl). It therefore appeared that the chirality at C-18' in the dimeric substances was important in terms of their antitumor activity.

A great deal of effort [30] was expended in evaluating various reaction parameters in the chloroindolenine method with the hope that such a study

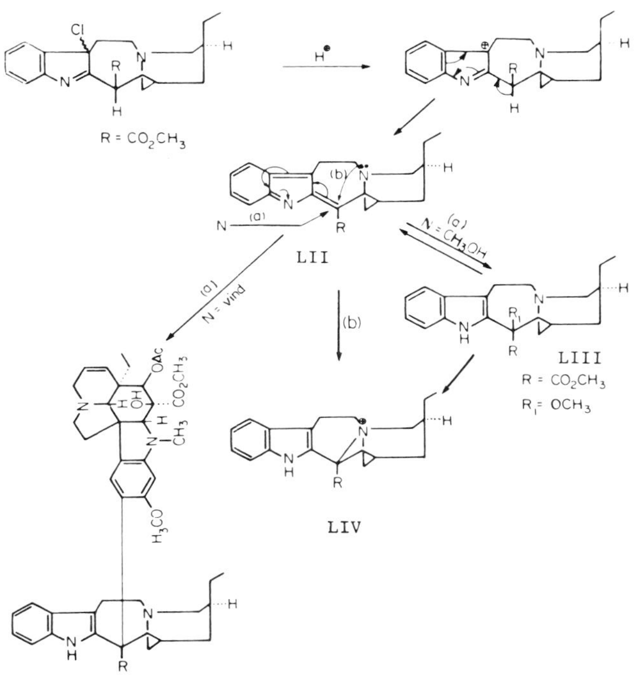

Fig. 16 A proposed mechanism for coupling and related reactions with chloroindolenine intermediates.

would allow a better understanding of the mechanism of this coupling reaction and, in turn, the opportunity for inverting the stereochemical course of this reaction. As a result of our numerous experiments, a mechanism (Fig. 16) could be proposed which accommodated not only our dimerization results but other related reactions with the chloroindolenine intermediates. We believe that the function of the acidic catalyst is to stimulate the cleavage of the C—Cl bond and generate the crucial intermediate **LII**. The stereochemical fate of the isolated products is then determined by the course of nucleophilic attack on **LII**. Molecular models reveal that, in **LII**, entry of the nucleophile N into the double bond is preferred from the β face of the molecule [pathway (a)]. Thus, reaction of **LII** with methanol would provide **LIII**, while with vindoline the isolated dimers would, as shown, possess the incorrect stereochemistry at C-18′. An alternative pathway [pathway (b)] is possible in which internal attack by the basic nitrogen atom would provide the quaternary salt **LIV**. Such a process has already been illustrated in the synthesis of 18β-carbomethoxycleavamine (Fig. 11). Clearly the reaction of **LIV** with vindoline would now provide the desired stereochemistry at C-18', but all of our attempts to perform this reaction proved to be unsuccessful.

The above-mentioned investigations involving the fragmentation approach involved the fragmentation of the catharanthine system to provide isolable and stable cleavamine derivatives. These substances were "activated" in a subsequent reaction to provide the desired intermediates for coupling with the dihydroindole unit. Another fragmentation approach that we generally term the "biogenetic approach" has been under investigation since 1973 and has led to some interesting results, which are discussed below.

In the "biogenetic approach" it was considered that the fragmentation of the catharanthine system could be generated *in situ*, and the "activated" intermediate thus formed would be allowed to react immediately with the dihydroindole unit to provide the dimeric product. It was felt that such a process might portray more closely the biosynthetic pathway in the plant and hopefully would allow the generation of synthetic dimers with the desired stereochemistry at C-18'. In considering such a process in synthetic and biosynthetic terms it appeared most appropriate to evaluate the consequences of electrophilic attack on the electron-rich catharanthine system, a pathway that would be feasible both chemically and enzymatically. Schemes 1–3. portray in general terms the expected modes of fragmentation that may be initiated by an electrophilic reagent ($E^{\oplus}$). Scheme 1 illustrates a fragmentation mode initiated by electrophilic attack at the β position of the indole ring in catharanthine (**XL**, R = CO_2CH_3). It is seen that the generated intermediate **LV** is reminiscent of that generated in the acid-catalyzed fragmentation of catharanthine (Fig. 12) and also in the coupling method involving chloro-indolenine intermediates (see **LII**, Fig. 16). Since the latter method provided

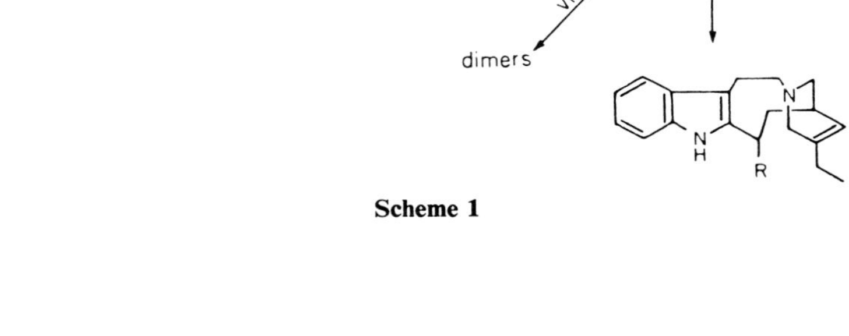

Scheme 1

dimers with incorrect stereochemistry at C-18′ the pathway proposed in this scheme was of little interest.

The alkene system in catharanthine provides another site of attack by the electrophilic reagent (Scheme 2). The carbonium ion intermediate expected from such a process could be considered either in classical (**LVI**) or non-classical (**LVII**) terms, and its reaction with the dihydroindole unit (vindoline) could provide the dimeric substances **LVIII**, **LIX**, or **LX**. Although studies involving a bicyclic quinuclidine system, present in catharanthine, are unavailable from the literature, the carbon analog, the bicyclo[2.2.2]octyl system, has been evaluated in considerable detail (for pertinent references see Bartlett [33]), and the evidence leads to the conclusion that the most likely dimeric product that would result is **LIX**.

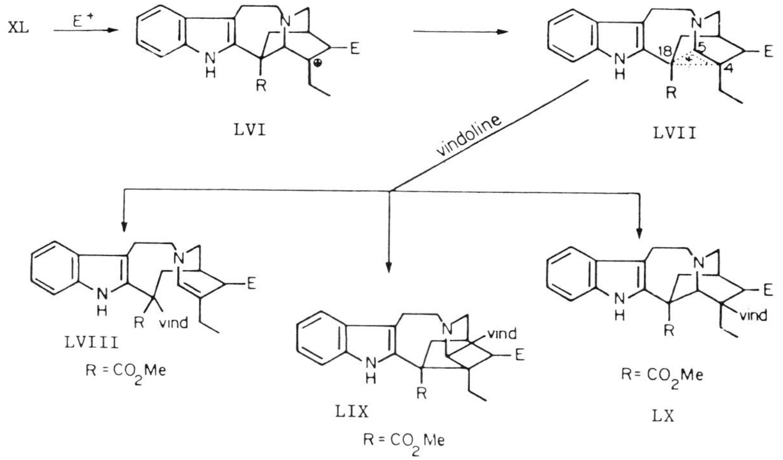

Scheme 2

Scheme 3 portrays attack at the basic nitrogen atom, for example, N-oxide formation, a well-known process in alkaloid chemistry. Fragmentation of N-oxides [34] is, in effect, a modification of the Polonovski reaction [35–40] and is well documented in the literature.* As Scheme 3 reveals, fragmentation in the direction **LXI → LXII → LXIII** is reminiscent of the previous studies in which intermediates generated from either the acid-catalyzed fragmentation of catharanthine (Fig. 12) or the chloroindolenine method (Fig. 16) lead to unnatural stereochemistry dimers (**LXIII**). On the other hand, a "concerted" process, shown by the pathway **LXI → LXIV**, in which the vindoline unit,

---

* For a general review, see Russell and Mikol [35].

acting as a nucleophile, displaces the C-18—C-5 bond of **LXI** in a *trans* coplanar fashion, would yield the desired stereochemistry, as shown in **LXIV**. A great deal of work has been performed in this area [41,42], and a summary of the pertinent results is provided below.

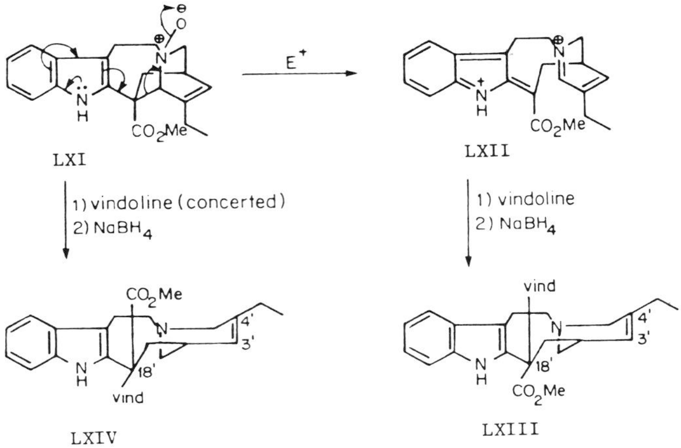

1) vindoline (concerted)
2) NaBH$_4$

1) vindoline
2) NaBH$_4$

Scheme 3

The earliest investigations (1973) of the "biogenetic approach" involved the reaction of catharanthine (or its hydrochloride salt) with *m*-chloro-perbenzoic acid and subsequent reaction of the intermediates thus generated with vindoline in the presence of methanolic hydrochloric acid, a reagent employed extensively in the earlier studies [27,30] already discussed. These experiments allowed the isolation of a dimeric product (**LXV**), which, although lacking the structural features of vinblastine, provided considerable stimulus for further experiments. It is clear that one plausible mode of formation of **LXV** would involve a Polonovski-type fragmentation, **LXI** → **LXII**, and subsequent attack of vindoline at the electrophilic iminium center

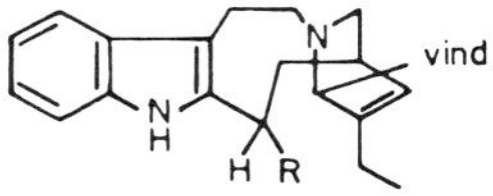

R = CO$_2$Me

LXV

of **LXII** to eventually provide, after sodium borohydride reduction, the isolated dimer. On this basis, fragmentation according to Scheme 3 was indeed occurring, but subsequent coupling with vindoline, in the methanolic hydrochloric acid medium, involved an inappropriate center.

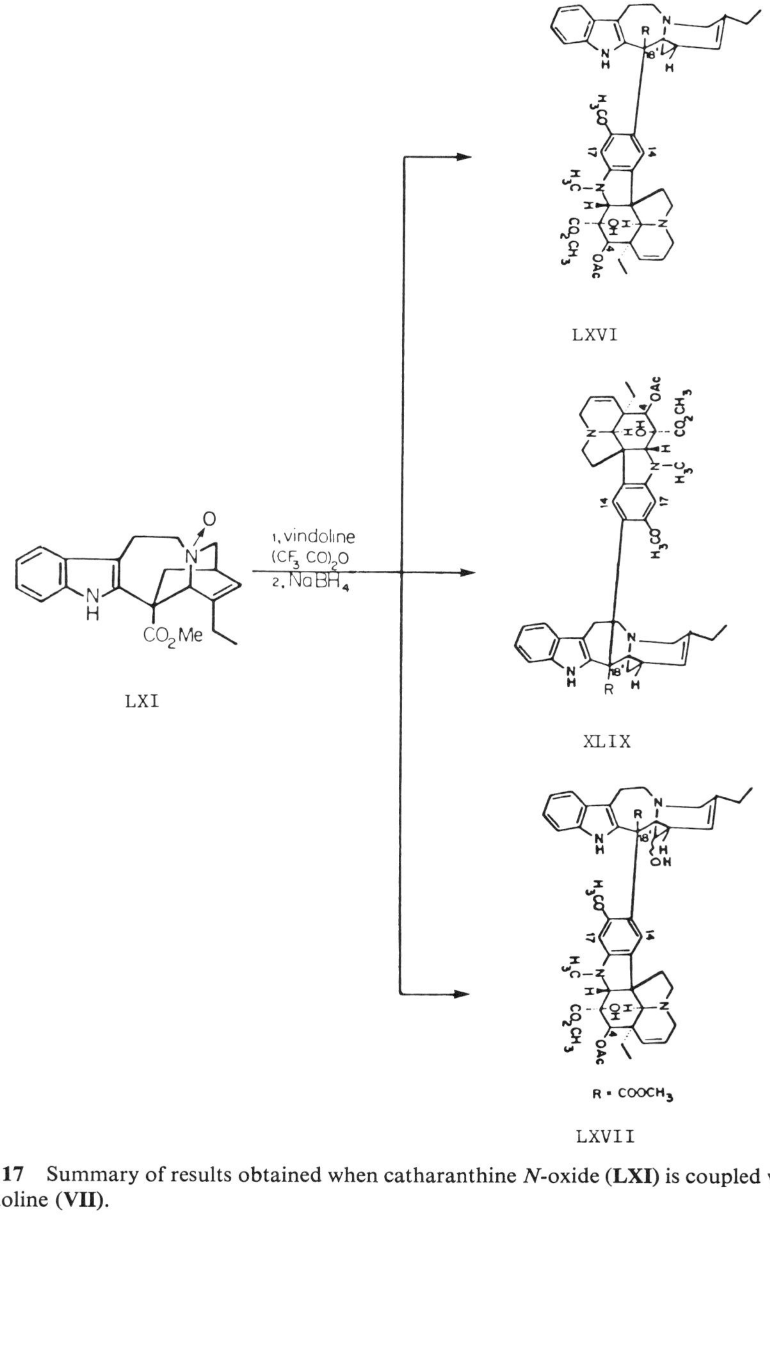

Fig. 17 Summary of results obtained when catharanthine N-oxide (**LXI**) is coupled with vindoline (**VII**).

Further investigations quickly revealed that the initially formed catharan-thine N-oxide or other intermediates were extremely sensitive to reaction conditions (temperature, solvent, reaction workup, molar ratio of reactants, etc.) and the resultant dimeric products isolated were markedly dependent on these factors. It is inappropriate to discuss the details of our many experiments here [42], so only the most salient features are presented.

Figure 17 summarizes the results obtained when catharanthine N-oxide generated in situ, is coupled with vindoline (VII) with trifluoroacetic anhydride as coupling reagent. As was deduced from later experiments, optimum reaction conditions [coupling with methylene chloride as solvent, $-50°C$, and $(CF_3CO)_2O$ as reagent] provided only dimeric products with natural stereochemistry at C-18'. In this study, 3',4'-dehydrovinblastine (LXVI, $R = CO_2CH_3$) was obtained consistently in about 50% yield, while the incorrect stereochemistry dimer (XLIX, $R = CO_2CH_3$) obtained earlier in the chloroindolenine studies was not observed in studies at low temperature. The carbinolamine dimer (LXVII, $R = CO_2CH_3$), a further oxidation product of LXVI ($R = CO_2CH_3$), was generally a minor product (8–18%) and, upon reaction with tin and hydrochloric acid, could be converted back to LXVI ($R = CO_2CH_3$).

Two reaction parameters, temperature and solvent, revealed rather dramatic differences in the yields and/or nature of the resultant products so they are briefly mentioned here. Table 1 reveals the effect of temperature, while Table 2 illustrates the different results obtained when various solvents were employed in the coupling process.

It is clear from Table 1 that low temperatures favor the formation of the desired C-18' stereochemistry dimers, while elevated temperatures provide predominantly the undesired products with unnatural stereochemistry at

TABLE 1

Effect of Temperature on Reaction of Catharanthine N-Oxide with Vindoline.

Experiment No.[a]	Temperature (°C)	Yield of dimer (%)		
		LXVI	XLIX	LXVII
1	−50 (standard)	50	0	8
2	−10	30	14	[b]
3	−4	38	18	[b]
4	42	17	29	[b]
5	61	0	34	[b]

[a] Solvent employed in experiments 1–4 was methylene chloride, while chloroform was used in experiment 5. All other reaction parameters were identical in these experiments.

[b] Not observed.

TABLE 2

**Effect of Solvent on Reaction of Catharanthine
N-Oxide with Vindoline.**

Experiment No.	Solvent	Yield of dimer (%)		
		LXVI	**XLIX**	**LXVII**
1	CH_2Cl_2	50	0	8
2	$CHCl_3$	0	0	23
3	Dimethylformamide	0	0	0
4	Tetrahydrofuran	0	0	0
5	CH_3CN	33	0	23
6	Toluene	0	0	0

C-18'. Table 2 shows that methylene chloride is the preferred solvent for the coupling reaction.

The versatility of the "biogenetic approach" was evaluated in terms of coupling various N-oxides of the catharanthine family with vindoline and/or vindoline derivatives. Again it is inappropriate to discuss the details of these numerous experiments, so only a summary of the most pertinent results is presented.

Figure 18 summarizes the results obtained when dihydrocatharanthine N-oxide (**LXVIII**) is coupled with vindoline. As in the previous studies, the number and nature of the dimeric products were dependent on reaction conditions. Thus, at a coupling temperature of $-10°C$ the four products **LXIX–LXXII** were isolated, while at lower temperatures ($-50°$ to $-30°C$) only the dimeric products **LXIX** and **LXX** bearing the natural stereochemistry at C-18' were obtained. In general, in all our studies in this series the yields of the dimeric products were significantly lower (5–20%), and we feel that this is probably due to the absence of the double bond which would be expected to stabilize the intermediate resulting from the Polonovski-type fragmentation.

Figure 19 summarizes the results obtained when decarbomethoxycatharanthine N-oxide (**LXXIII**, R = H) is coupled with vindoline at low temperature ($-15°$ to $-30°C$). The desired dimer **LXXIV** (R = H) is obtained in 27% yield, while the C-18' unnatural stereochemistry dimer **LXXV** (R = H), previously obtained in the chloroindolenine method, is isolated in 11% yield. An interesting feature of this study is the isolation of **LXXVI** (R = H, 32% yield), which is considered to arise from a normal Polonovski elimination occurring with **LXXIII** and the resulting iminium intermediate

$$\diagup C = \overset{\oplus}{N} \diagdown$$

then reacts with vindoline in the expected fashion.

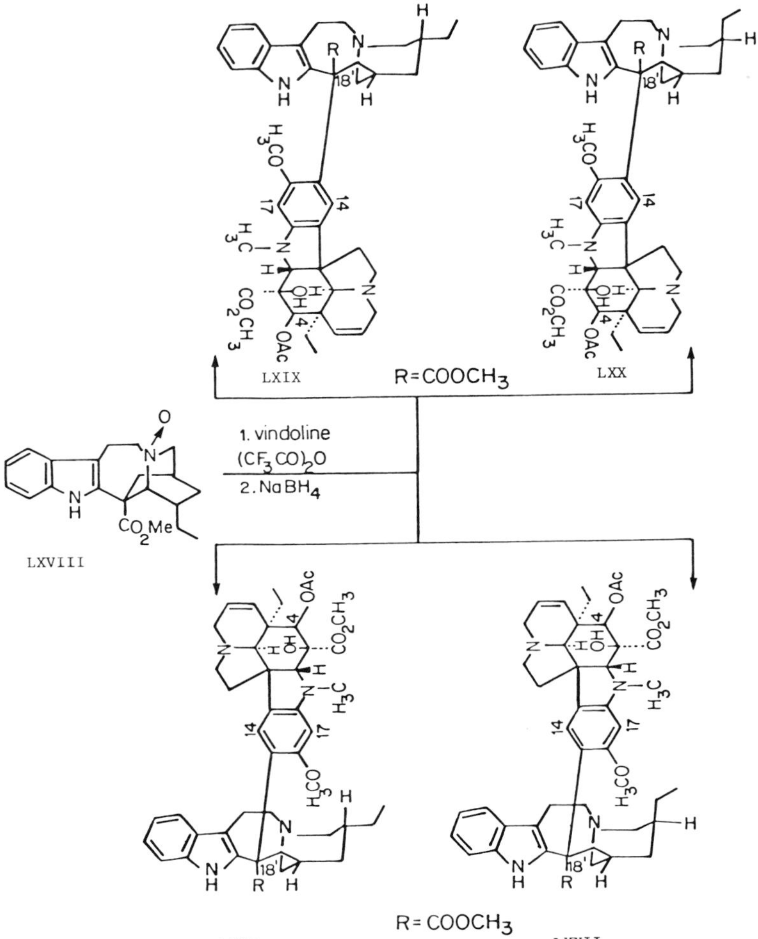

LXIX LXX LXVIII LXXI LXXII

Fig. 18 Summary of results obtained when dihydrocatharanthine *N*-oxide (**LXVIII**) is coupled with vindoline.

In another study, the coupling of catharanthine *N*-oxide and/or decarbomethoxycatharanthine *N*-oxide was considered with vindoline analogs to determine whether the functionality in the dihydroindole unit can be maintained during the coupling conditions. Figure 20 summarizes the results obtained with vindoline *N*-methylamide (**LXXVII**, $R_1 = CH_3$; $R_2 = Ac$). These studies were undertaken to provide a family of compounds (**LXXVIII**, $R = H$ or CO_2CH_3; $R_1 = CH_3$; $R_2 = Ac$; **LXXIX**, $R = H$ or CO_2CH_3; $R_1 = CH_3$; $R_2 = Ac$) structurally related to the vinblastine amides, the latter series having been shown to possess important biological activity [43].

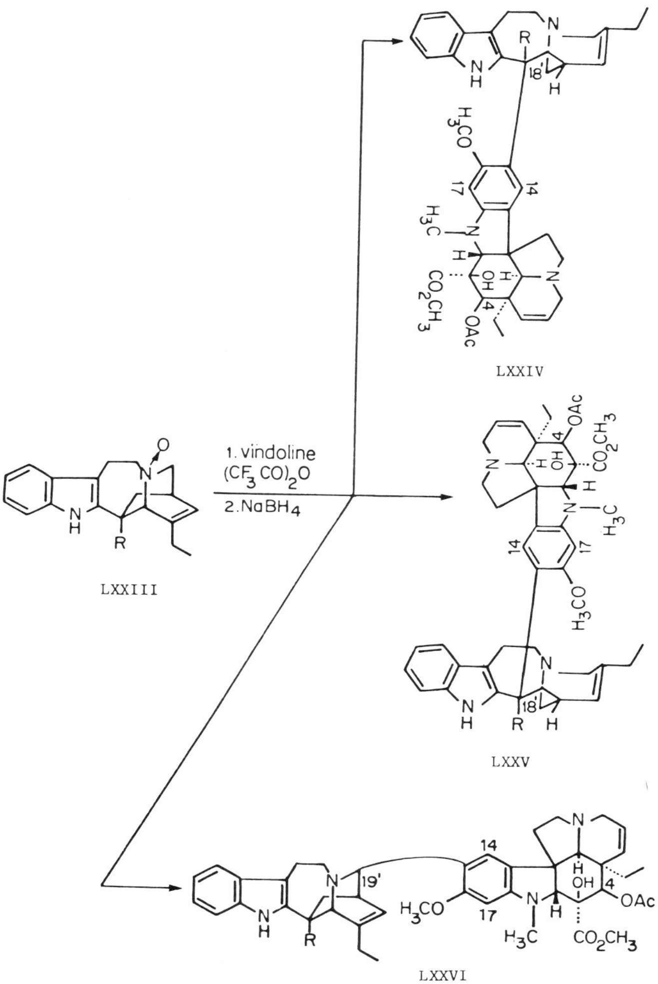

Fig. 19 Summary of results obtained when decarbomethoxycatharanthine N-oxide (**LXXIII**, R = H) is coupled with vindoline.

In summary, it is clear from the above discussion that the "biogenetic" approach" does provide an exciting synthetic entry into a series of bisindole systems closely related to the vinblastine family. It is possible to make the following generalizations with respect to these Polonovski-type fragmentation studies: (1) Catharanthine is the best substrate and generally leads to highest yields of dimers; (2) the absence of a double bond, for example, dihydro-

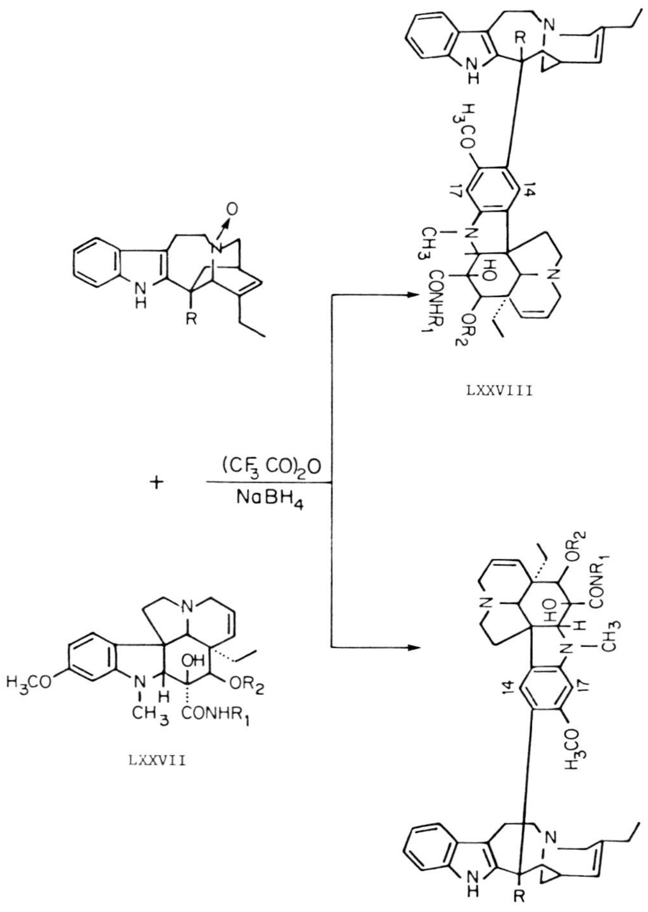

Fig. 20 Summary of results obtained when catharanthine N-oxide and/or decarbo-methoxycatharanthine N-oxide is coupled with vindoline N-methylamide (**LXXVII**), $R_1 = CH_3$; $R_2 = Ac$).

catharanthine, does not prevent fragmentation of the N-oxide derivative, but in general the products are obtained in lower yield and are isomeric at the ethyl-bearing center; (3) the absence of an ester group, as in decarbomethoxy-catharanthine, appears to reduce reactivity toward fragmentation and generally provides for lower yields of dimers; (4) the relative ratio of natural (C-18') vs. unnatural stereochemistry in the resultant dimers is markedly dependent on reaction conditions, particularly temperature. In all cases studied, the highest ratio of natural stereochemistry dimers is obtained when very low temperatures are employed.

It is possible to put forward a mechanistic rationale explaining the various results obtained in these studies. We feel that the Polonovski fragmentation process may proceed according to several routes: (1) a truly "concerted" or S_N2'-like mechanism in which there is a simultaneous fragmentation of the C-18—C-5 bond and formation of the C-18'—C-15 bond in the resulting dimer, (2) a stepwise process in which fragmentation of the C-18—C-5 bond generates an intermediate (**LXI** → **LXII**, Scheme 3) and the latter then reacts with vindoline to provide the dimer, and (3) both mechanisms occurring depending on the particular *Iboga* skeleton employed.

The "concerted" process, schematically represented by **LXXX**, must proceed in such a fashion that the vindoline unit enters from the α face of the indole system, thereby generating the desired stereochemistry at C-18'. On the other hand, this process cannot be invoked to explain the formation of the dimeric products with unnatural stereochemistry at C-18'.

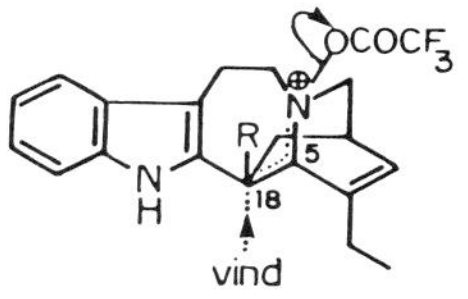

LXXX

The stepwise process can provide a rationalization for all of the obtained results. It is obvious that various conformational expressions can be written for the generated intermediate **LXII**, but only two (**LV** and **LXXXI**) are considered here since these are sufficient to provide the required explanation. Thus, if fragmentation of the rigid quinuclidine structure in the catharanthine system occurs in the indicated manner, the initially formed intermediate adopts what may be termed as the "*Iboga*" conformation **LV**. Under appropriate reaction conditions it is clear that **LV** can undergo conformational

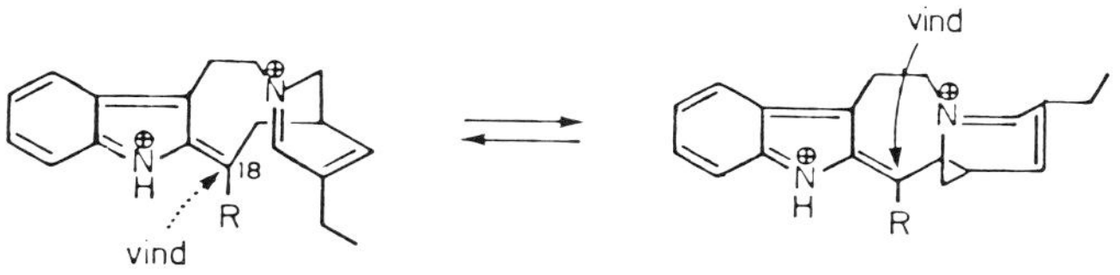

LV   LXXXI

alteration to **LXXXI**, the latter being the conformation revealed earlier in the X-ray studies of vinblastine and our synthetic dimers. From molecular models it is clear that in the "*Iboga*" conformation **LV** preferential attack of the vindoline molecule would occur from the α face, therefore providing dimers with natural stereochemistry at C-18'. On the other hand, in the alternative conformation **LXXXI**, approach of vindoline would occur from the β face of the original indole system to yield the opposite stereochemistry at C-18'. Obviously, reaction conditions such as temperature determine the conformational mobility between the two conformations already noted (as well as others). As Table 1 illustrates the relative proportion of unnatural stereochemistry dimer **XLIX** increases with temperature, and similar results are obtained in the dihydrocatharanthine and decarbomethoxycatharanthine series. It is therefore tempting to postulate that at low temperatures (for example, −50°C) the initially formed intermediate is held in the "*Iboga*" conformation **LV** and eventually provides the natural stereochemistry dimer **LXVI** while, at elevated temperatures, conformational inversion to **LXXXI** occurs and the dimer **XLIX** results.

An independent investigation of the Polonovski-type fragmentation to the synthesis of these types of dimeric substances has been recently reported by a French group [44].

An extension of the above studies toward the synthesis of vinblastine (**XLI**) and related bisindole alkaloids required methods for the introduction of oxygen functionality at the 3',4' positions in the synthetic dimers discussed above. Thus, vinblastine possesses a hydroxyl group at C-4, leurosidine has been recently shown [45] to be the C-4' hydroxy epimer of vinblastine, leurosine is known [45–47] to possess a C-3',4' epoxy group, and vincadioline [48] contains a diol system at these positions.

A successful approach to this problem employed the oxygen and/or hydroperoxide oxidation of the double bond, a reaction that has not been extensively employed by synthetic chemists [49–52]. Our initial investigations involved the N_a-carbomethoxy derivative of 18β-carbomethoxycleavamine (**LXXXII**), and the results are summarized in Fig. 21. The hydroperoxide

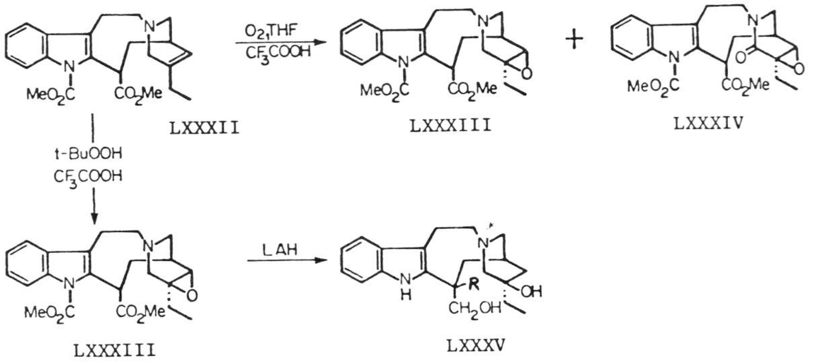

Fig. 21 Summary of results obtained when N_a-carbomethoxy-18β-carbomethoxy-cleavamine (**LXXXII**) is reacted with oxygen and/or *tert*-butyl hydroperoxide.

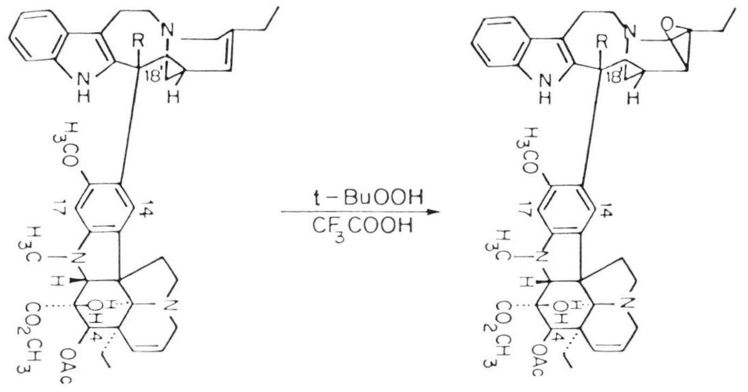

LXVI LXXXVI

Fig. 22 The synthesis of leurosine (**LXXXVI**, R = CO_2CH_3) from dimer **LXVI** (R = CO_2CH_3).

reaction provided higher yields (67%) of the desired epoxide (**LXXXIII**) than did the oxidation with oxygen (10–15%), where the predominant product (51% yield) was the lactam epoxide (**LXXXIV**).

Application of this reaction to the synthetic dimer **LXVI** (R = CO_2CH_3) (Fig. 22) obtained previously provided the first synthesis of the *Vinca* alkaloid leurosine (**LXXXVI**, R W CO_2CH_3). The β orientation of the epoxide ring shown for leurosine is assigned in a tentative manner since the structural elucidation studies mentioned earlier [45–47] did not allow a definitive assignment for this functionality.

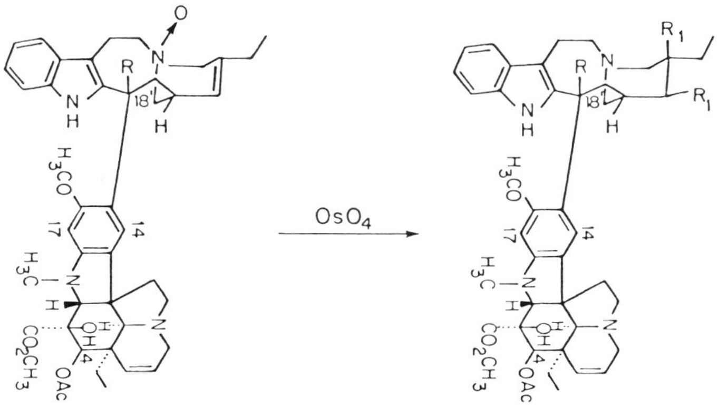

LXXXVII LXXXVIII

Fig. 23 The synthesis of hydroxyvinblastine 3′-(**LXXXVIII**, R = CO_2CH_3; R_1 = OH) from the *N*-oxide intermediate **LXXXVII** (R = CO_2CH_3).

In a related study concerned with the introduction of hydroxyl functionality at the C-3' and/or C-4' positions in the synthetic dimers, the N-oxide derivative of **LXVI** ($R = CO_2CH_3$) is reacted with osmium tetroxide, under controlled conditions, to provide 3'-hydroxyvinblastine (Fig. 23, **LXXXVIII**, $R = CO_2CH_3$; $R_1 = OH$). This synthetic material is a close relative of vincadioline [48] and a valuable intermediate in the synthesis of vinblastine. On the basis of the various studies presented, one can conclude that the "fragmentation approach" as it relates to the fragmentation of catharanthine and its various derivatives does provide a versatile synthetic entry into a whole series of bisindole alkaloids and novel synthetic dimers for biological evaluation. Studies in this direction are continuing.

In conclusion, this review summarizes some of our results in the program directed toward the synthesis of various indole and dihydroindole alkaloids of the *Aspidosperma*, *Iboga*, and *Vinca* alkaloids. Only those aspects of the synthetic investigations that arose from biogenetic considerations are emphasized here.

ACKNOWLEDGMENTS

A discussion of this work would not be complete without mention of the enthusiastic, persevering, and hard-working colleagues who made it possible. They are N. Abdurahman, J. Balsevich, J. Beck, G. Bokelman, T. Brocksom, R. T. Brown, U. Bunzli-Trepp, F. Bylsma, K. K. Chan, J. Cook, W. Cretney, A. Failli, J. Fromson, K. Fuji, Y. Fujise, D. Gregonis, C. Gletsos, J. Hadfield, T. Hibino, R. Imhof, I. Itoh, E. Jahngen, J. Katsube, P. Le Quesne, A. Leutwiler, B. McKague, V. Nelson, T. Okutani, E. Piers, A. Ratcliffe, M. Rohr, P. de Souza, A. Treasurywala, I. Vlattas, and S. Wunderly.

I would like to express my sincere thanks to Dr. N. Neuss, Dr. M. Gorman, and Dr. K. Gerzon, Lilly Research Laboratories, for numerous discussions and generous supplies of alkaloids which made these studies possible. Professor C. Djerassi, Stanford University, also kindly provided numerous comparison samples and/or data of various alkaloids.

Financial aid from the National Research Council of Canada, Medical Research Council of Canada, National Cancer Institute of Canada, and National Institutes of Health, Contract N01-CM-23223, Bethesda, Maryland, is gratefully acknowledged.

REFERENCES

1. J. P. Kutney, *Heterocycles* **4**, 169 and 429 (1976).
2. E. Wenkert, *J. Am. Chem. Soc.* **84**, 98 (1962).
3. J. P. Kutney and E. Piers, *J. Am. Chem. Soc.* **86**, 953 (1964).
4. J. P. Kutney, R. T. Brown, and E. Piers, *J. Am. Chem. Soc.* **86**, 2286 (1964).
5. J. P. Kutney, R. T. Brown, and E. Piers, *J. Am. Chem. Soc.* **86**, 2287 (1964).
6. J. P. Kutney, R. T. Brown, and E. Piers, *Lloydia* **27**, 447 (1964).

7. A. Camerman, N. Camerman, J. P. Kutney, E. Piers, and J. Trotter, *Tetrahedron Lett.* p. 637 (1965).
8. J. P. Kutney, N. Abdurahman, P. Le Quesne, E. Piers, and I. Vlattas, *J. Am. Chem. Soc.* **88**, 3656 (1966).
9. J. P. Kutney, W. J. Cretney, P. Le Quesne, B. McKague, and E. Piers, *J. Am. Chem. Soc.* **88**, 4756 (1966).
10. J. P. Kutney, K. K. Chan, A. Failli, J. M. Fromson, C. Gletsos, and V. R. Nelson, *J. Am. Chem. Soc.* **90**, 3891 (1968).
11. J. P. Kutney, E. Piers, and R. T. Brown, *J. Am. Chem. Soc.* **92**, 1700 (1970).
12. J. P. Kutney, R. T. Brown, E. Piers, and J. R. Hadfield, *J. Am. Chem. Soc.* **92**, 1708 (1970).
13. B. W. Bycroft, D. Schumann, M. B. Patel, and H. Schmid, *Helv. Chim. Acta* **47**, 1147 (1964).
14. J. Le Men and W. I. Taylor, *Experientia* **21**, 508 (1965).
14a. J. Trojanek and K. Blaha, *Lloydia* **29**, 149 (1966).
15. J. P. Kutney, N. Abdurahman, C. Gletsos, P. Le Quesne, E. Piers, and I. Vlattas, *J. Am. Chem. Soc.* **92**, 1727 (1970).
16. J. P. Kutney, K. K. Chan, A. Failli, J. M. Fromson, C. Gletsos, A. Leutwiler, V. R. Nelson, and J. P. de Souza, *Helv. Chim. Acta* **58**, 1648 (1975).
17. J. P. Kutney, W. J. Cretney, P. Le Quesne, B. McKague, and E. Piers, *J. Am. Chem. Soc.* **92**, 1712 (1970).
18. C. A. Grob, *Bull. Soc. Chim. Fr.* p. 1360 (1960).
19. M. F. Bartlett, R. Sklar, W. I. Taylor, E. Schlettler, R. L. S. Amai, P. Beak, N. V. Bringi, and E. Wenkert, *J. Am. Chem. Soc.* **84**, 622 (1962).
20. U. Renner, K. A. Jaeggi, and D. A. Prins, *Tetrahedron Lett.* p. 3697 (1965).
21. J. P. Kutney and F. Bylsma, *J. Am. Chem. Soc.* **92**, 6090 (1970).
22. J. P. Kutney and F. Bylsma, *Helv. Chim. Acta* **58**, 1672 (1975).
23. M. Gorman, N. Neuss, and N. J. Cone, *J. Am. Chem. Soc.* **87**, 93 (1965).
24. J. P. Kutney, R. T. Brown, and E. Piers, *Can. J. Chem.* **43**, 1545 (1965).
25. J. P. Kutney, W. J. Cretney, J. R. Hadfield, E. S. Hall, and V. R. Nelson, *J. Am. Chem. Soc.* **92**, 1704 (1970).
26. J. W. Moncrief and W. N. Lipscomb, *J. Am. Chem. Soc.* **87**, 4963 (1965).
27. J. P. Kutney, J. Beck, F. Bylsma, and W. J. Cretney, *J. Am. Chem. Soc.* **90**, 4504 (1968).
28. J. P. Kutney, J. Cook, K. Fuji, A. M. Treasurywala, J. Clardy, J. Fayos, and H. Wright, *Heterocycles* **3**, 205 (1975).
29. N. Neuss, M. Gorman, N. J. Cone, and L. L. Huckstep, *Tetrahedron Lett.* p. 783 (1968).
30. J. P. Kutney, J. Beck, F. Bylsma, J. Cook, W. J. Cretney, K. Fuji, R. Imhof, and A. M. Treasurywala, *Helv. Chim. Acta* **58**, 1690 (1975).
31. P. Bommer, W. McMurray, and K. Biemann, *J. Am. Chem. Soc.* **86**, 1439 (1964).
32. K. Biemann, *Lloydia* **27**, 397 (1964).
33. P. D. Bartlett, "Non-Classical Ions." Benjamin, New York, 1965.
34. A. Ahond, A. Cave, C. Kan-Fan, Y. Langlois, and P. Potier, *Chem. Commun.* p. 517 (1970), and references cited therein.
35. G. A. Russell and G. J. Mikol, *in* "Mechanisms of Molecular Migrations" (B. S. Thyagarajan, ed.), Vol. 1, p. 176. Wiley (Interscience), New York, 1968.
36. R. N. Renaud and L. C. Leitch, *Can. J. Chem.* **46**, 387 (1968).
37. R. Mitchelot, *Bull. Soc. Chim. Fr.* p. 4377 (1969).

38. R. T. Lalonde, E. Auer, C. F. Wong, and V. P. Muralidharan, *J. Am. Chem. Soc.* **93**, 2501 (1971).
39. C. A. Scherer, C. A. Dorschel, J. M. Cook, and P. Le Quesne, *J. Org. Chem.* **37**, 1083 (1972).
40. Y. Hayashi, Y. Nagano, S. Hongyo, and K. Teramura, *Tetrahedron Lett.* p. 1299 (1974).
41. J. P. Kutney, A. H. Ratcliffe, A. M. Treasurywala, and S. Wunderly, *Heterocycles* **3**, 639 (1975).
42. J. P. Kutney, T. Hibino, E. Jahngen, T. Okutani, A. H. Ratcliffe, A. M. Treasurywala, and S. Wunderly, to be published.
43. M. J. Sweeney, G. T. Cullivan, G. A. Poore, and K. Gerzon, *Proc. Annu. Meet., Am. Soc. Clin. Oncol., 10th*, Vol. 15 (1974).
44. P. Potier, N. Langlois, Y. Langlois, and F. Gueritte, *Chem. Commun.* p. 670 (1975).
45. E. Wenkert, E. W. Hagaman, B. Lal, G. E. Gutowski, A. S. Katner, J. C. Miller, and N. Neuss, *Helv. Chim. Acta* **58**, 1560 (1975).
46. N. Neuss, M. Gorman, N. J. Cone, and L. L. Huckstep, *Tetrahedron Lett.* p. 783 (1968).
47. D. J. Abraham and N. R. Farnsworth, *J. Pharm. Sci.* **58**, 694 (1969).
48. W. E. Jones and G. J. Cullivan, U.S. Patent 3,887,565 *Chem. Abstr.* **83**, 97687d (1975).
49. W. F. Brill, *J. Am. Chem. Soc.* **85**, 141 (1963).
50. W. F. Brill and N. Indicator, *J. Org. Chem.* **29**, 710 (1964).
51. N. Indicator and W. F. Brill, *J. Org. Chem.* **30**, 2074 (1965).
52. M. N. Sheng and J. G. Zajacek, *Adv. Chem. Ser.* **76**, 418 (1968).

Mechanisms of *cis-trans* Isomerization of Unsaturated Fatty Acids

Walter G. Niehaus, Jr.

INTRODUCTION

Although *trans*-isomers of unsaturated fatty acids are rather uncommon components of natural lipids, small amounts are found in animal tissues and milk [1,2], and large quantities of hydroxy and epoxy fatty acids with *trans* double bonds are found in a number of seed oils [3]. Deodorization [4] and partial hydrogenation [5] of vegetable oils result in a large content (20–50%) of *trans*-unsaturated fatty acids in commercial margarines and shortenings. The *trans*-octadecenoate isomers, which predominate, are principally the 10- and 11-isomers, with $10 > 11 > 9 > 8 > 12 > 13$ [5].

The biological origins of *trans*-unsaturated fatty acids are somewhat obscure, but in general they may be considered to arise from chemical or enzymatic isomerization of *cis* double bonds. Vaccenic acid (*trans*-11-octadecenoic acid) in beef fat and milk presumably arises from partial hydrogenation of linoleic and linolenic acids carried out by rumen micro-organisms [6]. The small content of elaidic acid (*trans*-9-octadecenoic acid) may arise by isomerization of oleic acid catalyzed by dietary mercaptans [7], which may account for the higher *trans*-fatty acid content of butter produced during the summer months [8]. Enzymatic action of soil microbes [9] may also contribute to the *trans*-fatty acid content of animal tissues. Although little is known about the biosynthesis of seed oils containing *trans*-unsaturated fatty acid derivatives, it seems safe to assume that all are derived via enzymatic modifications from typical *cis*-polyunsaturated fatty acids such as linoleic and α-linolenic acids. A number of plant seeds contain the enzymes lipoxidase and lipoperoxidase, whose sequential action on linoleic acid produces 13-hydroxy-10-oxo-*trans*-11-octadecenoic acid and 9-hydroxy-12-oxo-*trans*-10-octadecenoic acid [10].

METHODS OF DETECTING *trans* DOUBLE BONDS

The paucity of documented examples of naturally occurring fatty acids with isolated *trans* double bonds may be due in large part to the difficulty of identification of these acids. Until recently, *trans* unsaturation in fatty acids and glycerides was quantitated by infrared spectroscopy, based on the CH out-of-plane deformation observed at 10.35 μm [11]. This method has survived, partly because of the difficulty of separating *cis*- and *trans*-isomers by gas chromatography. Separation on open tubular columns was achieved in the early 1960's and has since been refined and extended [12–14]. Ackman and Hooper [15] have recently reported the separation of the methyl esters of *cis*- and *trans*-isomers of mono-, di-, and triethylenic fatty acids on open tubular columns coated with butanediol succinate, Silar-5CP, or Apiezon L. The open tubular or capillary column methodology is not routinely used by the average lipid chemist, however. A major advance has recently been achieved by Ottenstein, Bartley, and Supina [16], who have obtained complete separation of the *cis*- and *trans*-isomers of methyl octadec-9-enoate on packed columns of 15% SP 2340 or OV 275 on Chromosorb P AW DMCS. In my laboratory, this separation has been extended to the methyl esters of the *cis*- and *trans*-isomers of 9-tetradecenoic, 9-hexadecenoic, 6-octadecenoic, 11-octadecenoic, 13-docosenoic, and 15-tetracosenoic acids and to the mixture of *cis,cis*-, *cis,trans*- (*trans,cis*-), and *trans,trans*-9,12-octadecadienoic acids (see below).

Geometric isomers of esters of unsaturated fatty acids may also be separated analytically or preparatively by chromatography on silicic acid impregnated with silver nitrate, using either column [17] or thin-layer [18] methodology. Derivatives of monounsaturated fatty acids may also be prepared, which allows determination of both the position and configuration of the double bond. First, *cis*- or *trans*-unsaturated fatty acids are oxidized to the corresponding *erythro*- or *threo*-dihydroxy fatty acids using permanganate or osmic acid. The isopropylidene [19] or butyl boronate [20] derivatives of the dihydroxy fatty acid methyl esters are then analyzed by gas chromatography–mass spectrometry.

MECHANISMS OF *cis–trans* ISOMERIZATION

The isomerization of a double bond from the *cis* to the *trans* geometric configuration may occur with or without concomitant positional isomerization. In general, the nonenzymatic isomerization of double bonds of fatty acids, whether proceeding via an ionic or free-radical mechanism, occurs with retention of double-bond position and does not involve incorporation of hydrogen from the medium. A number of enzymes have been described which catalyze *cis–trans* isomerization of unsaturated fatty acids or fatty acyl thioesters. All these enzymatic isomerization reactions involve incorporation of hydrogen from the medium and concomitant migration of the double bond, usually to yield a product containing a conjugated *trans* double bond.

GEOMETRIC ISOMERIZATION WITHOUT BOND MIGRATION

Commonly used procedures for isomerization of unsaturated fatty acids from the *cis* to the *trans* configuration employ selenium or sulfur at high temperatures [21] or oxides of nitrogen at lower temperatures [22]. In general, isomerization proceeds with retention of double-bond position, but the selenium-catalyzed process when applied to polyunsaturated fatty acids may lead to side products and to double-bond migration [23]. Thus, for a number of years nitrous acid was the catalyst of choice for laboratory-scale isomerizations [22]. More recently, a variety of thiols have been employed as isomerization catalysts. Kircher [24] first demonstrated the conversion of methyl oleate to methyl elaidate by various thiols. Isomerization rate was shown to be dependent on thiol structure and concentration, the solvent employed, temperature, light intensity, and oxygen level. Thermodynamic equilibrium ($\sim 80\%$ *trans*) was attained and maintained starting with either the *cis*- or *trans*-isomer, indicating a reversible process. The thiol-catalyzed isomerization was independent of olefin concentration and was demonstrated with methyl oleate, methyl linoleate, methyl linolenate, and glyceryl trioleate (olive oil). Addition of thiol to the double bond was a competing reaction in the presence of oxygen but was suppressed in a nitrogen atmosphere. The mechanism of isomerization is proposed to involve reversible attack by a thiyl radical at an unsaturated carbon atom, rotation about the single bond so generated, followed by dissociation of the thiyl radical to generate primarily the more stable *trans*-isomer. Such a reaction mechanism was first proposed by Sivertz [25] and elaborated by Walling [26] in studies of the isomerization of butenes.

The thiyl radical catalysis of *cis–trans* isomerization of lipids was further studied by Gunstone and Ismail [27] and by Sgoutas and Kummerow [28]. The latter workers studied the *cis–trans* isomerization of monoenoic and dienoic fatty acid methyl esters catalyzed by thiyl or phosphinyl radicals that were generated by radical initiators in nonpolar organic solvents. An equilibrium mixture of *cis*- and *trans*-isomers was produced within 5 hr, with no detectable addition products or positional isomers. With longer reaction times and higher catalyst concentrations, overall yields of unsaturated fatty esters were lower and the formation of by-products was noted.

Although the procedures employed by Kircher [24] and by Sgoutas and Kummerow [28] involve essentially the same mechanism of isomerization, reversible radical addition to the double bond, they employ quite different reaction conditions. Kircher employed bifunctional thiols, e.g., 2-mercapto-propionic acid, glycol dimercaptoacetate, thiomalic acid, and 1,4-butane-dithiol, in a polar solvent such as ethanol and depended on sunlight to generate free radicals. Sgoutas and Kummerow used monofunctional thiols, thiophenol and dodecanethiol, or diphenylphosphine in a nonpolar solvent such as benzene, cyclohexane, or carbon tetrachloride and generated free radicals with azobis(isobutylnitrile). These methods are primarily of interest as means of preparing *trans*-unsaturated fatty acids and are probably unrelated to biological *cis–trans* isomerization, which is unlikely to proceed via a free-radical mechanism. Superoxide radical, frequently observed in biological systems, does not catalyze *cis–trans* isomerization of monoenoic fatty acids [29].

Thiol-catalyzed isomerization of double bonds of unsaturated fatty acids also occurs via an ionic mechanism. During a study in my laboratory of enzymatic transformations of oleic acid, Dr. C. E. Mortimer observed that thioglycol (2-mercaptoethanol) in neutral aqueous solution catalyzes non-enzymatic *cis–trans* isomerization. This process has been studied in some detail under a variety of incubation conditions with various thiol catalysts and fatty acid substrates [7]. Isomerization proceeds to equilibrium within 1–2 hr in neutral aqueous solution at moderate temperature and at relatively low thiol concentration (5–10 mEq/liter). Hydrogen from the medium is not incorporated into the product, and no double-bond migration occurs. The proposed mechanism of isomerization involves the formation of a mixed micelle of unsaturated fatty acid and thiol followed by nucleophilic attack by the thiol at the double bond, probably via a three-center mechanism. The resulting resonance-stabilized intermediate has significantly less double-bond character and is free to rotate about the former double bond. Subsequent elimination of thiol produces primarily the thermodynamically favored *trans*-isomer (Scheme 1). The general mechanism of isomerization is similar to that previously proposed by Sivertz [25], Walling [26], and Kircher [24], differing in that an ionic rather than a free-radical process is responsible for the reaction.

Evidence for this proposed mechanism is taken from my previous work [7] and current research and is summarized below.

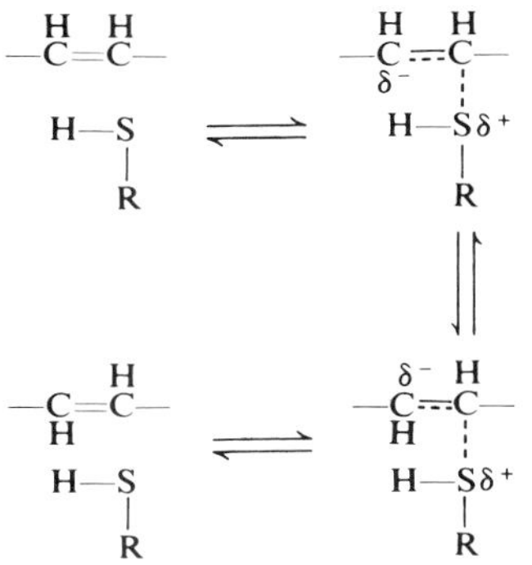

$$
\begin{array}{ccc}
\underset{\overset{\displaystyle |}{\text{H}}}{\overset{\text{H}}{\text{C}}}=\underset{\overset{\displaystyle |}{\text{H}}}{\overset{\text{H}}{\text{C}}} & & \overset{\text{H}}{\underset{\delta^-}{\text{C}}}=\cdots=\underset{}{\overset{\text{H}}{\text{C}}} \\
\end{array}
$$

Scheme 1

An ionic rather than radical mechanism is proposed since no radical-generating species is included in the reaction mixture, the rate of isomerization is independent of the presence of light or oxygen, and no free-radical signal was observed by electron paramagnetic resonance analysis of reaction mixtures [7]. Isomerization is proposed to occur in a mixed micelle composed of unsaturated fatty acid and thiol, in which the effective thiol concentration is much higher than that found in the surrounding medium. The ability of a particular thiol to effect the isomerization of oleic acid is related to its amphipathic character and hence its tendency to form mixed micelles with the strongly amphipathic oleic acid. Cysteine and glutathione, which are essentially insoluble in organic solvents, are ineffectual as catalysts, as are thiophenol and dodecanethiol, which are essentially insoluble in water. Dithiothreitol and thioglycol, with appreciable solubility in both aqueous and organic media, would be expected to partition into oleic acid micelles and are effective promoters of *cis–trans* isomerization. The effectiveness of a thiol catalyst is also related to its ionic nature and hence to its ability to interact electrostatically with the negatively charged oleic acid micelle. Thus, thioglycolic acid is totally ineffective due to electrostatic repulsion, whereas 2-mercaptoethylamine is by far the most efficient catalyst yet tested due to electrostatic attraction to the negatively charged oleic acid micelle [7]. The rate of isomerization of oleic acid is dependent on an intact micellar system. Disruption of the micelles by such agents as methanol, nonionic detergents, or serum albumin causes a parallel decrease in the turbidity of the system and the rate of isomerization [7]. High concentrations of 2-mercaptoethylamine (> 50 mEq/liter) also cause alteration of the micellar system, as evidenced by increased turbidity, with a concomitant decrease in isomerization rate.

Isomerization is expected to be related to the structural relationship between the thiol catalyst and the unsaturated fatty acid. The short aliphatic chain of 2-mercaptoethylamine would be expected to allow only minimal penetration of the SH group into the interior of the fatty acid micelle, with the NH_3^+ group remaining at the lipid–water interface, electrostatically bonded to a fatty acid carboxylate group. One would therefore expect greater ease of isomerization of double bonds in close proximity to the carboxyl group, in contrast to double bonds nearer to the hydrocarbon terminus of the fatty acid. This proposal was tested with the isomeric fatty acids cis-6-octadecenoic acid, cis-9-octadecenoic acid, and cis-11-octadecenoic acid. Fatty acid (0.5 mEq/liter) was incubated with 2-mercaptoethylamine (6 mEq/liter) in potassium phosphate buffer (0.1 M, pH 6.8) at 60°C for 30 min. The percentages of trans-isomer formed, assayed by gas chromatography [16], were 6-octadecenoic acid, 72%; 9-octadecenoic acid, 60%; 11-octadecenoic acid, 15%.

The overall chain length of the fatty acid also affects the rate of isomerization, presumably due to variations in micellar structure. Under these same incubation conditions, the extent of trans-isomer formation from cis-9-tetradecenoic acid, cis-9-hexadecenoic acid, and cis-9-octadecenoic acid was 2, 10, and 60%, respectively. Longer-chain fatty acids were isomerized to an extent intermediate between the 16- and 18-carbon acids: 13-docosenoic acid, 29%; and 15-tetracosenoic acid, 23%.

Alteration of the incubation conditions also affects the relative isomerization rates, presumably via changes in micellar structure. At pH 7.5 and 40 mEq/liter 2-mercaptoethylamine, one sees an enhancement of the rate of isomerization of tetradecenoic and hexadecenoic acids, and a reduced rate of isomerization of octadecenoic, docosenoic, and tetracosenoic acids.

The presence of an additional double bond in the fatty acid also affects the rate of isomerization. Linoleic acid (cis-9,cis-12-octadecadienoic acid) is isomerized much more slowly than is oleic acid (cis-9-octadecenoic acid), even though the overall carbon chain length and proximity of double bond to carboxyl group are identical. Incubation of linoleic acid with 2-mercaptoethylamine under the original incubation conditions results, after 60 min, in a mixture of 69% cis,cis-octadecadienoic acid, 24% cis,trans-octadecadienoic acid, and 6% trans,trans-octadecadienoic acid. Comparisons of these various isomerization rates emphasize the importance of micellar structure to isomerization rate. Experiments are currently underway to clarify these relationships between fatty acid structure, thiol structure, micellar structure, and isomerization rate.

No enzymes are currently known which catalyze cis–trans isomerization of fatty acids by a process analogous to this bioorganic system. The enzymatic isomerization of substrates containing a double bond conjugated to a dicarbonyl system does, however, involve such a mechanism. Three enzymes have been described which catalyze *cis–trans* isomerization without double-bond migration. Maleate isomerase from *Pseudomonas fluorescens* [30] has an absolute requirement for exogenous thiol, as does maleylacetone isomerase from *Vibrio 01* [31]. Maleate isomerase from *Alcaligenes faecalis* [32] does not require a thiol cofactor, but is inhibited by thiol-directed reagents, suggesting a catalytically essential cysteine residue at the active center. These reactions are all proposed to involve nucleophilic addition of thiol to a carbon–carbon double bond and enolization of the dicarbonyl system, with subsequent rotation and elimination of thiol.

GEOMETRIC ISOMERIZATION WITH CONCOMITANT BOND MIGRATION

Virtually all chemical procedures that effect *cis–trans* isomerization of unsaturated fatty acids proceed with retention of double-bond position. One notable exception is alkaline isomerization of polyunsaturated fatty acids to form *cis–trans* conjugated products [33]. Of the several enzymatic systems that are known to catalyze isomerization of unsaturated fatty acids or fatty acyl thioesters, all involve double-bond migration and all but one produce a product in which the *trans* double bond is conjugated to a carbonyl or aliphatic double bond.

The biosynthesis of unsaturated fatty acids by a number of bacteria, notably *Escherichia coli*, involves the enzyme β-hydroxydecanoylthioester dehydrase, which catalyzes the reversible dehydration of D(−)-β-hydroxy-decanoyl thioesters to a mixture of *trans*-2-decenoyl and *cis*-3-decenoyl thioesters and also effects a rapid interconversion of the two enoates. Studies of the mechanism of action of this enzyme have been reviewed by Bloch [34]. Two major questions posed by Bloch and his co-workers were: (a) What is the sequence of events leading from hydroxy acid to the two unsaturated products, the choices being formation of the *cis*- and *trans*-decenoates either by independent dehydration steps or by dehydration to one decenoate followed by isomerization to the other? (b) Does β-hydroxydecanoylthioester dehydrase have true isomerase activity and is this activity obligatory for dehydrase action? A combination of kinetic studies and incubations with deuterium- and tritium-labeled substrates led to the proposed mechanism (Fig. 1), in which enzyme-bound *trans*-2-decenoyl thioester is an obligatory intermediate linking all the substrates (products) of the reactions. The *cis*- and *trans*-decenoyl thioesters are interconverted by a direct isomerization of the double bond, presumably involving a carbanion intermediate. The proton transfers involved in the geometric and positional isomerization of the double bond are

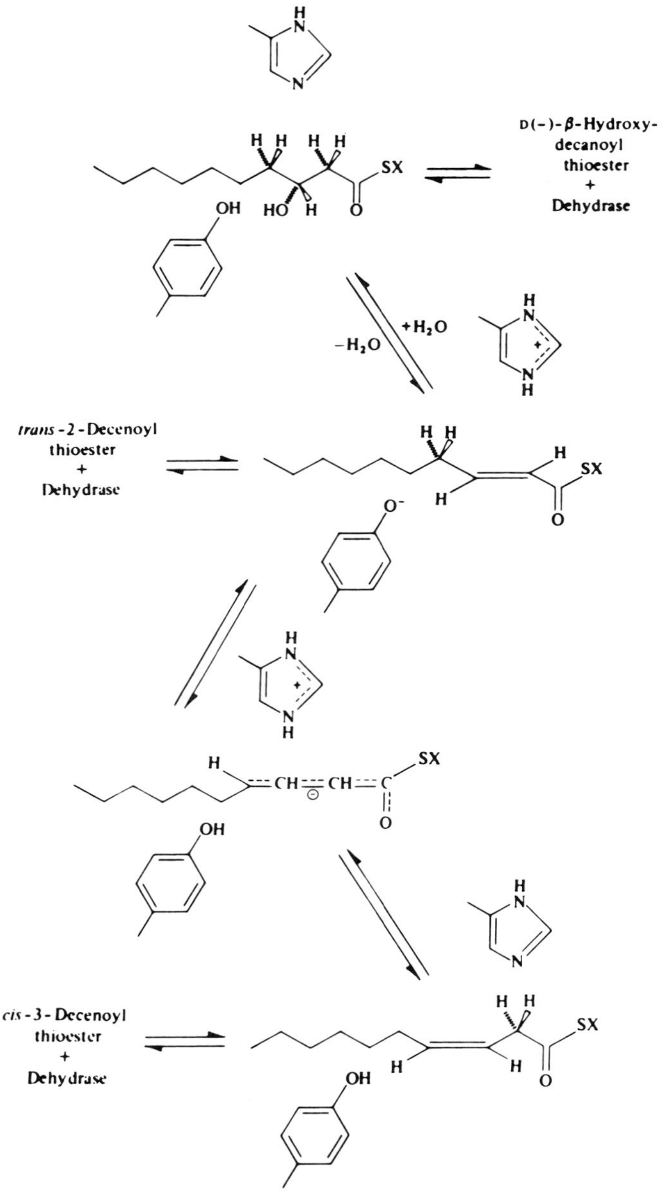

Fig. 1 Proposed mechanism of action of β-hydroxydecanoylthioster dehydrase.

mediated by a tyrosyl and a histidyl residue of the enzyme, acting as general acid–base catalysts. The enzyme also catalyzes the isomerization of the acetylenic substrate analog 3-decynoyl-*N*-acetylcysteamine to the allenic product (+)-2,3-decadienoyl-*N*-acetylcysteamine, which immediately inactivates the enzyme by covalent reaction with the catalytically essential histidine residue. The levorotatory isomer does not inhibit the enzyme [35].

Miesowicz and Bloch [36] have recently purified an enzyme from porcine liver which catalyzes the conversion of 3-acetylenic fatty acyl thioesters to (+)-2,3-dienoyl fatty acyl thioesters, which accumulate in the medium and do not inhibit this enzyme. The homogeneous acetylene–allene acyl thioester isomerase also converts *cis*-3- and *trans*-3-acyl coenzyme A thioesters to their *trans*-2-acyl coenzyme A isomers, the same reactions that are catalyzed by an enzyme described by Stoffel [37]. This enzyme is believed to be essential for the β-oxidation of oleic acid by liver mitochondria [38]. If these two enzymes should prove to be identical, it would be remarkable that the enzyme prefers an unnatural acetylenic substrate over a normal olefinic metabolite. Davidoff and Korn [39] have described a similar enzymatic activity from mitochondria which catalyzes the reversible isomerization of *trans*-2-hexadecenoyl coenzyme A to *cis*- and *trans*-3-hexadecenoyl coenzyme A. Their mitochondrial preparation also catalyzed the hydration of *trans*-2-hexadecenoyl coenzyme A, presumably by an enzyme distinct from the isomerase.

Tove and his associates have extensively characterized the enzymatic systems involved in the biohydrogenation of linoleic acid by rumen microorganisms [6,40–43]. The first step in the reaction sequence, conversion of linoleic acid to a *cis–trans* conjugated diene, is catalyzed by linoleic acid 12-*cis*,11-*trans*-isomerase, which has been partially purified from *Butyrivibrio fibrisolvens* [40]. The particulate enzyme has a strict requirement for a substrate with a free carboxyl group and a *cis*-9,*cis*-12-diene system and is highly specific for an ω chain length of six carbon atoms. Binding of substrate by enzyme thus involves interaction with the π system of the double bonds, hydrogen bonding with the substrate carboxyl group, and hydrophobic interaction with the hydrocarbon tail of the fatty acid. A direct isomerization rather than hydration–dehydration is seen to be responsible for the geometric and positional shift of the double bond, since neither (*R*)- nor (*S*)-12-hydroxyoctadec-9-enoic acid is converted to the conjugated acid, and no labeled hydroxyoctadecanoic acid product is detected when linoleic acid of high specific radioactivity is isomerized [42]. Linoleic acid isomerase does not catalyze the formation of an allene from *cis*-9-octadec-12-ynoic acid and is only slightly inhibited by the actylenic acid. A mechanism has been proposed [42] for the direct isomerization of linoleic acid to *cis*-9,*trans*-11-octadecadienoic acid, involving either a concerted protonation–deprotonation or an enzyme-bound carbanion intermediate that is protonated prior to dissociation from the enzyme surface (Fig. 2). Linoleic acid is proposed to bind to the enzyme in a loop conformation, stabilized by hydrophobic interactions between the enzyme and the fatty acid tail by π interactions between the cis-9,12-diene system and an electrophilic group (E) of the enzyme and by hydrogen bonding of the substrate carboxyl group with an electronegative group (B) of the enzyme. In a concerted fashion, a proton is transferred from carbon 11 of the substrate to the substrate carboxyl group, and another proton is transferred from the conjugate acid group BH of the enzyme to carbon 13 of the substrate. The product cis-9,trans-11-octadecadienoic acid then dissociates from the enzyme. Details of this mechanism are based on kinetic data obtained with various substrates and inhibitors. Binding of substrate in a loop conformation is suggested by the very similar Michaelis constants for binding of two substrates, linoleic acid (cis-9,cis-12-octadecadienoic acid) and γ-linolenic acid (cis-6,cis-9,cis-12-octadecatrienoic acid) and the inhibitor constants for cis-6-octadecenoic acid and cis-9-octadecenoic acid. In the extended conformation these four fatty acids have considerably different shapes, but their loop conformations are very similar. The substrate analogs, linoleyl alcohol, amine, and amide, also have inhibition constants that are essentially identical to the Michaelis constant for linoleic acid, indicating that the alignment of these substrates at the catalytic site of the enzyme is similar to, if not identical with, that of substrate. These substrate analogs, like linoleic acid, are also able to form a hydrogen bond with the enzyme. Thus, the carboxyl group of the substrate appears to be directly involved in the isomerization reaction and is proposed to function as a general acid–base catalyst for proton abstraction from carbon 11.

Garcia et al. [44] have recently reported that linoleic acid isomerase, prepared as described by Kepler and Tove [40], also catalyzes the conversion of cis-2,cis-5-octadecadienoic acid to trans-3,cis-5-octadecadienoic acid. No other cis,cis-methylene-interrupted isomers of octadecadienoic acid were isomerized by the enzyme. Nonenzymatic isomerization of cis-2,cis-5-octadecadienoic acid also occurred below pH 6. This additional reaction catalyzed by linoleic acid isomerase does not appear to be inconsistent with the proposed mechanism of action of the enzyme [45].

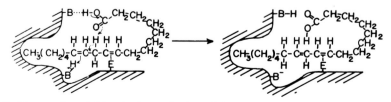

Fig. 2 Proposed model for the isomerization of linoleic acid by linoleic acid isomerase.

The enzymatic isomerization of long-chain monoenoic fatty acids is a major research interest of this author. A soluble enzyme from a pseudomonad was initially shown by Niehaus and Schroepfer [46,47] to catalyze the hydration and isomerization of oleic acid. The isomerization reaction was subsequently studied in greater detail by Mortimer and Niehaus [9]. This enzyme, oleate hydratase (isomerase), has several features in common with both β-hydroxydecanoylthioester dehydrase and linoleic acid isomerase. Like β-hydroxydecanoylthioester dehydrase it catalyzes both a reversible hydration and a direct isomerization of the unsaturated substrate. Like linoleic acid isomerase it requires a long-chain fatty acid with a free carboxyl group as substrate. The major distinction of oleate hydratase (isomerase) is that the reactive center of the substrate is an isolated double bond and that none of the substrates or products contain double bonds conjugated to carbonyl or olefinic double bonds.

Oleate hydratase (isomerase) is a multifunctional enzyme catalyzing the following reaction types:

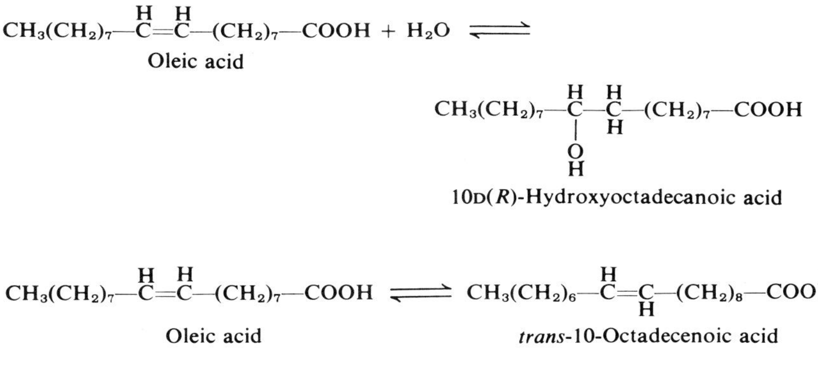

$$CH_3(CH_2)_7 \overset{H}{\underset{}{C}} = \overset{H}{\underset{}{C}} (CH_2)_7 - COOH + H_2O \rightleftharpoons$$
Oleic acid

$$CH_3(CH_2)_7 \overset{H}{\underset{|}{C}} - \overset{H}{\underset{H}{C}} (CH_2)_7 - COOH \qquad (1)$$
$$\overset{|}{\underset{}{O}} \overset{}{\underset{}{H}}$$

10D(R)-Hydroxyoctadecanoic acid

$$CH_3(CH_2)_7 \overset{H}{\underset{}{C}} = \overset{H}{\underset{}{C}} (CH_2)_7 - COOH \rightleftharpoons CH_3(CH_2)_6 \overset{H}{\underset{}{C}} = \overset{}{\underset{H}{C}} (CH_2)_8 - COOH$$
Oleic acid   *trans*-10-Octadecenoic acid

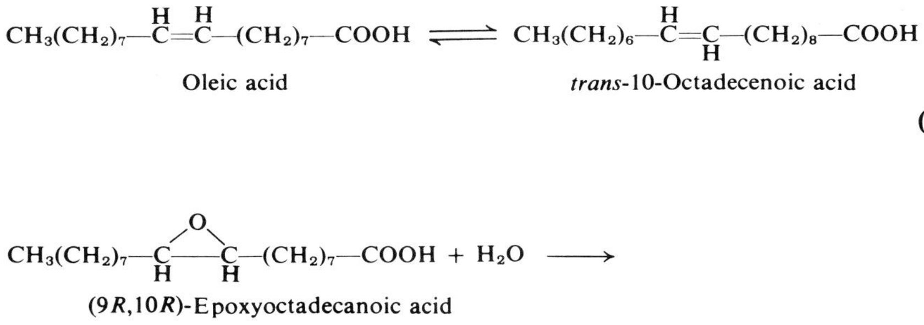

$$(2)$$

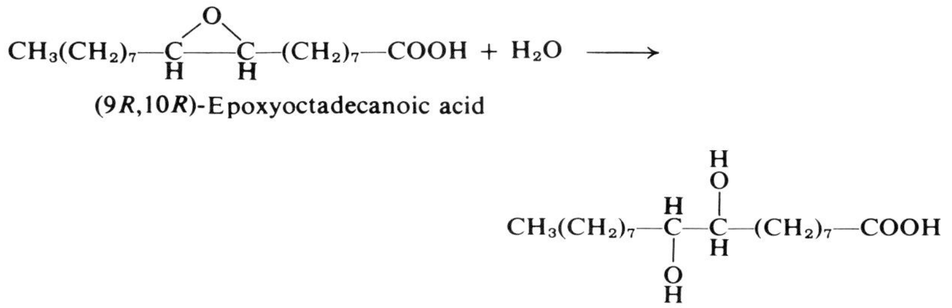

$$CH_3(CH_2)_7 \overset{O}{\underset{\overset{H}{C}----\overset{H}{C}}{}} (CH_2)_7 - COOH + H_2O \longrightarrow$$
(9R,10R)-Epoxyoctadecanoic acid

$$CH_3(CH_2)_7 \overset{H}{\underset{|}{\overset{O}{\underset{}{C}}}} - \overset{\overset{H}{O}}{\underset{H}{C}} (CH_2)_7 - COOH \qquad (3)$$
$$\overset{|}{\underset{}{O}} \overset{}{\underset{}{H}}$$

(9R,10R)-Dihydroxyoctadecanoic acid

$$CH_3(CH_2)_7-\overset{\overset{\displaystyle H}{|}}{\underset{\overset{\displaystyle |}{O}}{C}}\diagdown\hspace{-0.3em}\underset{O}{\diagup}\hspace{-0.3em}\overset{\displaystyle |}{\underset{\overset{\displaystyle |}{H}}{C}}-(CH_2)_7-COOH + H_2O \longrightarrow$$

(9R,10R)-Epoxyoctadecanoic acid

$$CH_3(CH_2)_7-\underset{\overset{\displaystyle |}{H}}{\overset{\overset{\displaystyle H}{\overset{\displaystyle |}{O}}}{C}}-\underset{\overset{\displaystyle |}{H}}{\overset{\overset{\displaystyle H}{\overset{\displaystyle |}{O}}}{C}}-(CH_2)_7-COOH \qquad (4)$$

(9R,10S)-Dihydroxyoctadecanoic acid

The enzyme has stringent specificity with regard to geometric and positional isomers of substrate. Thus, hydration occurs with cis-9-unsaturated fatty acids of chain length 15–20 [9], with linoleic acid [48], and in the intact organism with linolenic acid and with ricinoleic acid [49]. Fatty acids with trans-9 unsaturation, with cis-10 or cis-11 unsaturation, and cis-9-unsaturated fatty acids shorter than 14 carbons are not substrates for hydration. The only exception to this specificity found to date is the conversion of cis-8-heptadecenoic acid to 9-hydroxyheptadecanoic acid [9]. Other cis-8-unsaturated fatty acids have not yet been studied.

The cis–trans isomerization has more stringent substrate specificity and occurs only with cis-9-fatty acids containing 16, 17, 18, or 19 carbon atoms [9]. The isomerization occurs with stereospecific incorporation of hydrogen from the medium, involving the 9-pro-R hydrogen of trans-10-octadecenoic acid and the 11-pro-S hydrogen of cis-9-octadecenoic acid [9]. Two lines of evidence have been employed to demonstrate that a true isomerization is involved rather than an asymmetric hydration–dehydration reaction sequence. Upon incubation of enzyme with a mixture of [14C]oleic acid and [3H]-hydroxyoctadecanoic acid, the 3H/14C ratio of the product trans-10-octadecenoic acid parallels that of the oleic acid, not of the hydroxyoctadecanoic acid. Also, the conversion of trans-10-octadecenoic acid to 10-hydroxyoctadecanoic acid in a medium enriched in 2H2O involves the incorporation of two atoms of deuterium, one at carbon 9 and one at carbon 11, thus implicating oleic acid as an obligatory intermediate [9].

Although the enzyme has not yet been purified to homogeneity, all observations made thus far strongly support the proposal that hydration and isomerization are catalyzed by a single enzyme at a unique catalytic site.

1. All activities are induced simultaneously when oleic acid is added to the organism's growth medium.

2. Oleate hydratase and epoxyoctadecanoate hydratase activities [reactions (1) and (3)] undergo marked thermal inactivation at identical rates [47].

3. Oleate hydratase and oleate isomerase activities [reactions (1) and (2)]

are also heat inactivated at the same rate and, in addition, are comparably inhibited by low concentrations of several detergents [50].

4. No separation of activities has been achieved during analytical-scale salt fractionation, ion-exchange chromatography, density gradient centrifugation, polyacrylamide gel electrophoresis, or isoelectric focusing [50].

5. Recently, oleate hydratase has been extensively purified using adsorption onto hydroxylapatite, ammonium sulfate fractionation, DEAE-cellulose chromatography, and gel filtration on Sephadex G-200. The final preparation retained activities (1)–(4) in essentially the same proportions that were present in the initial crude extract [50].

6. The substrate specificities and stereospecificity of the various reactions are also consistent with a single binding site for the various substrates and products of reactions (1)–(4). Hydration of oleic acid and of epoxyoctadecanoic acid involves addition of OH to carbon 10 from the same side of the fatty acid, and hydration and isomerization both involve addition and abstraction of hydrogen at carbon 9 from the same side of the fatty acid.

Other data contributing to the development of this model include the pH profile for hydration (maximum pH 7) and isomerization (maximum pH 5) [9], the more restricted substrate specificity for isomerization with respect to

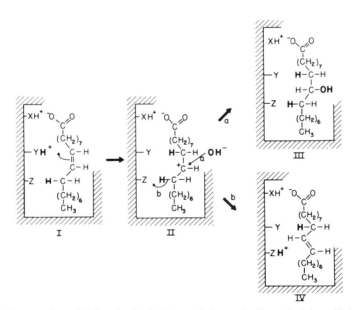

Fig. 3 Proposed model for the hydration and isomerization of oleic acid by oleate hydratase.

the hydrocarbon tail [9], the strong discrimination against tritium in the addition of the 11L(S) hydrogen but not in the addition of the 9L(R) hydrogen [reaction (2)], the primary kinetic isotope effect for the removal of the 9L(R) and 11L(R) hydrogens [reaction (2)] [9], and the strong depression in ^2H_2O of the rate of isomerization of *trans*-10-octadecenoic acid to oleic acid with no depression of the rate of the reverse reaction [9].

These data provide the basis for a working model in which reactions (1)–(4) occur at a single catalytic site according to the mechanism [51] shown in Fig. 3. The binding site on the enzyme contains a positively charged group, designated XH^+, which forms an electrostatic interaction with the carboxylate group of the substrate, and a hydrophobic pocket in which the hydrocarbon tail of the fatty acid is bound by nonpolar interaction. The specificity of binding is such that a *cis*-9 double bond is in appropriate juxtaposition to a proton-donating group of the enzyme, designated in Fig. 3 as YH^+; proton transfer to the L(S) side of carbon 9 results in a carbonium ion intermediate (II) bound to the enzyme so as to prevent rotation about the C-9—C-10 bond. This carbonium ion may be neutralized in either of two ways: (a) nucleophilic attack at the D(R) side of carbon 10 by a hydroxyl group, which may be free or protein bound, resulting in the formation of (10R)-hydroxyoctadecanoic acid (III), or (b) transfer of the 11-pro-S hydrogen to an acceptor group of the enzyme, Z, resulting in the formation of *trans*-10-octadecenoic acid (IV). The proper juxtaposition of the 11-pro-S hydrogen and the hydrogen acceptor Z is proposed to be extremely sensitive to the fit of the hydrocarbon tail of the substrate in the hydrophobic pocket, thus explaining the marked decrease in the rate of the isomerization reaction when the number of methylene groups in the tail is changed. A conformational change in the enzyme could occur in ^2H_2O affecting the binding of the hydrocarbon tail in this pocket, resulting in the lower rate of isomerization of *trans*-10-octadecenoic acid to *cis*-9-octadecenoic acid. The large isotope discrimination against tritium in the addition of the 11L hydrogen [10] is consistent with the proton-donating group Z being in an extremely hydrophobic environment, as postulated in the model, since in this environment it would exchange its proton with water very slowly.

This model is also consistent with the pH optima for hydration and isomerization. As the pH of the medium is lowered from 7 to 5, the rate of hydration (reaction a) decreases in direct response to the decrease in hydroxyl ion concentration. Thus, at pH 5, the rate of reaction b, the abstraction of a proton from carbon 11, increases due to the decreased rate of the competing nucleophilic attack by hydroxyl ion. The decrease in the pH of the medium need not decrease the effectiveness of group Z in proton abstraction if, as proposed above, it exists in a very hydrophobic environment. For example, a catalytically active lysyl residue in acetoacetate decarboxylase has an apparent pK_a of 5.9 [52]. This same mechanism can also be used to explain the hydration of *cis-* and *trans*-9,10-epoxyoctadecanoic acids. Morris [53] deduced the stereochemistry of these reactions from optical rotation data published by Niehaus and Schroepfer [54].

CONCLUSIONS

The fact that relatively few enzymes are currently known which catalyze *cis–trans* isomerization of unsaturated fatty acids may be due in large part to the difficulty in assaying this reaction. With the recent development of a routine gas chromatographic separation of *cis-* and *trans*-isomers of monoenoic and polyenoic fatty acids [16] it is highly likely that additional fatty-acid isomerases will be discovered. A number of enteric bacteria have been reported to convert oleic acid to hydroxyoctadecanoic acid [55,56]. Their enzymes, like the pseudomonad oleate hydratase [9], may also catalyze *cis–trans* isomerization of the double bond. Considering the ease with which a thiol nucleophile catalyzes isomerization of an isolated double bond, requiring only about 9 kcal/mole activation energy [7], it would be surprising if enzymes do not exist which catalyze the thiol-dependent isomerization of fatty acids. Finally, one should recognize the *cis–trans* isomerization of hydrophobic unsaturated biomolecules other than fatty acids. Thiols and dihydroflavins catalyze the nonenzymatic isomerization of retinal in neutral aqueous solution [57], and liver alcohol dehydrogenase catalyzes the *cis–trans* isomerization of farnesol [58].

ACKNOWLEDGMENTS

Original investigations in the author's laboratory, cited in this chapter, were supported in part by USPHS Grants MH 15788 and HD 04505.

REFERENCES

1. D. G. Cornwell, R. Backderf, C. L. Wilson, and J. B. Brown, *Arch. Biochem. Biophys.* **46**, 364–375 (1953).
2. D. Swern, H. B. Knight, and C. R. Eddy, *J. Am. Oil Chem. Soc.* **31**, 44–46 (1952).
3. C. R. Smith, *Prog. Chem. Fats Other Lipids* **11**, Part 1, 137–150 (1974).
4. R. G. Ackman, S. N. Hooper, and D. L. Hooper, *J. Am. Oil Chem. Soc.* **51**, 42–49 (1974).
5. C. R. Scholfield, V. L. Davison, and H. J. Dutton, *J. Am. Oil Chem. Soc.* **44**, 649–651 (1967).
6. C. E. Polan, J. J. McNeill, and S. B. Tove, *J. Bacteriol.* **88**, 1056–1064 (1964).
7. W. G. Niehaus, *Biorg. Chem.* **3**, 302–310 (1974).

8. J. Boer, B. C. P. Jansen, and A. Kentie, *J. Nutr.* **33**, 339–358 (1947).
9. C. E. Mortimer and W. G. Niehaus, *J. Biol. Chem.* **249**, 2833–2842 (1974).
10. W. J. Esselman and C. O. Clagett, *J. Lipid Res.* **15**, 173–178 (1974).
11. O. D. Sreve, M. R. Heeter, H. B. Knight, and D. Swern, *Anal. Chem.* **22**, 1261–1270 (1950).
12. C. Litchfield, A. F. Isbell, and R. Reiser, *J. Am. Oil Chem. Soc.* **39**, 330–346 (1962).
13. C. R. Scholfield, *J. Am. Oil Chem. Soc.* **49**, 583–590 (1972).
14. J. Hrivnak, L. Sojak, J. Krupcik, and Y. P. Duchesne, *J. Am. Oil Chem. Soc.* **50**, 68–71 (1973).
15. R. G. Ackman and S. N. Hooper, *J. Chromatogr. Sci.* **12**, 131–138 (1974).
16. D. M. Ottenstein, D. A. Bartley, and W. R. Supina, *J. Chromatogr.* **119**, 401–407 (1976).
17. R. A. Stein and V. Slawson, *Anal. Chem.* **40**, 2017–2020 (1968).
18. L. Morris, *Chem. Ind. (London)* pp. 1238–1240 (1962).
19. J. A. McCloskey and M. J. McClelland, *J. Am. Chem. Soc.* **87**, 5090–5093 (1965).
20. C. J. W. Brooks and I. MacLean, *J. Chromatogr. Sci.* **9**, 18–23 (1971).
21. J. D. Fitzpatrick and M. Orchin, *J. Am. Chem. Soc.* **79**, 4765–4771 (1957).
22. C. Litchfield, R. Harlow, A. Isbell, and R. Reiser, *J. Am. Oil Chem. Soc.* **42**, 73–76 (1965).
23. V. V. R. Subrahmanyam and F. W. Quackenbush, *J. Am. Oil Chem. Soc.* **41**, 275–282 (1964).
24. H. W. Kircher, *J. Am. Oil Chem. Soc.* **41**, 351–354 (1964).
25. C. Sivertz, *J. Phys. Chem.* **63**, 34–38 (1959).
26. C. Walling and W. Helmreich, *J. Am. Chem. Soc.* **81**, 1144–1148 (1959).
27. F. D. Gunstone and I. A. Ismail, *Chem. Phys. Lipids* **1**, 264–270 (1967).
28. D. S. Sgoutas and F. A. Kummerow, *Lipids* **4**, 283–287 (1969).
29. W. G. Niehaus and E. M. Gregory, unpublished observations (1976).
30. Y. Takamara, T. Takamura, M. Soejima, and T. Vomura, *Agric. Biol. Chem.* **333**, 718–728 (1967).
31. S. Seltzer, *J. Biol. Chem.* **248**, 215–222 (1973).
32. W. Scher and W. B. Jakoby, *J. Biol. Chem.* **244**, 1878–1882 (1969).
33. B. A. Bruce, M. L. Sivain, S. F. Herb, P. L. Nichols, and R. W. Riemenschneider, *J. Am. Oil Chem. Soc.* **29**, 279–287 (1952).
34. K. Bloch, *in* "The Enzymes" (P. D. Boyer, ed.), 3rd ed., Vol. 5, pp. 441–464. Academic Press, New York, 1971.
35. M. Morisaki and K. Bloch, *Biorg. Chem.* **1**, 188–193 (1972).
36. F. M. Miesowicz and K. Bloch, *Biochem. Biophys. Res. Commun.* **65**, 331–335 (1975).
37. W. Stoffel, R. Ditzer, and H. Caesar, *Hoppe-Seyler's Z. Physiol. Chem.* **339**, 167–181 (1964).
38. W. Stoffel and H. Caesar, *Hoppe-Seyler's Z. Physiol. Chem.* **341**, 76–83 (1965).
39. F. Davidoff and E. D. Korn, *J. Biol. Chem.* **240**, 1549–1558 (1965).
40. C. R. Kepler and S. B. Tove, *J. Biol. Chem.* **242**, 5686–5692 (1967).
41. C. R. Kepler, W. P. Tucker, and S. B. Tove, *J. Biol. Chem.* **245**, 3612–3620 (1970).
42. C. R. Kepler, W. P. Tucker, and S. B. Tove, *J. Biol. Chem.* **246**, 2765–2771 (1971).
43. I. S. Rosenfeld and S. B. Tove, *J. Biol. Chem.* **246**, 5025–5030 (1971).
44. P. T. Garcia, W. W. Christie, H. M. Jenkin, L. Anderson, and R. T. Holman, *Biochim. Biophys. Acta* **424**, 296–302 (1976).
45. S. B. Tove, personal communication (1976).
46. W. G. Niehaus and G. J. Schroepfer, *Biochem. Biophys. Res. Commun.* **21**, 271–275 (1965).

47. W. G. Niehaus, A. Kisic, A. Torkelson, D. J. Bednarczyk, and G. J. Schroepfer, *J. Biol. Chem.* **245**, 3790–3797 (1970).
48. G. J. Schroepfer, W. G. Niehaus, and J. McCloskey, *J. Biol. Chem.* **245**, 3798–3801 (1970).
49. L. L. Wallen, E. N. Davis, Y. U. Wu, and W. K. Rohwedder, *Lipids* **6**, 745–750 (1971).
50. W. G. Niehaus, unpublished observations (1975).
51. W. G. Niehaus and C. E. Mortimer, *Abstr. Int. Congr. Biochem., 9th, 1973* p. 404 (1973).
52. D. E. Schmidt and F. H. Westheimer, *Biochemistry* **10**, 1249–1253 (1971).
53. L. J. Morris, *Biochem. J.* **118**, 681–693 (1970).
54. W. G. Niehaus, A. Kisic, A. Torkelson, D. J. Bednarczyk, and G. J. Schroepfer, *J. Biol. Chem.* **245**, 3802–3809 (1970).
55. P. J. Thomas, *Gastroenterology* **62**, 430–435 (1972).
56. J. R. Pearson, H. S. Wiggins, and B. S. Drasar, *J. Med. Microbiol.* **7**, 265–275 (1974).
57. S. Futterman and M. Rollins, *J. Biol. Chem.* **248**, 7773–7779 (1973).
58. C. Capellini, A. Corbella, P. Gariboldi, and G. Jommi, *Bioorg. Chem.* **5**, 129–136 (1976).

The Synthesis and Metabolism of Chirally Labeled α-Amino Acids

Ronald J. Parry

INTRODUCTION

Organic chemists have had a long-standing interest in the nature of biosynthetic processes. This interest began with general speculations regarding the types of chemical reactions that might be involved in the formation of those secondary metabolites called natural products and led to attempts to mimic biosynthetic processes *in vitro* [1]. With the advent of radioactive tracers, it became possible to put these speculations to the test by appropriate experiments with living systems. As the result of many such tracer experiments, a picture of natural-product biosynthesis has developed in which the outlines of the major biosynthetic pathways are clear. The emergence of this picture has coincided with an increased interest in the mechanistic details of individual biosynthetic reactions.

A chemist who chooses to investigate reaction mechanisms has a number of tools at his disposal. One of the most powerful of these tools is stereochemistry. A genuine understanding of the stereochemical subtleties of enzymatic reactions began in 1947 with Ogston's classic paper on the metabolism of citric acid [2]. This paper laid the foundation for the concept of prochirality and provided the impetus for investigation of the stereochemistry of enzymatic reactions occurring at prochiral centers. Two recent surveys [3,4] provide a general introduction to this rapidly developing field of research. This review concerns itself with the synthesis and metabolism of α-amino acids in which a chiral center has been created by the stereospecific introduction of a hydrogen isotope at a prochiral carbon atom. The literature has been reviewed through February 1976.

GLYCINE

The simplest of the α-amino acids, glycine (1), has been prepared in chirally labeled form by two methods, both of which are enzymatic. One method utilizes the pyridoxal-dependent enzyme serine transhydroxymethylase, which catalyzes the interconversion of glycine and L-serine (2):

$$H_2NCH_2COOH \quad + \quad HCHO \quad \rightleftarrows \quad HOCH_2\underset{\underset{NH_2}{|}}{C}HCOOH$$

$$1 \hspace{6cm} 2$$

It was reported [5] in 1964 that serine transhydroxymethylase catalyzes an exchange between the α-hydrogen atoms of glycine and tritiated water, and it was observed that the rate of exchange is greatly increased in the presence of tetrahydrofolate. Subsequently, this exchange reaction was utilized [6,7] to prepare stereospecifically tritiated samples of glycine. Incubation of $(2R,2S)$-[2-³H]glycine with tetrahydrofolate in the absence of formaldehyde and in an unlabeled medium led to a stereospecific exchange of one tritium atom for protium with formation of $(2R)$-[2-³H]glycine. When unlabeled glycine was incubated with the enzyme in the presence of tritiated water, $(2S)$-[2-³H]-glycine was produced. The assignment of absolute configuration to these two stereospecifically tritiated forms of glycine was accomplished by using D-amino-acid oxidase. This enzyme converts D-alanine (3) to pyruvate (4), but it also accepts glycine as a substrate, transforming the latter into glyoxylate (5). Since the conversion of D-alanine to pyruvate by D-amino-acid oxidase results in removal of the α-hydrogen atom of the amino acid, it was assumed that the α-hydrogen atom of glycine with the pro-S configuration would be removed in the formation of glyoxylate by the enzyme (Scheme 1). When stereospecifically tritiated glycine obtained from the incubation of non-radioactive glycine with serine transhydroxymethylase and tritiated water was oxidized with D-amino-acid oxidase, most of the tritium was absent from the glyoxylate produced, showing that the labeled glycine had the (S) configuration. In a parallel experiment, glycine obtained by the enzymatic equilibration of $(2R,2S)$-[2-³H]glycine with water was incubated with D-amino-acid oxidase. The glyoxylic acid produced in this experiment retained most of the tritium label, proving that this sample of glycine had the (R) configuration. As anticipated, the oxidation of $(2R,2S)$-[2-³H]glycine with D-amino-acid oxidase yielded glyoxylate containing 50% of the original tritium label.

Stereospecifically tritiated glycine has also been prepared [7] by carrying

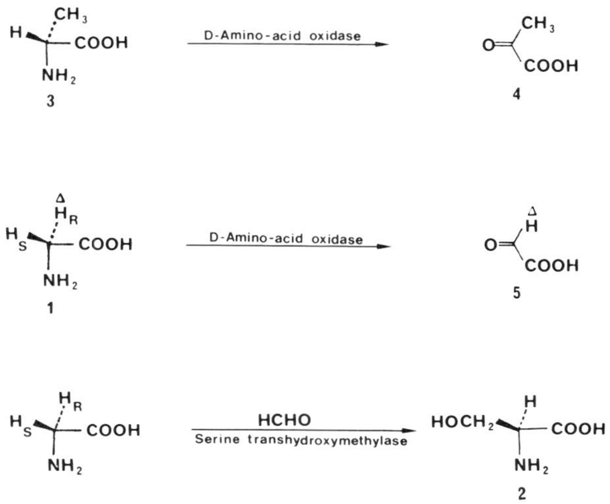

$$\text{3} \quad \xrightarrow{\text{D-Amino-acid oxidase}} \quad \text{4}$$

$$\text{1} \quad \xrightarrow{\text{D-Amino-acid oxidase}} \quad \text{5}$$

$$\text{2} \quad \xleftarrow[\text{Serine transhydroxymethylase}]{\text{HCHO}}$$

Scheme 1

out a transamination reaction between L-aspartic acid (6) and glyoxylic acid in the presence of tritiated water. When this conversion was catalyzed by an L-specific transaminase from rat liver, the resulting tritiated glycine had the (R) configuration as shown by the production of labeled glyoxylate on treatment with D-amino-acid oxidase (Scheme 2).

The first application of chirally labeled glycine to the investigation of the stereochemistry of enzymatic reactions was with respect to serine transhydroxymethylase [6,7]. In the presence of formaldehyde, glycine is converted by this enzyme to L-serine (2) with loss of the pro-S hydrogen atom. The enzymatic replacement of the pro-S hydrogen atom of glycine with a hydroxymethyl group therefore proceeds with retention of configuration (Scheme 1). Serine transhydroxymethylase also catalyzes the condensation of glycine with acetaldehyde to give a mixture of threonine and allothreonine; this reaction has also been shown to proceed with loss of the pro-S hydrogen atom of glycine [6].

The stereospecificity of the decarboxylation of aminomalonic acid (7) has been investigated with the aid of chiral glycine [8]. L-Aspartate β-decarboxylase from *Alcaligenes faecalis* is a pyridoxal-dependent multifunctional enzyme that catalyzes a number of reactions in addition to the β-decarboxylation of L-aspartic acid (6). These reactions include transaminations between several pairs of L-amino acids and the decarboxylation of aminomalonic acid to glycine. When the latter reaction was carried out in tritiated water, $(2S)$-[2-^{3}H]glycine was produced. The chirality of the labeled glycine was determined in the usual manner using D-amino-acid oxidase. The formation

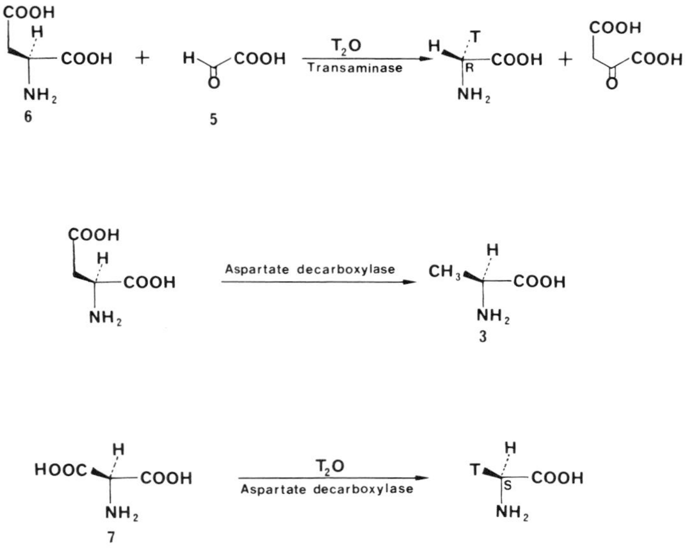

Scheme 2

of (2S)-[2-^{3}H]glycine in the decarboxylation reaction suggests that L-aspartate β-decarboxylase acts selectively on the pro-R carboxyl group of amino-malonate, which corresponds to the β-carboxyl group of aspartate, and that the replacement of the carboxyl group by a proton occurs with retention of configuration (Scheme 2). However, the less likely possibility that the decarboxylation of aminomalonate proceeds with loss of the pro-S carboxyl group followed by protonation with inversion cannot be ruled out.

The enzyme 5-aminolevulinic acid synthetase is involved in the initial stages of porphyrin biosynthesis. The enzyme, which requires pyridoxal phosphate as a cofactor, catalyzes the condensation of glycine (1) with succinyl–CoA (8) to give 5-aminolevulinic acid (9):

$$\text{H}_2\text{NCH}_2\text{COOH} \ + \ \text{HOOCCH}_2\text{CH}_2\text{COSCoA} \ \longrightarrow \ \text{H}_2\overset{5}{\text{N}}\text{CH}_2\text{COCH}_2\text{CH}_2\overset{1}{\text{COOH}} \ + \text{CO}_2 \ + \text{CoASH}$$
$$\quad\quad 1 \quad\quad\quad\quad\quad\quad 8 \quad\quad\quad\quad\quad\quad\quad\quad\quad\quad\quad 9$$

Insight into the mechanism of this reaction has been gained by investigating the fate of the prochiral hydrogen atoms of glycine when this amino acid is transformed into 5-aminolevulinate [9,10]. Incubation of (2R,2S)-[2-^{3}H,2-^{14}C]glycine with a highly purified preparation of 5-aminolevulinate synthetase from *Rhodopseudomonas spheroides* yielded radioactive 5-amino-levulinic acid which was reduced with sodium borohydride to 5-amino-4-hydroxypentanoic acid. The latter compound was then cleaved with periodate. The C-5 carbon atom of 9 liberated as formaldehyde was trapped as its crystalline dimedone adduct. The radioactivity of the dimedone–formaldehyde showed that the biosynthesis of 5-aminolevulinate involved the loss of about 50% of the original radioactivity. The stereochemistry associated with the loss was then examined. (2R)-[2-³H,2-¹⁴C]Glycine and (2S)-[2-³H,2-¹⁴C]glycine prepared using serine transhydroxymethylase were separately incubated with purified 5-aminolevulinate synthetase, and the radioactive **9** obtained was degraded to isolate C-5. These experiments demonstrated that the conversion of glycine to 5-aminolevulinic acid proceeds with complete loss of the pro-*R* hydrogen atom but complete retention of the pro-*S* hydrogen.

The nature of the stereochemical events occurring at position 5 of 5-amino-levulinic acid remained to be determined. Because of the ready exchange of

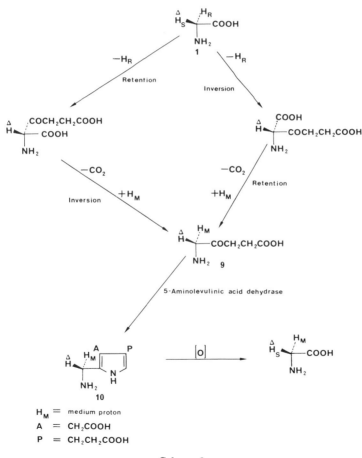

Scheme 3

the C-5 hydrogens of **9** with the medium, a system was developed whereby the labeled **9** produced by 5-aminolevulinate synthetase was immediately converted enzymatically to porphobilinogen (**10**) (Scheme 3). The biosynthesized porphobilinogen was N-acetylated and then oxidized to yield N-acetylglycine, which was hydrolyzed to glycine. The chirality of the tritium label present in the glycine so produced was analyzed by carrying out a stereospecific exchange of the pro-S hydrogen atom using serine transhydroxymethylase. The exchange process resulted in nearly complete removal of tritium from the glycine obtained by degradation of labeled porphobilinogen. The 2-pro-S hydrogen atom of glycine therefore occupies the pro-S position at C-5 of 5-aminolevulinic acid. These results, when combined with the observation that the 2-pro-R hydrogen atom of glycine is lost during 5-aminolevulinic acid formation, allow some interesting conclusions to be drawn about the mechanism of conversion of glycine to **9**. This conversion requires two bond-forming events: the replacement of the glycine carboxyl group by a proton and the formation of a carbon–carbon single bond between succinyl–CoA and C-2 of glycine. The stereochemistry observed for the 5-aminolevulinate synthetase reaction requires that one of the two steps proceed with retention of configuration and the other with inversion (Scheme 3).

Another unusual metabolic pathway for glycine occurs in the anaerobic bacterium *Clostridium sticklandii*. This organism contains a multicomponent enzyme system which links the reduction of glycine to the phosphorylation of ADP to ATP:

$$H_2NCH_2COOH + 2H + ADP + P_i \longrightarrow H{-}CH_2COOH + NH_3 + ATP$$

One of the components of the system is a low molecular weight selenium-containing protein called protein A, while the remaining components have been collectively termed glycine reductase. The stereochemistry of glycine reduction by this *Clostridium* enzyme system has recently been elucidated [11].

Enantiomeric, stereospecifically doubly labeled forms of glycine were prepared using serine transhydroxymethylase. $(2S)$-[2-^{3}H,2-^{2}H$_1$]Glycine (**11**) was prepared by enzyme-catalyzed exchange of [2-^{2}H$_2$]glycine in tritiated water, while $(2R)$-[2-^{3}H,2-^{2}H$_1$]glycine (**12**) was obtained by exchange of $(2R,2S)$-[2-^{3}H]glycine in deuterated water. [^{14}C]Glycine was added to each sample of [2-^{3}H,2-^{2}H$_1$]glycine, and the resulting multiply labeled forms of glycine were incubated with glycine reductase to yield chirally labeled forms of acetic acid (Scheme 4). Configurational analysis of the two samples of acetic acid was accomplished by conversion to L-malate followed by equilibration with fumarase. The malate (**13**) derived from $(2S)$-[2-^{3}H,2-^{2}H$_1$]glycine (**11**) lost the majority of its tritium label; the malate (**14**) derived from $(2R)$-[2-^{3}H,2-^{2}H$_1$]glycine (**12**) retained its tritium. Since it is known that

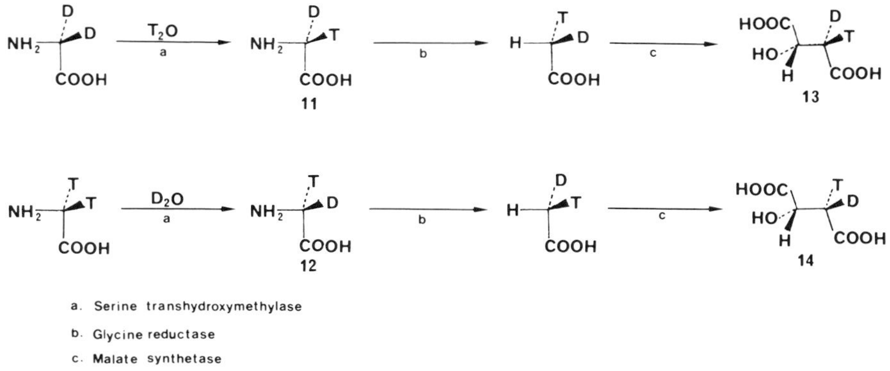

a. Serine transhydroxymethylase

b. Glycine reductase

c. Malate synthetase

Scheme 4

fumarase exchanges the 3-pro-*R* hydrogen of malate [12] and that the malate synthetase reaction proceeds with inversion of configuration at the methyl group of acetic acid [13,14], it follows that the glycine reductase reaction proceeds with inversion of configuration at the methylene carbon of glycine (Scheme 4).

CYSTEINE

One synthesis of chirally labeled cysteine has been reported [15]. The reaction sequence that was utilized to prepare (2*R*,3*R*)-[2,3-³H]- and (2*R*,3*S*)-[3-³H]cysteine is outlined in Scheme 5. Pyruvic acid (15, X = H) or [³H]pyruvic acid (15, X = ³H) was transformed in the manner shown to the thiazoline 16. Catalytic reduction of the labeled thiazoline (16, X = ³H) yielded the racemic thiazolidine 17 (X = ³H; Y = H), which was hydrolyzed to the corresponding acid and resolved with strychnine to afford the (2*R*,3*S*)-thiazolidine (18, X = ³H; Y = H). Hydrolysis of 18 (X = ³H; Y = H) yielded (2*R*,3*S*)-[3-³H]cysteine (19, X = ³H; Y = H). Catalytic tritiation of 16 (X = H) generated the racemic thiazolidine 17 (X = H; Y = ³H), which was resolved to give the (2*R*,3*R*)-[2,3-³H]thiazolidine (18, X = H; Y = ³H); from this, (2*R*,3*R*)-[2,3-³H]cysteine (19, X = H; Y = ³H) was obtained.

Chirally labeled cysteine has been used [15] to examine the fate of the hydrogen atoms attached to C-3 of the amino acid when it is incorporated into penicillin G (20). The two stereospecifically labeled forms of cysteine whose preparation has just been discussed were each administered to *Penicillium chrysogenum* in conjunction with L-[*U*-¹⁴C]cysteine as an internal reference. L-[3,3'-³H,*U*-¹⁴C]Cystine was also administered to *P. chrysogenum* as a control experiment. The results indicated a low retention of tritium with

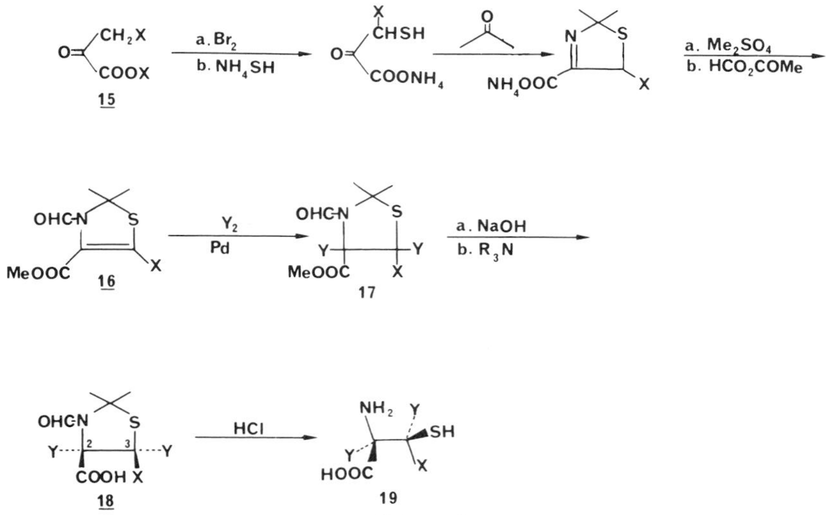

Scheme 5

(2R,3S)-[2-³H]cysteine, the expected retention of ca. 50% of the tritium with labeled cystine, and a higher tritium retention with (2R,3R)-[2,3-³H]-cysteine. Unfortunately, the interpretation of the data was complicated by the fact that one of the labeled forms of cysteine was tritiated at C-2 as well as at C-3 and that some tritium loss appeared to take place from C-2 of the amino acid during its conversion to **20**. Nevertheless, the authors appear to be justified in concluding that cysteine is incorporated into penicillin G (**20**) with retention of the 3-pro-R hydrogen atom of the amino acid. This result indicates that the incorporation of C-3 of cysteine into **20** proceeds with overall retention of stereochemistry:

SERINE

Chirally labeled serine has been prepared enzymatically in conjunction with an examination of the stereochemistry of the pyridoxal phosphate-dependent enzymes tryptophan synthetase and tryptophanase [16,17]. (2S,3R)- and (2S,3S)-[3-³H]phosphoglyceric acid (**21**) can be obtained enzymatically from [1-³H]glucose and [1-³H]mannose, respectively [18]. Incubation of the

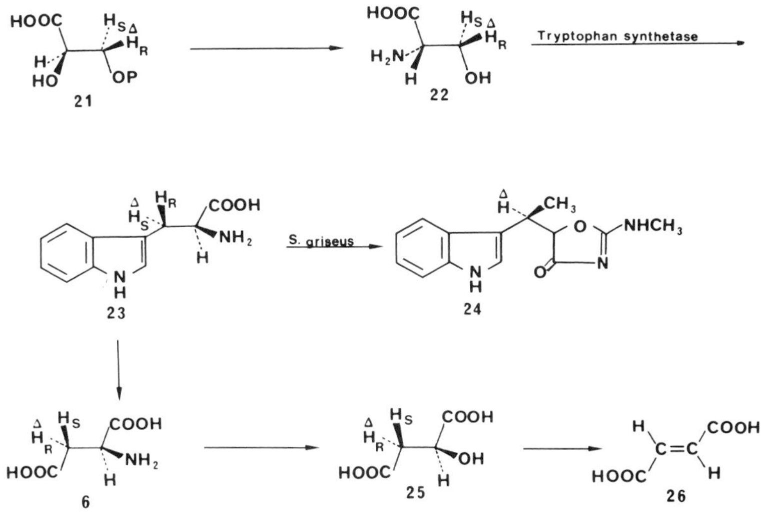

Scheme 6

specifically labeled forms of **21** with an enzyme preparation from *Escherichia coli* followed by treatment of the products with alkaline phosphatase gives (2S,3R)- and (2S,3S)-[3-³H]serine (**22**) (Scheme 6).

The two chirally tritiated forms of L-serine were each mixed with [U-¹⁴C]-serine and incubated in the presence of indole with purified, native tryptophan synthetase from *Neurospora crassa*. The formation of tryptophan (**23**) from **22** proceeded without tritium loss in each case. The tritium label in each sample of tryptophan was shown to be stereospecific by administration of both samples to *Streptomyces griseus* and isolation of indolmycin (**24**). The labeled tryptophan obtained from (2S,3R)-[3-³H]serine (**22**, H$_R$ = ³H) gave indolmycin nearly devoid of tritium, while the labeled tryptophan produced from (2S,3S)-[3-³H]serine (**22**, H$_S$ = ³H) led to indolmycin that retained all of the tritium label [16,19]. The configuration at the chirally labeled center in the two samples of tryptophan was analyzed by reduction to give 4,7-dihydro-tryptophan followed by ozonolysis with an oxidative workup to yield L-aspartic acid (**6**). The chirality of the labeled samples of aspartic acid was determined by conversion to L-malic acid (**25**) and incubation of the labeled malic acids with fumarase (Scheme 6). Since it had been previously established that the conversion of malic acid to fumaric acid (**26**) by fumarase results in the loss of the 3-pro-*R* hydrogen atom from malate [12], these experiments defined the chirality of the two isotopically labeled samples of tryptophan: The tryptophan obtained from (2S,3R)-[3-³H]serine has the (3S) configuration, while that obtained from (2S,3S)-[3-³H]serine has the (3R) configuration.

From this, one can conclude that the methylation of C-3 of tryptophan as a result of its conversion to indolmycin proceeds with loss of the 3-pro-*R* hydrogen atom [19], and the tryptophan synthetase reaction catalyzed by the *Neurospora* enzyme takes place with retention of configuration at C-3 of serine (Scheme 6). The same approach has been utilized to show that the formation of L-tryptophan from indole and L-serine catalyzed by the *E. coli* enzyme tryptophanase also occurs with retention of configuration at C-3 of serine [17]. These two enzymes therefore appear to fit the stereochemical pattern that is emerging for pyridoxal phosphate enzymes, namely, that the reactions take place on only one face of the amino acid–pyridoxal phosphate complex (see the discussion below as well as the sections on glycine and tyrosine).

The investigation of the mechanism of the reactions catalyzed by tryptophanase and tryptophan synthetase has been extended to an analysis of the steric course of the α,β-elimination reactions catalyzed by these two enzymes [17]. The deamination of L-serine to pyruvate and ammonia by *E. coli* tryptophanase and by tryptophan synthetase β_2 protein from *E. coli* mutant A2/F′A2 was examined using (2*S*,3*R*)- and (2*S*,3*S*)-[3-³H,¹⁴C]serine. The deamination of L-tryptophan to indole, pyruvate, and ammonia by tryptophanase was studied with (2*S*,3*R*)- and (2*S*,3*S*)-[3-³H,3-¹⁴C]tryptophan. The reactions were carried out in D_2O, and the resulting pyruvate was trapped as lactate by using NADH and an excess of lactate dehydrogenase. The lactate samples were oxidized to acetic acid, and the chirality of the methyl group in the acetate samples was then determined by conversion to malate with malate synthetase [13,14] followed by treatment with fumarase [12]. The results of these experiments showed that the two deamination reactions catalyzed by tryptophanase have the same stereochemical consequence: Protonation at C-3 of the amino acid side chain occurs with retention of configuration. This finding is in harmony with published observations on the deamination of D-threonine by D-serine dehydratase [20] and of L-threonine by L-serine dehydratase [21] since both of these reactions also proceed with retention of configuration. In contrast, an investigation of the stereochemistry of the deamination of chirally labeled serine catalyzed by tryptophan synthetase β_2 protein revealed that the replacement of the hydroxyl group by hydrogen is, in this case, nonstereospecific. The authors suggested that these results may reflect a subtle difference between those enzymes primarily catalyzing α,β-eliminations and those primarily catalyzing β-replacement reactions. Both types of enzyme are presumed to operate via the same α-aminoacrylate–pyridoxal phosphate Schiff base intermediate, but the enzymes catalyzing α,β-eliminations apparently have a base that can transfer a proton to C-3 of this intermediate, while the enzymes catalyzing β-replacement reactions do not.

HOMOSERINE

The synthesis of homoserine stereospecifically labeled at C-3 and C-4 has recently been reported [22]. The methods used to prepare $(3R)$-[3-^{2}H$_1$]- and $(3S)$-[3-^{2}H$_1$]homoserine are shown in Scheme 7. Addition of hydroxylamine in ethanol to (E)-[2-^{2}H$_1$]cinnamic acid (27) afforded the enantiomeric 3-amino-3-phenyl[2-^{2}H$_1$]propionic acids 28 and 29. Ozonolysis of the mixture of 28 and 29 gave *erythro*-DL-[3-^{2}H$_1$]aspartic acid, thereby establishing that the addition of the nitrogen nucleophile across the double bond of 28 occurred in *cis* fashion. Resolution of the mixture of 28 and 29 gave 28 $(2R,3S)$, which was reduced with lithium aluminum hydride to the alcohol 30. Acetylation of 30 followed by ozonolysis, oxidative workup, and acid hydrolysis produced $(2S,3R)$-[3-^{2}H$_1$]homoserine (31). $(2S,3S)$-[3-^{2}H$_1$]Homoserine (32) was obtained via the same reaction sequence by beginning with the addition of hydroxyl amine to unlabeled (E)-cinnamic acid in a deuterated medium (Scheme 7).

Homoserine stereospecifically labeled in the terminal methylene group was prepared from $(1S)$-3-phenyl[1-^{2}H$_1$]propanol (33), which is available by reduction of [*formyl*-^{2}H$_1$]cinnamaldehyde with fermenting yeast [23]. Acetylation of 33 followed by bromination and treatment of the bromoester with sodium azide gave the azide 34. Catalytic reduction of the azide function in 34 and ozonolysis of the resulting amine yielded $(4S)$-DL-[4-^{2}H$_1$]homoserine (35) (Scheme 8). Conversion of $(1S)$-3-phenyl[1-^{2}H$_1$]propanol (33) to $(1R)$-3-phenyl[1-^{2}H$_1$]propanol (36) and repetition of the same sequence of reactions produced $(4R)$-DL-[4-^{2}H$_1$]homoserine (37).

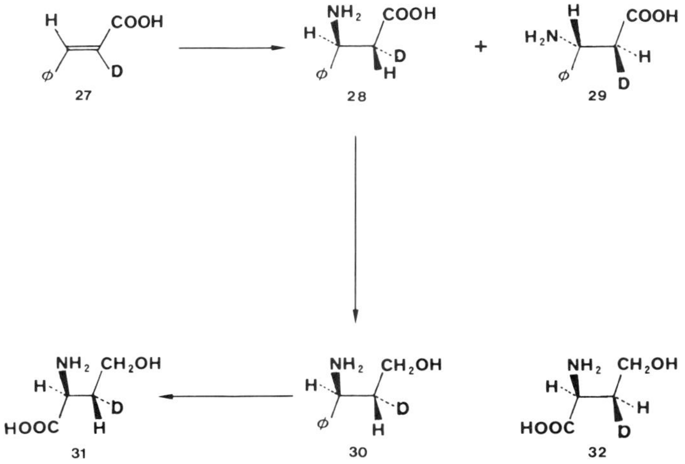

Scheme 7

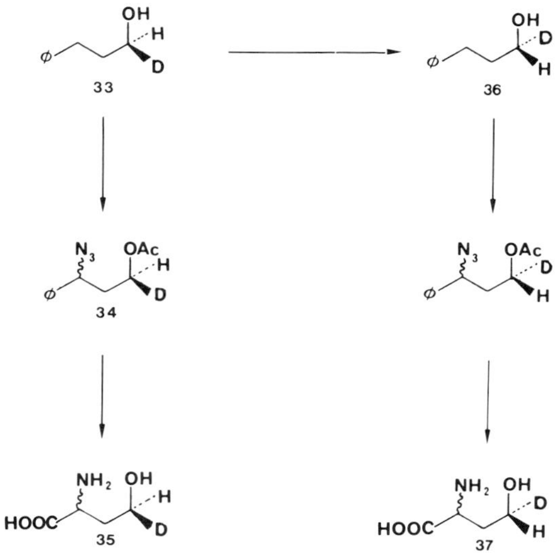

Scheme 8

ASPARTIC ACID

The enzyme L-aspartate ammonia-lyase (aspartase) catalyzes the reversible elimination of ammonia from L-aspartic acid (**6**) to give fumaric acid (**27**):

Chirally labeled aspartic acid has been prepared enzymatically in conjunction with a study of the stereochemistry of this ammonia-lyase reaction. Incubation of aspartate and fumarate in D_2O with aspartase obtained from *Bacillus cadaveris* [24] or *Proteus vulgaris* [25] gave unlabeled fumarate plus aspartate containing one deuterium atom at C-3 per mole. The configuration of the deuterium label in the aspartic acid was shown to be (*R*) in the following way. Treatment of the L-[3-²H₁]aspartate (**38**) with either nitrous acid or a combination of transaminase and malate dehydrogenase gave L-[3-²H₁]malate (**39**). Treatment of the labeled malate with fumarate hydratase gave fumaric acid that did not contain deuterium (Scheme 9) [24]. When the deuterated malate **39** obtained from the labeled aspartate **38** was examined by nmr spectrometry, the coupling constant between the protons at C-2 and C-3 of **39** indicated that it had the *erythro* configuration [25]. Further evidence in favor of this conclusion was obtained by conversion of the labeled aspartate **38** to

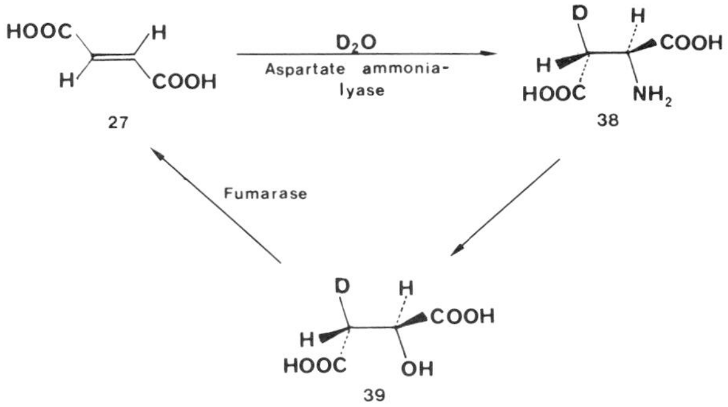

Scheme 9

D-[3-^{2}H$_1$]malate, whose nmr spectrum agreed with that of the *threo* compound [25]. Hence, aspartate ammonia-lyase catalyzes a *trans* elimination of ammonia from L-aspartic acid [26].

GLUTAMIC ACID

The formation of chirally labeled L-glutamic acid has been reported as part of a study of the enzyme methylaspartate mutase. This enzyme, which is isolated from *Clostridium tetanomorphum*, catalyzes the reversible isomerization of L-glutamic acid (**40**) to (3S)-3-methyl-L-aspartic acid (**41**):

The enzyme requires the participation of a cobamide coenzyme, and labeling studies have established that the isomerization proceeds by migration of C-1 and C-2 of glutamate from C-3 to C-4. In the course of this rearrangement,

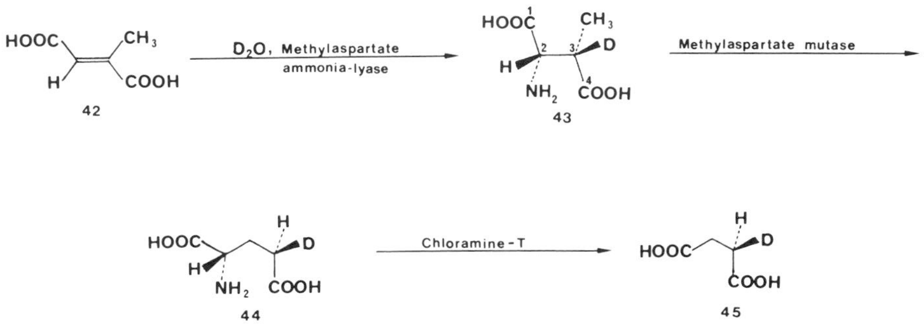

Scheme 10

the migrating group is replaced by a proton at C-3. This proton is derived from C-4 of glutamate [27,28]. The stereochemistry of the migration has been analyzed in the following way [29]. The addition of ammonia to mesaconic acid (42) catalyzed by methylaspartate ammonia-lyase gave (3S)-3-methyl-L-[3-^{2}H$_1$]aspartate (43) when the reaction was run in D$_2$O (Scheme 10). On treatment with methylaspartate mutase, 43 yielded (4R)-L-[4-^{2}H$_1$]glutamic acid (44), whose configuration at C-4 was defined by oxidation to (2R)-[2-^{2}H$_1$]succinic acid (45) with chloramine-T. The configuration of the labeled succinic acid followed from optical rotatory dispersion (ORD) measurements. This stereochemical information requires that the rearrangement proceed with a net inversion of configuration at C-3.

PROLINE

The chemical synthesis of (4R)- and (4S)-L-[4-^{3}H]proline has been accomplished [30]. The preparation of these labeled forms of proline was carried

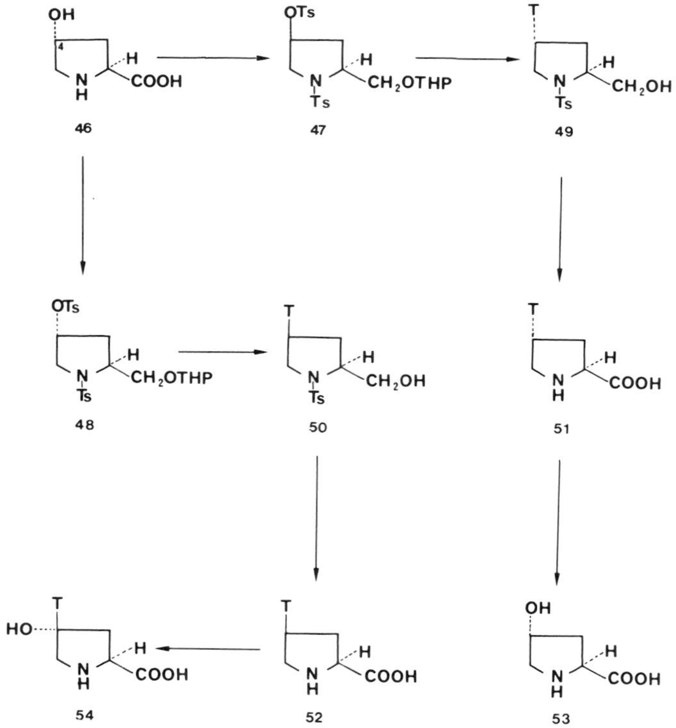

Scheme 11

out according to Scheme 11. (4R)-4-Hydroxy-L-proline (46) was converted in several steps to the two tetrahydropyranyl ethers 47 and 48, which are epimeric at C-4. Reduction of each of these compounds with tritiated lithium aluminum hydride resulted in displacement of the p-toluenesulfonyl ester group from C-4 and introduction of tritium with inversion of configuration. Acid hydrolysis of the reduction products gave (4R)-L-[4-³H]prolinol (49) and (4S)-L-[4-³H]prolinol (50). The stereochemical homogeneity of each of these compounds was ensured by preparation of the corresponding deuterated prolinols and examination of their nmr spectra. Chromic acid oxidation of 49 and 50 followed by removal of the N-p-toluenesulfonyl groups with hydrogen bromide in acetic acid afforded (4R)-L-[4-³H]proline (51) and (4S)-L-[4-³H]proline (52).

The chirally tritiated prolines 51 and 52 have been used to study the mechanism of the conversion of L-proline to (4R)-4-hydroxy-L-proline (53) in chick embryo [30] and in *Streptomyces antibioticus* [31]. In both cases, hydroxylation of (4R)-L-[4-³H]proline (51) proceeded with complete tritium loss to give 53, while hydroxylation of (4S)-L-[4-³H]proline (52) resulted in complete tritium retention to give 54. Consequently, the hydroxylation of proline at C-4 takes place with retention of configuration.

HISTIDINE

The reversible elimination of ammonia from L-histidine (55) to give urocanic acid (56) is catalyzed by the enzyme L-histidine ammonia-lyase (histidase):

When this transformation is carried out in a tritiated medium using a highly purified enzyme from *Pseudomonas putida*, (3R)-L-[3-³H]histidine (57) is obtained [32]. The chirality of the isotopic label in 57 was established by means of the degradative sequence found in Scheme 12. Esterification of the radioactive histidine followed by treatment with benzoyl chloride led to the ring-opened tribenzoate 58. Ozonolysis of 58 with oxidative workup generated tritiated N,N'-dibenzoyl-L-asparagine methyl ester (59), which was hydrolyzed to (3R)-L-[3-³H]aspartic acid (60). The stereochemistry of the tritium label in 60 was analyzed by treatment with aspartate ammonia-lyase. The enzyme-catalyzed loss of ammonia from 60 gave radioinactive fumaric acid. The tritium label in 60 therefore has the (R) configuration [24,25]. These results established that the histidine ammonia-lyase reaction is mechanistically

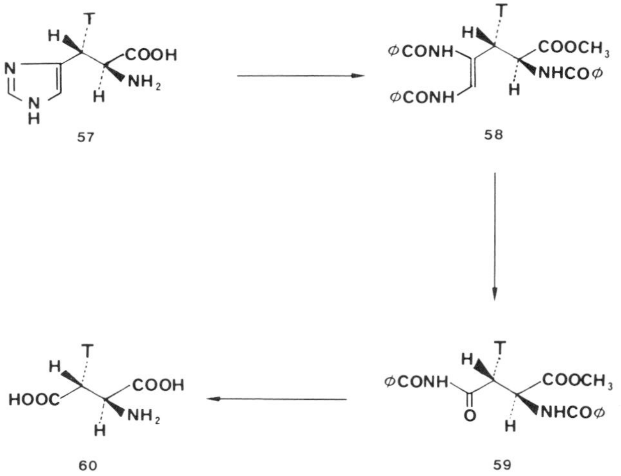

Scheme 12

similar to the aspartate ammonia-lyase reaction; both enzymes catalyze *trans* eliminations in which the 3-pro-*R* hydrogen atom of the amino acid is lost.

PHENYLALANINE AND TYROSINE

Two groups have devised methods for the synthesis of phenylalanine (**64**, X = R = H) and tyrosine (**64**, X = H; R = OH) in chirally labeled form [33–36]. Both groups employed the synthetic sequence outlined in Scheme 13. A labeled benzaldehyde (**61**, X = ^{2}H or ^{3}H; R = H, OMe, or OBz) was condensed with *N*-benzoylglycine to give a labeled α-benzoylaminocinnamic acid (**63**, X = ^{2}H or ^{3}H; R = H, OMe, or OBz). The (*Z*) configuration assigned to **62** and **63** followed from the chemical correlations shown in Scheme 13. Catalytic reduction of **63** proceeded in a *cis* fashion to give a mixture of *N*-benzoyl-α-amino acids, which was debenzoylated to afford a racemic mixture of the (2*S*,3*R*)-amino acid (**64**, X = ^{2}H or ^{3}H; R = H or OH) and the (2*R*,3*S*)-amino acid (**65**, X = ^{2}H or ^{3}H; R = H or OH). The racemic mixture was resolved by preparation of the *N*-chloroacetyl derivatives of **64** and **65** followed by hydrolysis with hog renal acylase I for the phenylalanine derivatives (R = H) or carboxypeptidase A for the tyrosine derivatives (R = OH). This technique gave the (3*R*)-L-amino acids (**64**) and the *N*-chloroacetyl derivatives of the (3*S*)-D-amino acids; the free (3*S*)-D-amino acids (**65**) were obtained from their *N*-chloroacetyl derivatives by

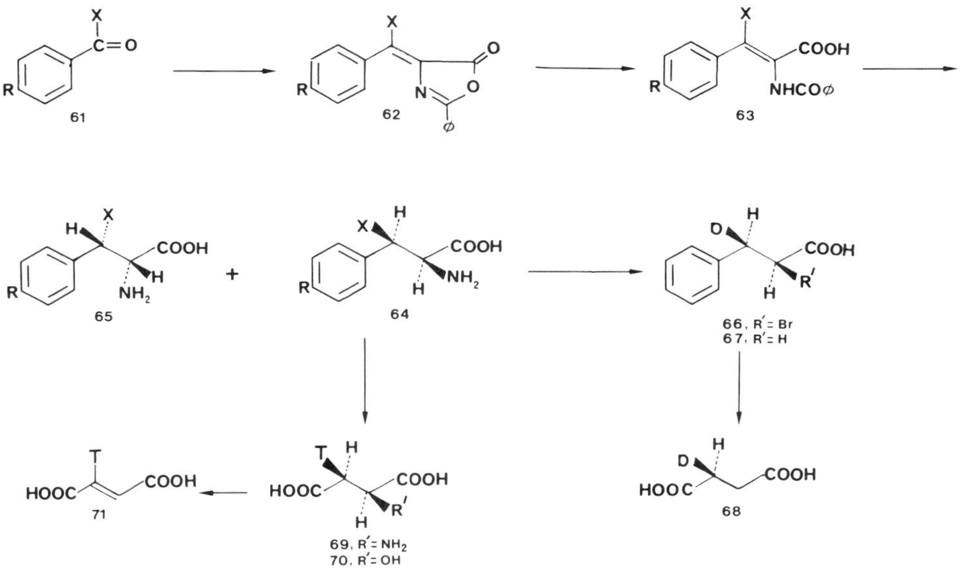

Scheme 13

chemical hydrolysis. The samples of chirally labeled L- and D-amino acids were then chemically racemized at C-2 in order to permit parallel experiments with (3R)- and (3S)-labeled materials having identical configurations at C-2.

The configuration of the label present at C-3 of **64** and **65** was determined in the manner shown in Scheme 13. Treatment of **64** (R = H; X = ^{2}H) with nitrous acid and hydrogen bromide gave (2S,3R)-2-bromo-3-phenyl[3-^{2}H$_1$]-propionic acid (**66**), which was catalytically reduced to (3S)-3-phenyl-[3-^{2}H$_1$]propionic acid (**67**). Destruction of the aromatic nucleus with ozone followed by an oxidative workup gave (2S)-[2-^{2}H$_1$]succinic acid (**68**), whose configuration was apparent from ORD measurements. A similar degradation of **65** (R = H; X = ^{2}H) gave (2R)-[2-^{2}H$_1$]succinic acid. In the tyrosine series, the configuration of the label in **64** (R = OH; X = ^{3}H) was checked by ozonolysis to yield (3S)-L-[3-^{3}H]aspartic acid (**69**), which was converted to (3S)-L-[3-^{3}H]malic acid (**70**) by nitrous acid. The configuration at the labeled carbon atom of **70** was shown to be (S) by virtue of the fact that most of the radioactivity was retained in the fumaric acid (**71**) obtained by the action of fumarase [12].

An alternative approach to the synthesis of stereospecifically deuterated phenylalanine has also been devised [33,37]. [formyl-^{2}H$_1$]Benzaldehyde was reduced with the enzyme liver alcohol dehydrogenase in the presence of an excess of ethanol and a catalytic quantity of NAD to give (S)-[methylene-^{2}H$_1$]-

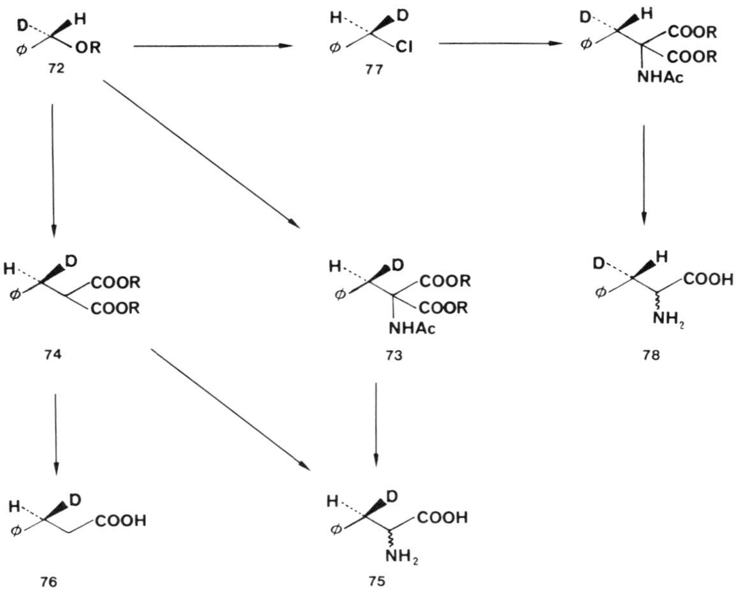

Scheme 14

benzyl alcohol (**72**, R = H) (Scheme 14) [38]. The corresponding *p*-toluene-sulfonyl ester (**72**, R = Ts) reacted with the sodium salt of ethyl acetamido-malonate or ethyl malonate to yield the inverted [3-^{2}H$_1$]malonic ester derivatives **73** and **74**, respectively. Hydrolysis of **73** gave (3*R*)-DL-[3-^{2}H$_1$]-phenylalanine (**75**). Hydrolysis of **74** with subsequent bromination, decarboxylation, and ammonolysis likewise gave **75**. Decarboxylation of the malonic acid **74** (R = H) afforded (3*S*)-3-phenyl[3-^{2}H$_1$]propionic acid (**76**), which was degraded to (2*S*)-[2-^{2}H$_1$]succinic acid, thereby establishing the absolute configuration at C-3 of the labeled phenylalanine **75** as (*R*). (3*S*)-DL-[3-^{2}H$_1$]Phenylalanine (**78**) was obtained from the (*S*)-[*methylene*-^{2}H$_1$]-benzyl alcohol (**72**, R = H) by conversion of the alcohol to the (*R*)-[*methylene*-^{2}H$_1$]chloride **77** with phosphorus trichloride and pyridine. Subsequent alkylation with ethyl acetamidomalonate and hydrolysis yielded **78**.

The chirally labeled forms of phenylalanine and tyrosine prepared by the methods just described have been used to explore the stereochemistry of a number of enzymatic processes involving these amino acids. One group of reactions that has been examined are those catalyzed by ammonia-lyases. L-Phenylalanine ammonia-lyase occurs widely in higher plants and in some fungi where it catalyzes the elimination of ammonia from L-phenylalanine (**79**) to give *trans*-cinnamic acid (**80**):

The stereochemistry of this elimination process has been elucidated [33,37] for an enzyme obtained from potato tubers. A number of experiments with stereospecifically deuterated and tritiated forms of phenylalanine established that the conversion of **79** to **80** proceeds with loss of the 3-pro-*S* hydrogen atom. The formation of cinnamic acid from phenylalanine in *Colchicum* plants was also found to occur with loss of the 3-pro-*S* hydrogen atom of the amino acid [33]. The phenylalanine ammonia-lyases therefore utilize a *trans* elimination mechanism, and the stereochemical course of this ammonia-lyase reaction is the same as that observed for aspartate and histidine ammonia-lyases (*vide supra*) [39].

Various plants contain an ammonia-lyase that converts L-tyrosine (**81**) to *trans*-*p*-hydroxycinnamic acid (**82**):

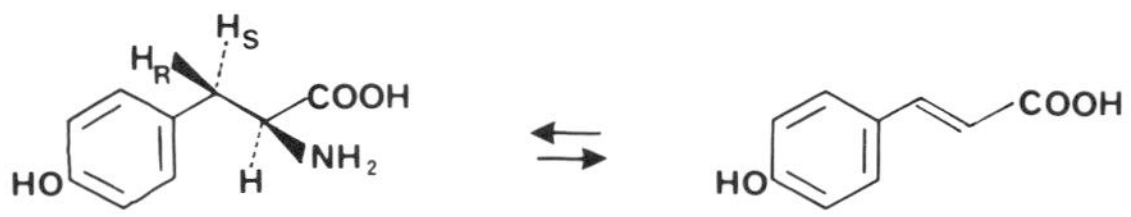

An enzyme that is isolable from maize (*Zea mays* L.) catalyzes this conversion as well as the conversion of L-phenylalanine to *trans*-cinnamic acid. In contrast, phenylalanine ammonia-lyase from potatoes possesses almost no tyrosine ammonia-lyase activity. The stereochemistry of the elimination process catalyzed by the maize enzyme has been probed [36] with chirally tritiated tyrosine and phenylalanine. The findings were consistent with the information available concerning other ammonia-lyase reactions: The maize enzyme catalyzes a *trans* elimination of ammonia from tyrosine or phenylalanine in which the 3-pro-*S* hydrogen atom is lost from the amino acid side chain.

Another reaction in the metabolism of aromatic amino acids that involves loss of one of the C-3 hydrogen atoms is a desaturation process. The stereochemistry of two formal desaturations of tyrosine has been determined. Administration [40] of the (3*R*), (3*S*), and (3*R*,3*S*) forms of DL-[3-³H,3-¹⁴C]-tyrosine to cultures of *Penicillium griseofulvum* gave three doubly labeled samples of mycelianamide (**81a**) (Scheme 15). The tritium to carbon-14 ratios in the three samples indicated nearly complete tritium loss from (3*S*)-[3-^{3}H]-tyrosine, little tritium loss from (3*R*)-[3-^{3}H]tyrosine, and ca. 50% tritium loss from (3*R*,3*S*)-[3-^{3}H]tyrosine. These results are consistent with a stereospecific removal of the 3-pro-*S* hydrogen atom from tyrosine during its conversion to **81a**.

L-Tyrosine has been shown [41] to be the precursor of carbon atoms 6–13 in the alkaloid securinine (**82**). Administration of (3*R*,3*S*)-, (3*R*)-, and (3*S*)-DL-[3-^{3}H,3-^{14}C]tyrosine to young *Securinega suffruticosa* plants [42] gave results similar to those observed for mycelianamide biosynthesis. The tritium to carbon-14 ratios in the labeled securinine isolated in each feeding experiment indicated that the 3-pro-*S* hydrogen atom of tyrosine is lost during the conversion of the amino acid to the alkaloid (Scheme 15). It is of interest that the 3-pro-*S* hydrogen atom is removed from tyrosine in both of the desaturation processes shown in Scheme 15. Since the identical stereochemistry has been observed in desaturation processes involving tryptophan (*vide infra*), there may be a stereochemical uniformity associated with reactions of this type.

The stereochemistry of hydrogen loss from C-3 of tyrosine as a result of hydroxylation has been examined [34]. Tyrosine provides biosynthetically the hydroaromatic (C_6–C_2) but not the aromatic (C_6–C_1) unit of the Amaryllidaceae alkaloid haemanthamine (**83**) (Scheme 15) [43]. During conversion of tyrosine to haemanthamine, one C-3 hydrogen atom is replaced by a hydroxyl group having the (*R*) configuration [44]. Feeding experiments with (3*R*)- and (3*S*)-DL-[3-^{3}H,2-^{14}C]tyrosine using flowering "Texas" daffodils established that the 3-pro-*R* hydrogen atom of tyrosine is lost as a result of the hydroxyla-

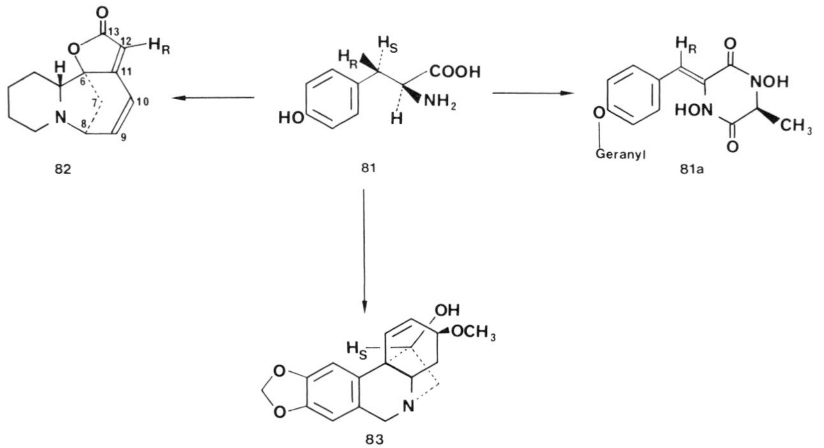

Scheme 15

tion process. Consequently, the hydroxylation reaction takes place with retention of configuration (Scheme 15).

An unexpected nonoxidative loss of hydrogen from C-3 of phenylalanine was encountered [45] during an investigation of the incorporation of chirally labeled phenylalanine into gliotoxin (**84**). Feeding experiments with the (3*R*) and (3*S*) forms of DL-[3-²H₁]- and DL-[3-³H]phenylalanine indicated that the biosynthesis of gliotoxin in *Trichoderma viride* occurs with loss of the 3-pro-*R* hydrogen atom of the amino acid precursor. Nuclear magnetic resonance analysis of monodeuteriogliotoxin formed from DL-[3-²H₂]phenylalanine allowed assignment of the (*S*) configuration to the new deuterium-bearing chiral center; this corresponds to incorporation of the original 3-pro-*S* deuterium atom with retention of configuration:

It was also observed that some dideuterio species accompanied the mono-deuteriogliotoxin derived from [3-²H₂]phenylalanine. This suggested that hydrogen loss from the methylene group of the precursor could not be an obligatory step in the biosynthesis of **84**, and, in fact, the incorporation of [3-³H]phenylalanine into mycelial protein was found to involve extensive loss of the 3-pro-*R* tritium atom. Thus, it appears that, in *T. viride*, phenylalanine undergoes an exchange reaction involving replacement of the pro-*R* methylene proton by an external proton with overall retention of configuration. This exchange apparently proceeds substantially faster than incorporation of the amino acid into gliotoxin or protein, but it does not seem to be an obligatory preliminary step for either process.

β-Replacement reactions are another class of enzymatic transformations associated with the metabolism of aromatic amino acids. The stereochemistry of two of these reactions, catalyzed by tryptophanase and tryptophan synthetase, has already been discussed (see the section on serine). The enzyme tyrosine phenol-lyase is a pyridoxal-dependent multifunctional enzyme that catalyzes a number of α,β-elimination, racemization, and β-replacement reactions. The stereochemistry of one of the β-replacement reactions has been elucidated [46] by studying the reaction of tyrosine with resorcinol to give 2,4-dihydroxy-L-phenylalanine (**85**):

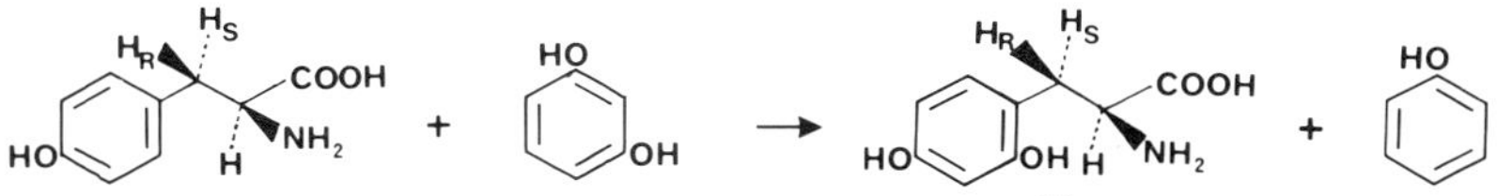

The $(3R)$ and $(3S)$ forms of [3-^{2}H$_1$]tyrosine and 2,4-dihydroxyphenylalanine were synthesized by the methods outlined in Scheme 13. Reaction of each of the stereospecifically deuterated samples of tyrosine with resorcinol, catalyzed by purified tyrosine phenol-lyase from *Escherichia intermedia*, gave two stereospecifically deuterated specimens of **85**. The configuration of the deuterium label in these specimens was elucidated by comparison of their nmr spectra with the nmr spectrum of synthetic chirally deuterated 2,4-dihydroxy-phenylalanine. The results showed that the β-replacement reaction occurs with complete retention of configuration at C-3. Related work has shown that at least one other β-replacement reaction catalyzed by tyrosine phenol-lyase proceeds with an identical stereochemical course; the enzyme-catalyzed formation of tyrosine from L-serine and phenol takes place with retention of configuration at C-3 of L-serine [47]. Hence, the reactions promoted by tyrosine phenol-lyase exhibit the same stereospecificity as those catalyzed by related pyridoxal-dependent enzymes such as tryptophanase and tryptophan synthetase.

TRYPTOPHAN

The synthesis of tryptophan stereospecifically labeled at C-3 with deuterium or tritium has been achieved via the oxazolinone method (Scheme 16). Treatment of anilinium 3-indolylglyoxylate (**86**, X = H) with deuterated or tritiated water followed by pyrolysis gave labeled 3-formylindole (**87**, X = ^{2}H or ^{3}H). *N*-Acetylation and subsequent condensation with *N*-acetylglycine gave the (*Z*)-oxazolinone **88**, which was hydrolyzed to the indolylacrylic acid **89**. Catalytic reduction of **89** produced a labeled racemic mixture composed of **90** and **91**. The racemate was resolved by formation of diastereomeric salts with (−)-1-phenylethylamine. The resolved samples were α-epimerized and then deacetylated to give the $(3R)$ and $(3S)$ forms of DL-tryptophan (**92** and **93**). The stereochemistry assigned to the isotopic labels in **92** and **93** was confirmed by degradation of the tritiated compounds (X = ^{3}H) to aspartic acid. The configuration of the label in the aspartic acid samples was analyzed by conversion to malic acid and dehydration with fumarase. The configurational purity at C-3 was shown to be ca. 94% by biosynthetic incorporation of the labeled tryptophan into indolmycin (**24**, Scheme 6).

Stereospecifically labeled tryptophan prepared in the manner described has been used to examine two biosynthetic processes that result in the loss of a hydrogen atom from C-3 of the amino acid. The toxic metabolite sporodesmin A (**94**) (Scheme 17) produced by the fungus *Pithomyces chartarum* appears to be derived biosynthetically from tryptophan, alanine, and methionine [48]. Incorporation experiments [49] with $(3R)$- and $(3S)$-DL-]3-^{3}H,^{14}C]tryptophan

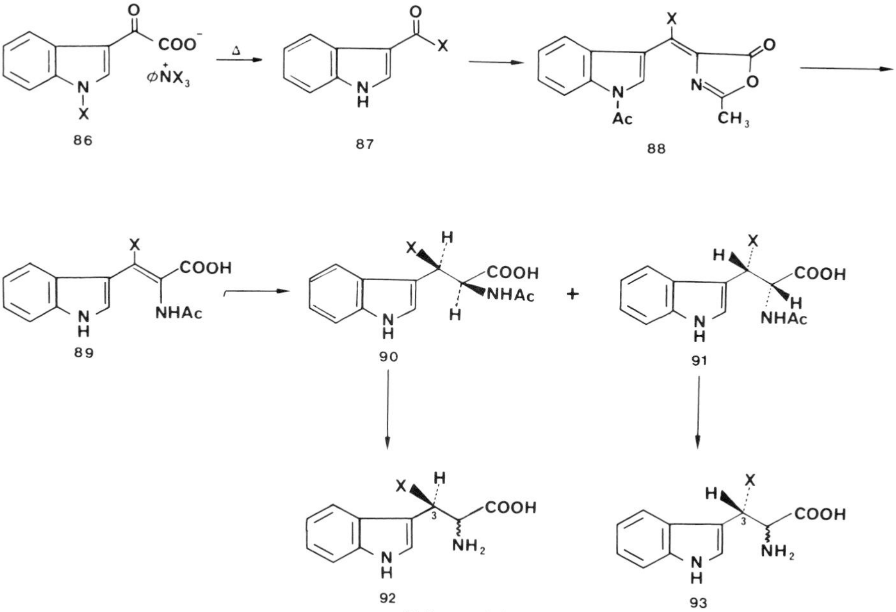

Scheme 16

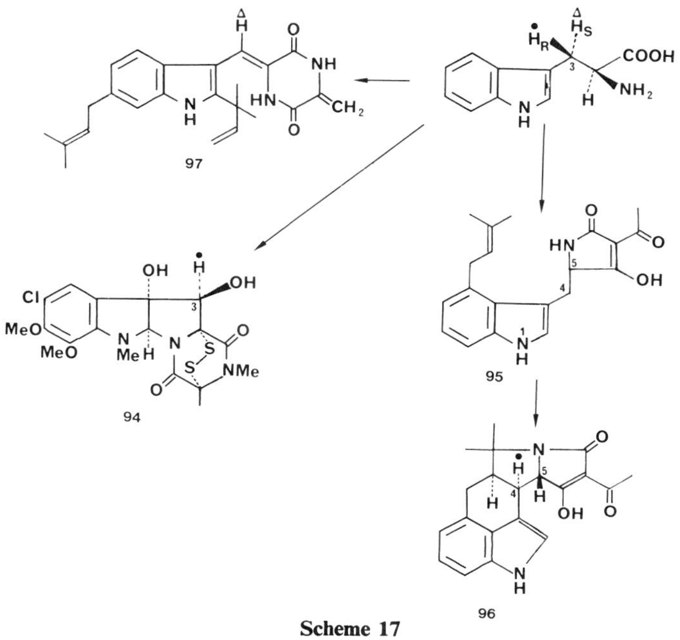

Scheme 17

gave tritium to carbon-14 ratios in the isolated sporodesmin A that were consistent with replacement of a methylene hydrogen (or tritium) at C-3 by OH with retention of configuration.

α-Cyclopiazonic acid (96) (Scheme 17) is a fungal metabolite formed biosynthetically from tryptophan via β-cyclopiazonic acid (95) [50,51]. The stereochemistry of the final, oxidative cyclization reaction (95 → 96) has been elucidated [52]. Administration of stereoselectively labeled DL-[3-³H,3-¹⁴C]-tryptophan to cultures of *Penicillium cyclopium* gave labeled samples of α-cyclopiazonic acid (96) with tritium to carbon-14 ratios indicating loss of the 3-pro-*S* hydrogen atom of tryptophan in the cyclization step; no tritium loss was observed in the β-cyclopiazonic acid (95) produced in these experiments. It had been found previously that β-cyclopiazonic acid derived from [2-³H]tryptophan is incorporated into α-cyclopiazonic acid with complete retention of tritium. This finding, taken in conjunction with the results obtained using [3-³H]tryptophan, is consistent with cyclization via a 1,4- but not a 4,5-dehydro derivative of 95 with C—C bond formation at C-4 occurring from the opposite side of the molecule to proton removal.

L-Tryptophan stereospecifically labeled with tritium at C-3 has also been prepared enzymatically using stereospecifically labeled forms of [3-³H]serine and the enzyme tryptophan synthetase [16,17,53]. Chirally labeled tryptophan obtained in this fashion has been used [53] to examine the steric course of a desaturation process associated with the incorporation of the amino acid into cryptoechinuline A (97) (Scheme 17). When (3R)- and (3S)-L-[3-³H,3-¹⁴C]-tryptophan were fed to cultures of *Aspergillus amstelodami*, the labeled cryptoechinuline A retained the tritium of (3R)-[3-³H]tryptophan and lost the tritium present in the (3S)-[3-³H]amino acid. The formal desaturation of the tryptophan side chain therefore takes place with loss of the 3-pro-*S* hydrogen atom. The same stereochemical course is exhibited when L-tyrosine is incorporated into mycelianamide [40] and securinine [42] (Scheme 15).

REFERENCES

1. Sir R. Robinson, *J. Chem. Soc.* pp. 762 and 876 (1917); "The Structural Relations of Natural Products" Oxford Univ. Press (Clarendon), London and New York, 1955; C. Schöpf, *Justus Liebigs Ann. Chem.* **497**, 1 (1923); C. Schöpf, G. Lehman, W. Arnold, K. Kach, H. Bayerle, K. Falk, F. Oechler, and H. Stever, *Angew. Chem.* **50**, 779 and 797 (1937); L. Ruzicka, *Experientia* **9**, 357 (1953); J. N. Collie, *J. Chem. Soc.* pp. 63, 122, and 329 (1893); pp. 787 and 1806 (1907); A. J. Birch and F. W. Donovan, *Aust. J. Chem.* **6**, 369 (1953).
2. A. G. Ogston, *Nature (London)*, **162**, 963 (1948).
3. W. L. Alworth, "Stereochemistry and Its Application in Biochemistry." Wiley (Interscience), New York, 1972.

4. R. Bentley, "Molecular Asymmetry in Biology," Vols. I and II. Academic Press, New York, 1969 and 1970.
5. L. Schirch and W. T. Jenkins, *J. Biol. Chem.* **239**, 3801 (1964).
6. P. M. Jordan and M. Akhtar, *Biochem. J.* **116**, 277 (1970).
7. D. Wellner, *Biochemistry* **9**, 2307 (1970).
8. A. G. Palekar, S. S. Tate, and A. Meister, *Biochemistry* **9**, 2310 (1970).
9. Z. Zaman, P. M. Jordan, and M. Akhtar, *Biochem. J.* **135**, 257 (1973).
10. M. M. Abboud, P. M. Jordan, and M. Akhtar, *J. Chem. Soc., Chem. Commun.* p. 643 (1974).
11. G. Barnard and M. Akhtar, *J. Chem. Soc., Chem. Commun.* p. 980 (1975).
12. F. A. L. Anet, *J. Am. Chem. Soc.* **82**, 994 (1960); O. Gawron and T. P. Fondy, *ibid.* **81**, 6333 (1959).
13. J. W. Cornforth, J. W. Redmond, H. Eggerer, W. Buckel, and C. Gutschow, *Nature (London)* **221**, 1212 (1969).
14. J. Luthy, J. Retey, and D. Arigoni, *Nature (London)* **221**, 1213 (1969).
15. D. J. Morecombe and D. W. Young, *J. Chem. Soc., Chem. Commun.* p. 198 (1975).
16. G. E. Skye, R. Potts, and H. G. Floss, *J. Am. Chem. Soc.* **96**, 1593 (1974).
17. E. Schleicher, K. Mascaro, R. Potts, D. R. Mann, and H. G. Floss, *J. Am. Chem. Soc.* **98**, 1043 (1976).
18. H. G. Floss, O. K. Onderka, and M. Carroll, *J. Biol. Chem.* **247**, 736 (1972).
19. L. Zee, U. Hornemann, and H. G. Floss, *Biochem. Physiol. Pflanz.* **168**, 19 (1975).
20. I. Y. Yang, Y. Z. Huang, and E. E. Snell, *Fed. Proc., Fed. Am. Soc. Exp. Biol.* **34**, 496 (1975).
21. G. Kapke, Paper presented at conference on vitamin B₆, Ames, Iowa, October 10–11, 1975. (Cited in reference [17].)
22. D. Coggiola, C. Fuganti, D. Ghiringhelli, and P. Grasselli, *J. Chem. Soc., Chem. Commun.* p. 143 (1976).
23. C. Fuganti, D. Ghiringhelli, and P. Grasselli, *J. Chem. Soc., Chem. Commun.* p. 846 (1975).
24. S. Englard, *J. Biol. Chem.* **233**, 1003 (1958).
25. A. I. Krasna, *J. Biol. Chem.* **233**, 1010 (1958).
26. The stereochemistry of the elimination was originally concluded to be *cis* on the basis of an early, incorrect assignment of configuration to the L-[3-²H₁]malate formed in the fumarate hydratase reaction.
27. A. A. Iodice and H. A. Barker, *J. Biol. Chem.* **238**, 2094 (1963).
28. R. G. Eagar, B. G. Baltimore, M. M. Herbst, H. A. Barker, and J. H. Richards, *Biochemistry* **11**, 253 (1972).
29. M. Sprecher, R. L. Switzer, and D. B. Sprinson, *J. Biol. Chem.* **241**, 864 (1966).
30. Y. Fujita, A. Gottlieb, B. Peterkofsky, S. Udenfriend, and B. Witkop, *J. Am. Chem. Soc.* **86**, 4709 (1964).
31. L. A. Salzman, H. Weissbach, and E. Katz, *Proc. Natl. Acad. Sci. U.S.A.* **54**, 542 (1965).
32. I. L. Givot, T. A. Smith, and R. H. Abeles, *J. Biol. Chem.* **244**, 6341 (1969); J. Retey, H. Fierz, and W. Zeylemaker, *FEBS Lett.* **6**, 203 (1970).
33. R. H. Wightman, J. Staunton, A. R. Battersby, and K. R. Hanson, *J. Chem. Soc., Perkin Trans. 1* p. 2355 (1972).
34. G. W. Kirby and J. Michael, *J. Chem. Soc., Perkin Trans. 1* p. 115 (1973).
35. G. W. Kirby, S. Narayanaswami, and P. S. Rao, *J. Chem. Soc., Perkin Trans. 1* p. 645 (1975).
36. P. G. Strange, J. Staunton, H. R. Wiltshire, A. R. Battersby, K. R. Hanson, and

E. A. Havir, *J. Chem. Soc., Perkin Trans. 1* p. 2364 (1972); B. E. Ellis, M. H. Zenk, G. W. Kirby, J. Michael, and H. G. Floss, *Phytochemistry* **12**, 1057 (1973).

37. R. Ife and E. Haslam, *J. Chem. Soc. C* p. 2818 (1971).
38. Cf. A. Streitwieser, J. R. Wolfe, and W. D. Schaeffer, *Tetrahedron* **6**, 338 (1959); V. E. Althouse, D. M. Feigl, W. A. Sanderson, and H. S. Mosher, *J. Am. Chem. Soc.* **88**, 3595 (1966); A. Horeau and H. Nouaille, *Tetrahedron Lett.* p. 3953 (1966).
39. Note that, although the aspartate and histidine ammonia-lyase reactions each proceed with loss of a 3-*pro*-R hydrogen atom, the hydrogen atom lost from phenylalanine is the same hydrogen atom in an absolute sense. Because of the sequence rules, the (*R*)–(*S*) designations of the hydrogen atoms at C-3 of phenylalanine are reversed from those of the C-3 hydrogen atoms of aspartate and histidine.
40. G. W. Kirby and S. Narayanaswami, *J. Chem. Soc., Chem. Commun.* p. 322 (1973).
41. R. J. Parry, *Tetrahedron Lett.* p. 307 (1974); U. Sankawa, K. Yamasaki, and Ebinzuka, *ibid.* p. 1867.
42. R. J. Parry, *J. Chem. Soc., Chem. Commun.* p. 144 (1975).
43. W. C. Wildman, *Alkaloids (N.Y.)* **11**, 308 (1968).
44. J. Clardy, F. M. Hauser, D. Dahm, R. A. Jacobson, and W. C. Wildman, *J. Am. Chem. Soc.* **92**, 6337 (1970).
45. N. Johns, G. W. Kirby, J. D. Bu'Lock, and A. P. Ryles, *J. Chem. Soc., Perkin Trans. 1* p. 383 (1975).
46. S. Sadawa, H. Kumagai, H. Yamada, and R. K. Hill, *J. Am. Chem. Soc.* **97**, 4334 (1975).
47. C. Fuganti, D. Ghiringhelli, D. Giangrasso, and P. Grasselli, *J. Chem. Soc., Chem. Commun.* p. 726 (1974).
48. N. R. Towers and D. E. Wright, *N.Z. J. Agric. Res.* **12**, 275 (1969).
49. G. W. Kirby and M. J. Varley, *J. Chem. Soc., Chem. Commun.* p. 833 (1974).
50. C. W. Holzapfel and D. C. Wilkins, *Phytochemistry* **10**, 351 (1971).
51. J. C. Schabort, D. C. Wilkins, C. W. Holzapfel, D. J. J. Potgieter, and A. W. Neitz, *Biochem. Biophys. Acta* **250**, 311 (1971); J. C. Schabort and D. J. J. Potgieter, *ibid.* p. 329.
52. P. S. Steyn, R. Vleggaar, N. P. Ferreira, G. W. Kirby, and M. J. Varley, *J. Chem. Soc., Chem. Commun.* p. 465 (1975).
53. R. Cardillo, C. Fuganti, D. Ghiringhelli, P. Grasselli, and G. Gatti, *J. Chem. Soc., Chem. Commun.* p. 778 (1975).

Reactions of Sulfur Nucleophiles with Halogenated Pyrimidines

Eugene G. Sander

INTRODUCTION

The halogenated pyrimidines and their related nucleosides and nucleotides are an important class of antiviral [1–3] and antitumor [4] agents. The exact nature of these important pharmacological activities is not completely understood; however, 5-iodo-2'-deoxyuridine, an antiviral agent used for treatment of herpes simplex infections of the eye, is thought to act by forming less stable DNA [1], while 5-fluorouracil, an antitumor drug, may exert its activity by blocking DNA synthesis via inhibition of thymidylate synthetase [5].

Little is known about the biochemical events associated with *in vivo* dehalogenation of the 5-halopyrimidines at the level of either the bases, nucleosides, or nucleotides. Several groups, using both halogenated uracil and halogenated dihydrouracil derivatives, have shown that halide ion and uracil are among the urinary excretion products [6–11]; however, no evidence is available for the dehalogenation of either 5-fluorouracil or any of its nucleosides or nucleotides. In 1970, Cooper and Greer demonstrated the existence of enzyme systems in soluble rat liver preparations that catalyze the release of iodide ion from ^{125}I-labeled 5-iodouracil [12]. This enzyme activity had an absolute requirement for reduced nicotinamide adenine dinucleotide phosphate (NADP) and was inhibited by both 5-cyanouracil and 5-diazouracil. The former inhibitor had previously been shown to inhibit dihydrouracil dehydrogenase, a reduced pyridine nucleotide-linked enzyme responsible for the reversible reduction of the 5,6 double bond of both uracil and thymine [13]. It was also known from the earlier work of Barrett and West [11] that incubation of both 5-bromo-5,6-dihydrouracil and 5-bromo-6-methyl-5,6-dihydrouracil in isotonic phosphate, pH 7.4, 37°C, resulted in the very slow appearance of uracil. Thus, it was postulated that *in vivo* dehalogenation of both 5-iodo- and 5-bromouracil proceeds via NADP-linked, dihydrouracil dehydrogenase-catalyzed reduction of the 5,6 double bond of the halouracil [Eq. (1)] followed by nonenzymatically catalyzed E2 elimination of HX from the halodihydrouracil intermediate [Eq. (2)] to yield uracil as a product that could be detected in the urine [4,5,9,10]. Recently, Wataya and Santi [14]

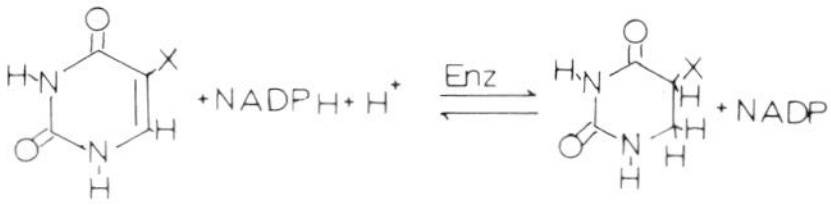

$$\text{(1)}$$

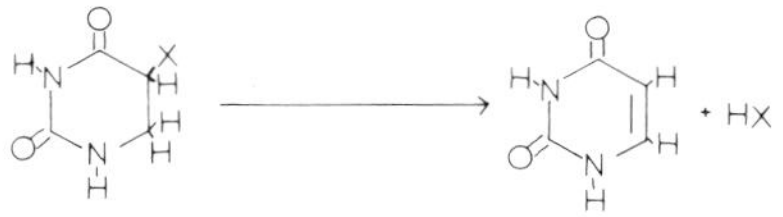

$$\text{(2)}$$

X= —I, —Br

showed that both 5-bromo- and 5-iodo-2′-deoxyuridylate can be enzymatically dehalogenated in the presence of dithiothreitol by thymidylate synthetase from amethopterin-resistant *Lactobacillus casei*. Thus, the halogenated pyrimidines may also be physiologically dehalogenated at the deoxynucleotide level.

The objectives of this review are to summarize fairly recent work on model systems for both halopyrimidine and halodihydropyrimidine dehalogenation and to relate these models to what is known about the enzymatic process.

DEHALOGENATION OF HALOPYRIMIDINES BY BISULFITE BUFFERS

Sodium bisulfite is a common reducing agent which, upon loss of a proton, yields sulfite, a powerful "α-effect" nucleophile that can add to a wide variety of carbonyl compounds as well as to carbon–carbon double bonds to form sulfonic acids. Reactions of the bisulfite buffer system with nucleic acid components were essentially ignored until 1967, when Notari demonstrated that sodium bisulfite dramatically increased the rate of arabinosylcytosine deamination in phosphate buffers, pH 6.9, 70°C [15]. This work supported the mechanism of nucleophile-catalyzed deamination of cytosine, previously proposed by Shapiro and Klein [16], and fairly rapidly led to the elucidation of the reactions shown in Eqs. (3) and (4) for the reversible addition of sulfite

$$(3)$$

$$(4)$$

and a proton to the 5,6 double bond of both uracil and cytosine [17–19]. These reactions have been recently reviewed by Hayatsu [20] and are not covered here except as they relate to the mechanisms of halopyrimidine dehalogenation promoted by sulfur nucleophiles. Pogolotti and Santi have also briefly reviewed halopyrimidine dehalogenation in Chapter 12 of Volume I.

In 1972, work from both our laboratory [21] and Fourrey's [22] demonstrated that dilute sodium bisulfite buffers of near neutral pH could cause the facile dehalogenation of 5-bromo-, 5-iodo-, and 5-chlorouracil at room temperature. Under the same conditions, 5-fluorouracil was not dehalogenated. Our conclusions were based principally on ultraviolet difference

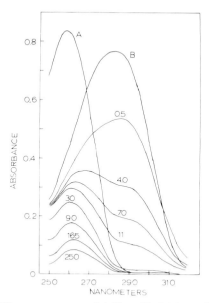

Fig. 1 Ultraviolet difference spectra of 1.01×10^{-4} M uracil in H_2O (A), 1.25×10^{-4} M 5-iodouracil in H_2O (B), and 1.25×10^{-4} M 5-iodouracil in 0.10 M sodium bisulfite, pH 7.10. Numbers on curves refer to the time, in minutes, at which the spectra of the 5-iodouracil–bisulfite reaction mixtures were recorded [21].

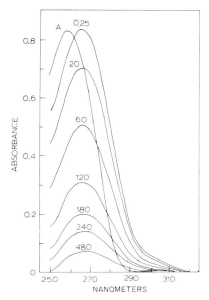

Fig. 2 Ultraviolet difference spectra of $1.01 \times 10^{-4}\ M$ uracil in H_2O (A) and $1.32 \times 10^{-4}\ M$ 5-fluorouracil in $0.10\ M$ sodium bisulfite, pH 7.15. Numbers on curves indicate the time at which the 5-fluorouracil–bisulfite spectra were recorded [21].

spectra of $0.10\ M$ sodium bisulfite buffer (pH 7.10) reacting with approximately $1.0 \times 10^{-4}\ M$ concentrations of the various halogenated pyrimidines measured as a function of time. Figures 1 and 2 show typical spectra for 5-iodouracil and 5-fluorouracil, respectively. The spectra for 5-iodouracil reacting with sodium bisulfite buffer (Fig. 1) show a fairly rapid decrease in 5-iodouracil absorbance at 283 nm concomitant with the appearance of a uracil peak at 259 nm. This absorbance peak then more slowly decays to yield an equilibrium mixture of uracil and 5,6-dihydrouracil 6-sulfonate as the final products of the reaction. In the case of 5-fluorouracil, which does not dehalogenate, no intermediate uracil absorption peak could be detected (Fig. 2). Product identification after treatment with sodium hydroxide, which promotes the elimination of sulfite from 5,6-dihydrouracil 6-sulfonate, showed that the final dehalogenation product of 5-bromo-, 5-iodo-, and 5-chlorouracil was uracil. In the case of 5-fluorouracil, which does not dehalogenate, base-treated reaction mixtures yielded the starting pyrimidine, 5-fluorouracil. On the basis of these data, we proposed the following reaction pathway for the bisulfite-promoted dehalogenation of the 5-halouracils [Eqs. (5)–(7)]:

$$\tag{5}$$

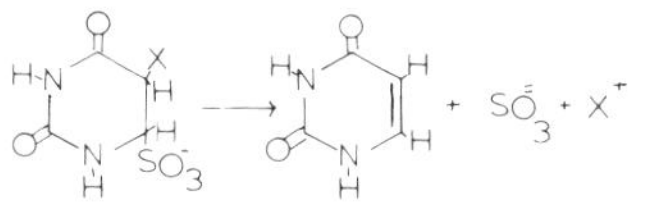

$$\text{(6)}$$

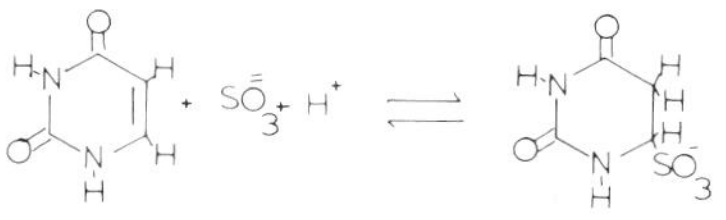

$$\text{(7)}$$

$$X = -Br, -Cl, -I$$

Fourrey [22], studying reactions of 5-bromouridine with sodium bisulfite using nuclear magnetic resonance techniques, could not, however, detect the intermediacy of either 5-bromo-5,6-dihydrouridine 6-sulfonate or uridine as intermediates in the formation of the final 5,6-dihydrouridine 6-sulfonate products. Thus, he proposed a reaction pathway for bisulfite-mediated 5-bromouridine dehalogenation which yielded 5,6-dihydrouridine 6-sulfonate without the formation of a uracil intermediate [Eq. (8)]. Detailed kinetic

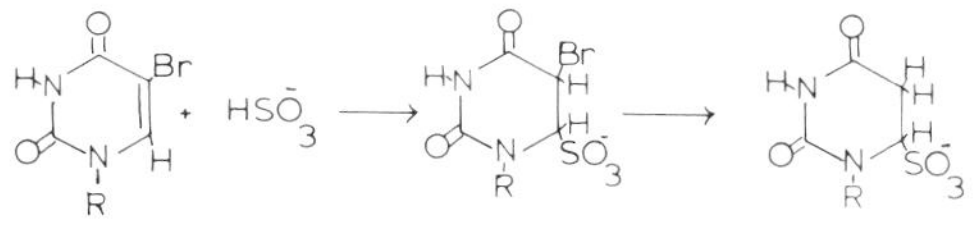

$$\text{(8)}$$

analysis of bisulfite-promoted dehalogenation of 5-bromouracil, 5-iodouracil, 5-bromo-6-methoxy-5,6-dihydrothymine, and 5-bromo-6-methoxy-5,6-dihydrouracil by Rork and Pitman [23,24] and Sedor et al. [25] showed that both of these earlier pathways have some validity in that, in the dehalogenation of the 5-bromo- and 5-iodouracils, the 5-halo-5,6-dihydrouracil 6-sulfonate intermediate does not accumulate but, instead, is present only in steady-state concentrations. Furthermore, both groups showed that uracil, while formed in the reaction, need not be an obligatory intermediate in the formation of 5,6-dihydrouracil 6-sulfonate. These results are discussed more fully later.

The reaction pathway shown in Eqs. (5)–(7) indicates the release of a positively charged halonium ion species, which was considered to be analogous to the mechanisms proposed by Garrett et al. [26,27] for the pH-independent dehalogenation of 5-iodouracil to yield uracil and hypoiodous acid [Eq. (9)].

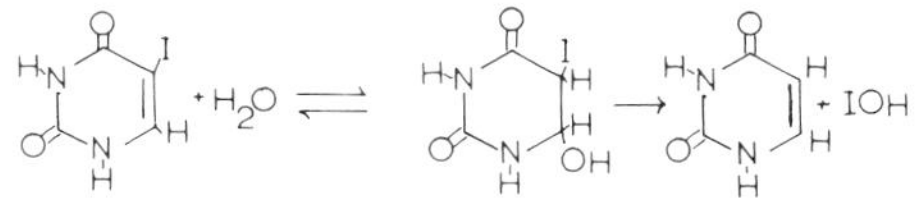

$$\text{(9)}$$

Since halonium ions are electron-deficient species, the reaction of 5-bromouracil was further investigated by Sedor and Sander [28] to determine any additional role for bisulfite buffers other than the formation of 5-halo-5,6-dihydrouracil 6-sulfonate, shown in Eq. (5). Kinetic examination of this

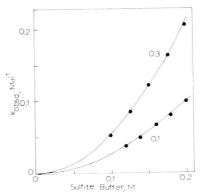

Fig. 3 Relationship between the observed pseudo first-order rate constants for the dehalogenation of 5-bromouracil and increasing bisulfite buffer concentration, 25°C, ionic strength 1.0 M. Numbers on curves refer to the fraction bisulfite in the buffers [28].

reaction showed that the pseudo first-order rate constants for the reaction of bisulfite buffers with 5-bromouracil had a second-order dependence on bisulfite buffer, thus indicating that 2 moles of either sulfite or bisulfite were involved in the overall dehalogenation reaction (Fig. 3). These kinetic observations were substantiated by iodometric titrations of bisulfite buffer utilization which, along with measurement of 5,6-dihydrouracil, uracil, and sulfate formation, showed that, in addition to the mole of sulfite required to form the 5,6-dihydrouracil 6-sulfonate product, another mole of sulfite was converted to sulfate during the course of the bisulfite-mediated 5-bromouracil dehalogenation. Thus, the overall stoichiometry for the reaction of 5-bromouracil with bisulfite buffers could be formulated as shown in Eq. (10). The

$$\text{(structure)} + 2\mathrm{S\bar{O}_3} + \mathrm{H_2O} \;=\; \text{(structure)} + \mathrm{S\bar{O}_4} + \mathrm{Br^-} \tag{10}$$

sulfite molecule converted to sulfate during the reaction was believed to act either directly [Eq. (11)] or via an intervening water molecule [Eq. (12)] by attacking the halogen atom on the 5-bromo-5,6-dihydrouracil to yield either bromosulfonic acid or hypobromous acid as an intermediate. Either of these

$$\text{(structure)} \longrightarrow \text{(structure)} + \mathrm{BrS\bar{O}_3} + \mathrm{S\bar{O}_3} \tag{11}$$

$$\text{(structure)} \longrightarrow \text{(structure)} + \mathrm{BrOH} + \mathrm{HS\bar{O}_3} + \mathrm{S\bar{O}_3} \tag{12}$$

intermediates reacts further by known reactions to yield bromide and sulfate as final products, according to the stoichiometry shown in Eq. (10).

In all of the work described above, it was assumed that 5-halo-5,6-dihydrouracil 6-sulfonate is a discrete intermediate in the overall dehalogenation of 5-iodo-, 5-bromo-, and 5-chlorouracil. This assumption was reasonable based on the fact that 5,6-dihydrouracil 6-sulfonate is an isolable product from reaction mixtures containing bisulfite buffers and uracil; however, this important point was not proved until Rork and Pitman [23] showed the accumulation of 5-chloro-5,6-dihydrouracil 6-sulfonate by measuring the spectral changes associated with the first several minutes of the reaction of 5-chlorouracil with bisulfite buffers (Fig. 4). Thus, by analogy, this work indicates the existence of similar intermediates in the dehalogenation of 5-bromo- and 5-iodouracil.

The kinetics of the overall dehalogenation of 5-iodo-, 5-chloro-, and 5-bromouracil by bisulfite buffers are difficult to interpret in terms of detailed mechanism because of the multitude of steps that can control the rate of the reaction under differing experimental conditions. Work in Pitman's laboratory and in ours attempted to circumvent these problems by detailed studies of parts of the reaction, namely, the addition of bisulfite buffers to uracils [29,30] and to 5-fluorouracil [31], which cannot dehalogenate, and the

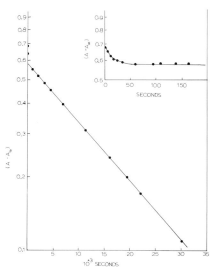

Fig. 4 Semilogarithmic relationship between extent reaction ($A - A_\infty$, 295 nm) and time for the reaction of 5-chlorouracil with 0.45 M sodium bisulfite, pH 6.50, 25°C, ionic strength 1.0 M [23]. Reprinted with permission from *J. Am. Chem. Soc.* **97**, 5559 (1975). Copyright by the American Chemical Society.

bisulfite buffer-mediated dehalogenation of a series of halodihydrouracils [24], which included 5-bromo-6-methoxy-5,6-dihydrouracil, a compound that can be considered a model for the 5-halo-5,6-dihydrouracil 6-sulfonate thought to be an intermediate in halopyrimidine dehalogenation.

The addition of sulfite and protons to the uracil ring system is a reversible reaction. Hence, catalysis of the process can be kinetically studied in either direction. In 1972, Erickson and Sander [29] showed that the elimination of sulfite from 1,3-dimethyl-5,6-dihydrouracil 6-sulfonate, at 25°C, $\mu = 1.0\ M$, was subject to general base catalysis of proton transfer by a series of amines ($\beta = 0.87$). Furthermore, the reaction was subject to a kinetic hydrogen–deuterium isotope effect of 4.1, which indicated that proton abstraction was from a carbon atom, such as C-5, of the dihydropyrimidine ring system rather than from an electronegative atom, such as either oxygen or nitrogen. Rork and Pitman [30] further studied this reaction over a far wider amine buffer concentration range than that employed by Erickson and Sander [29] and thus made the important observation that the observed pseudo first-order rate constants for the general-base-catalyzed elimination of sulfite from both the 6-sulfonates of uracil and 1,3-dimethyluracil increase and then become invariant as a function of increasing morpholine buffer concentration (Fig. 5). Such behavior implies that sulfite elimination from the 6-sulfonates of uracil

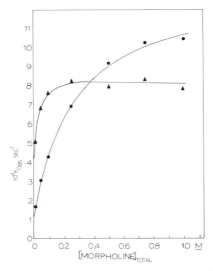

Fig. 5 Nonlinear relationship between the observed first-order rate constants for sulfite elimination from 6-sulfonate adducts of uracil (▲) and 1,3-dimethyluracil (●) and increasing morpholine buffer concentration, pH 8.77, 25°C, ionic strength 1.0 M [30]. Reprinted with permission from *J. Am. Chem. Soc.* **96**, 4654 (1974). Copyright by the American Chemical Society.

and 1,3-dimethyluracil is a multistep process with proton abstraction by general bases occurring in a step that is discrete from the elimination of SO_3^{2-} from an enolate anion, such as that shown in Eq. (13). Sedor *et al.* [31] confirmed this observation by showing similar behavior for the general-acid-catalyzed addition of sulfite to 5-fluorouracil measured as a function of increasing bisulfite and bis(2-hydroxyethyl)iminotris(hydroxymethyl)methane hydrochloride concentrations. Thus, the mechanism of the addition of sulfite and protons across the 5,6 double bond of the uracils is a multistep rather than a concerted process, which can be formulated as shown in Eq. (13).

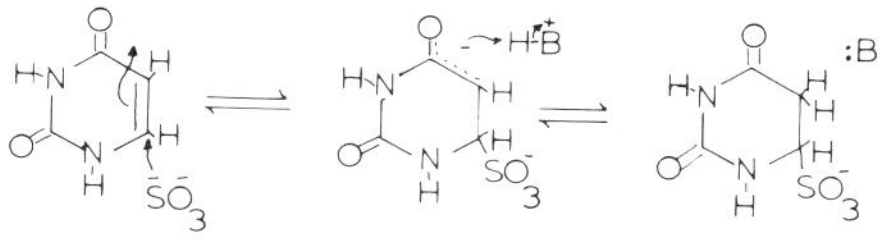

$$\text{(13)}$$

While this mechanism satisfies the buffer catalysis data [29–31], the site of general catalyzed proton donation and abstraction to the enolate anion, shown in Eq. (13), is not clear. In the elimination of sulfite from both the 6-sulfonates of 1,3-dimethyluracil [29] and from 5-fluorouracil [31], a hydrogen–deuterium isotope effect of about 4 is observed using ethanolamine buffers, a result that argues for proton transfer to C-5 rather than to the oxygen atom on C-4.

In contrast to the elimination reaction, a small solvent hydrogen–deuterium isotope effect of about 1.1 is observed for the bisulfite-catalyzed addition of sulfite to both 5-fluorouracil [31] and 5-bromouracil [25], a result that could argue for proton transfer to the oxygen at C-4 followed by a rapid tautomerization to yield the final 5,6-dihydrouracil 6-sulfonate product. Clearly, more research involving the measurement of isotope effects for the reaction in both directions, using the same buffer system and experimental conditions, is required to gain further insight into this problem.

The second major step in the bisulfite-mediated dehalogenation of the 5-halouracils involves the role of sulfite in the actual dehalogenation of the 5-halo-5,6-dihydrouracil 6-sulfonates [Eqs. (11) and (12)]. Rork and Pitman [24] carefully examined the reactions of bisulfite buffers with a series of halogenated 5,6-dihydrouracils, shown below (**I–VII**):

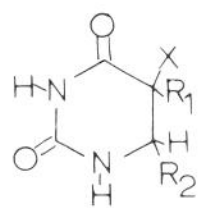

Compound	X	R_1	R_2
I	—Br	—H	—OCH$_3$
II	—Br	—CH$_3$	—OCH$_3$
III	—Cl	—H	—OCH$_3$
IV	—Cl	—CH$_3$	—OCH$_3$
V	—Br	—H	—H
VI	—I	—H	—H
VII	—Br	—H	—CH$_3$

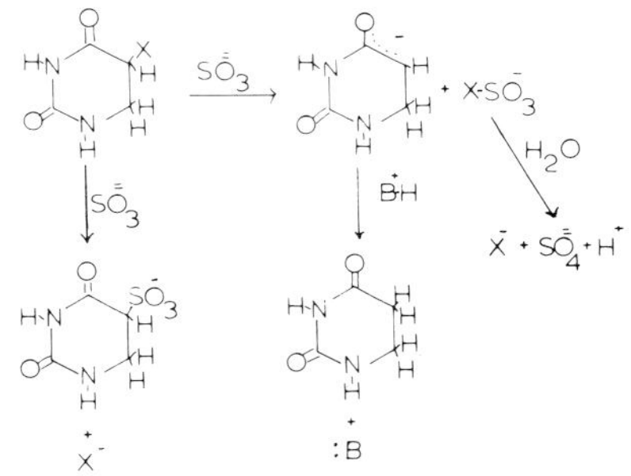

Scheme 1 X = —Br, —I.

These important studies illustrate the following points. First, the dehalogena-tion of 5-bromo-6-methoxy-5,6-dihydrouracil (**I**) by bisulfite buffers yields uracil and 6-methoxy-5,6-dihydrouracil as initial products, along with methanol, bromide, sulfate, and protons. Second, the product ratio (uracil: 6-methoxy-5,6-dihydrouracil) of the reaction of 5-bromo-6-methoxy-5,6-dihydrouracil (**I**) with bisulfite is dependent on the concentration of general acids, with uracil being the dominant product at low buffer concentrations. In this respect, similar results have been demonstrated for the dehalogenation of the 5-halouracils by bisulfite in that the final yield of 5,6-dihydrouracil 6-sulfonate is increased relative to uracil as a direct function of potential general acid concentration [23,25]. Third, the reaction of the simple 5-halo-5,6-dihydrouracils (**V** and **VI**), which do not have a methoxy leaving group at C-6, with bisulfite buffers gives both 5,6-dihydrouracil and 5,6-dihydro-uracil 5-sulfonate along with halide, sulfate, and protons as products.

These results allowed Rork and Pitman to conclude that two potential mechanisms were available for the dehalogenation of the 5-halo-5,6-dihydro-uracils by bisulfite. The mechanisms differ depending on the presence of a leaving group at C-6 of the dihydropyrimidine. The first of these mechanisms, shown in Scheme 1, relates to the 5-halo-5,6-dihydrouracils (**V** and **VI**) in which part of the reaction goes via an S_N2 displacement of halide to yield 5,6-dihydrouracil 5-sulfonate as a product, while in another reaction SO_3^{2-} attacks the halogen atom directly to yield a halosulfonic acid and the enolate anion of 5,6-dihydrouracil, which subsequently adds a proton to give the final dihydrouracil product. The nature of the halogen atom at C-5 appears to have an effect on the relative contribution of these two pathways as the S_N2 pathway decreases in significance in going from a bromo substituent to an iodo substituent, a result that may be due, in part, to greater ease of reduction of the iodo compound and ring conformation [24]. The second of these mechanisms, shown in Scheme 2, applies to the dehalogenation of halo-dihydropyrimidines which have a methoxy leaving group at C-6. In this

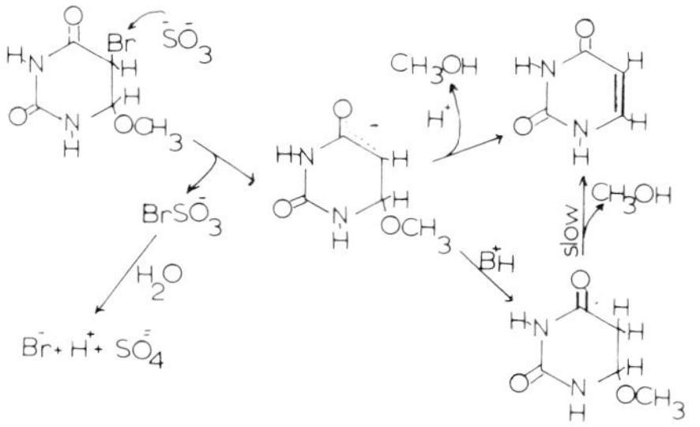

Scheme 2

mechanism, sulfite attacks the halogen directly to yield as intermediates a halosulfonic acid and the enolate anion of 6-methoxy-5,6-dihydrouracil. The enolate anion then either eliminates the methoxy group at C-6 to yield uracil or, in a buffer-dependent step, adds a proton to yield 6-methoxydihydrouracil, thus accounting for the fact that the product ratio (uracil:6-methoxy-5,6-dihydrouracil) is dependent on the concentration of general acids. These workers [24] concluded that the mechanism shown in Scheme 2 for the bisulfite-mediated dehalogenation of 5-bromo-6-methoxy-5,6-dihydrouracil was most applicable to the dehalogenation of 5-halo-5,6-dihydrouracil 6-sulfonate, the intermediate involved in halouracil dehalogenation, because the large sulfite group at C-6 would both sterically inhibit the attack of another sulfite at C-5 (S_N2 process) and promote the attack of sulfite on the C-5 halogen by an inductive effect.

Further work by Rork and Pitman [23] and from our laboratory [25] helped to promote a better understanding of the bisulfite-mediated dehalogenation of the 5-halouracils. First, only in the case of 5-chlorouracil did an appreciable amount of the 5-halo-5,6-dihydrouracil 6-sulfonate intermediate accumulate [23]. Second, the relationship between the pseudo first-order rate constants for 5-bromouracil dehalogenation changes from a second-order to an almost first-order dependence on total bisulfite buffer concentrations as a function of increasing total bisulfite [25]. Third, in the bisulfite buffer concentration range where the reaction of 5-bromouracil is second order with respect to total bisulfite, the relative contributions of two different reactions to the overall dehalogenation rate are influenced by fraction bisulfite (α) in the various bisulfite buffers employed. This was illustrated by the fact that the reaction exhibits a small kinetic hydrogen–deuterium isotope effect of 1.23 at $\alpha = 0.20$, but not at $\alpha = 0.80$, that sensitivity to added general acids (imidazolium ion) increases with decreasing fraction bisulfite, and that, relative to perchlorate ion, chloride, bromide, and sulfate enhance the reaction rate maximally at high values of fraction bisulfite [25]. This effect of rate enhancements caused by specific anions, such as chloride and perchlorate, was also observed by Rork and Pitman, who ascribed these rate differences to changes in the structure of water and hence to changes in the activity coefficients of the reactants [23,24]. Consequently, reactions similar to that proposed by Wang [32] for the halide ion- and proton-promoted dehalogenation of several 5,5-dibromo-6-hydroxy-5,6-dihydrouracils [Eq. (14)] probably do not affect the bisulfite-mediated dehalogenation of the 5-halouracils.

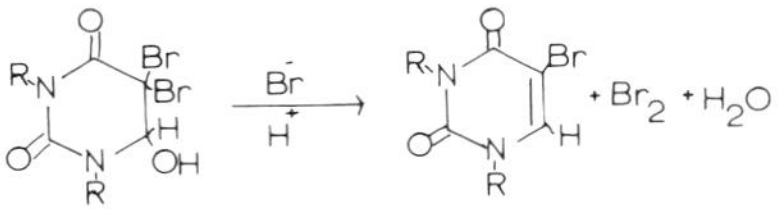

$$\tag{14}$$

Thus, based principally on the work outlined above from our laboratory and from Pitman's laboratory, the more detailed reaction pathway shown in Scheme 3 can be formulated for the bisulfite-mediated dehalogenation of the 5-halouracils.

Recently, Hayatsu et al. [33] reported that the bisulfite-mediated dehalogenation of N-1-substituted 5-bromouracils, such as 5-bromo-2'-deoxyuridine and 1-methyl-5-bromouracil, is several orders of magnitude slower than the reaction of 5-bromouracil under similar conditions. To explain the faster reaction of the unsubstituted 5-bromouracil, they have proposed a mechanism involving sulfite elimination across the N-1—C-6 bond of 5-bromo-5,6-dihydrouracil 6-sulfonate prior to debromination (Scheme 4), a mechanism not available to the N-1-substituted uracils because of the difficulty associated with quaternization of the nitrogen at position 1 of the ring system. This pathway may have some validity; however, it must be noted that the steric effects associated with N-1 substitution of the 5-bromouracil could be invoked to explain the rate differences observed by these workers without

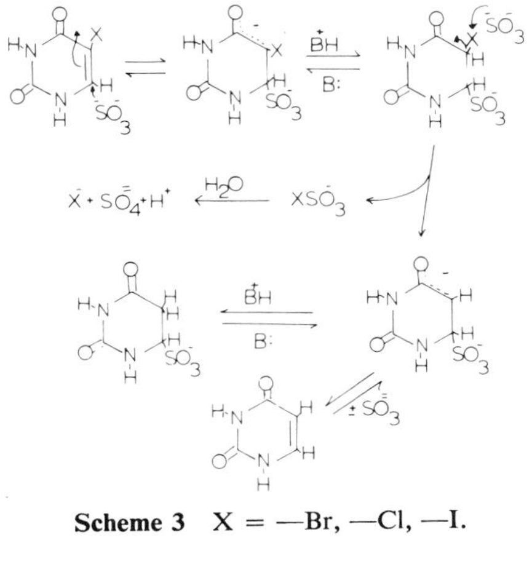

Scheme 3   X = —Br, —Cl, —I.

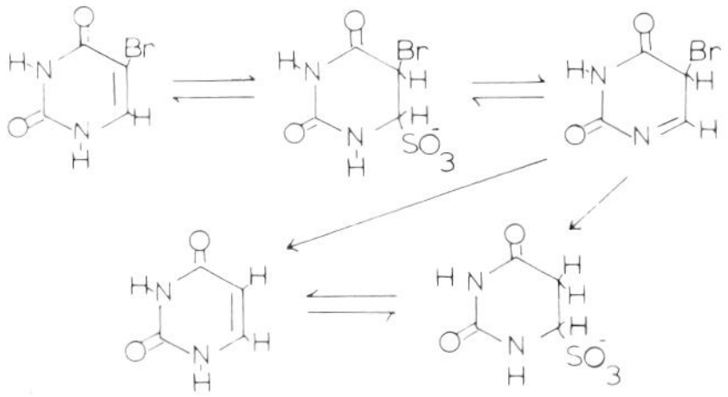

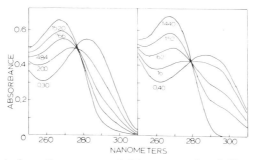

Scheme 4 From Hayatsu *et al.*[33].

invoking a separate pathway of dehalogenation. Obviously, this point needs
further clarification.

Both 5-iodo- and 5-bromocytosine are dehalogenated by dilute bisulfite
buffers [34] under conditions in which bisulfite-mediated cytosine deamination
is minimal [35,36]. Figure 6 shows the change in ultraviolet absorption spectra

Fig. 6 Ultraviolet absorption spectra of 5-bromocytosine (left) and 5-iodocytosine
(right) reacting with 0.05 *M* sodium bisulfite in 0.05 *M* sodium phosphate buffer,
pH 7.18, 22°–24°C, ionic strength 1.0 *M* [34].

observed when either 5-iodo- or 5-bromocytosine is allowed to react at room
temperature (22°–24°C) with 0.05 *M* bisulfite buffers, pH 7.18. Under these
conditions, cytosine is the major pyrimidine product, and 1 mole of sulfite is
consumed to form sulfate per mole of 5-halocytosine dehalogenated. The
kinetic behavior of the reactions varies depending on the 5-halo substituent;
however, a pathway similar to that shown in Scheme 3 for bisulfite-promoted
halouracil dehalogenation can be supported (Scheme 5). With 5-bromo-
cytosine, but not with 5-iodocytosine, extrapolation of semilogarithmic plots
of extent reaction vs. time indicates the bisulfite buffer concentration-
dependent formation of an intermediate, most likely the conjugate acid of
5-bromo-5,6-dihydrocytosine 6-sulfonate, which subsequently reacts to
control the overall rate of 5-bromocytosine dehalogenation. The kinetics of
disappearance of both of the 5-halocytosines have a second-order dependence
on bisulfite buffer concentration at pH 7.22, and the pH optimum of the

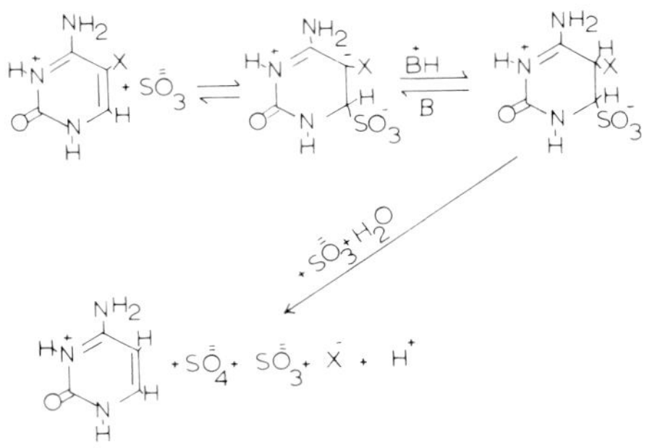

Scheme 5 X = —Br, —I.

reaction of about 4.5 argues for the participation of the conjugate acid of cytosine rather than neutral cytosine, a conclusion similar to that drawn for the case of bisulfite-mediated cytosine deamination [36]. The dehalogenation of 5-bromocytosine is insensitive to added general acids, such as imidazolium ion and acetic acid, indicating that proton transfer is not a kinetically significant feature of the rate-determining step of this dehalogenation reaction. Thus, the dehalogenation of the conjugate acid of 5-bromo-5,6-dihydrocytosine 6-sulfonate probably controls the rate of this reaction. In the case of 5-iodocytosine dehalogenation, the reaction has a marked sensitivity to the addition of both imidazole and acetate buffers. At pH 6.86 and 7.21, the equality of slopes of plots of k_{obsd} against increasing imidazolium ion indicates that the reaction of 5-iodocytosine with sulfite, like the 5-halouracils, is subject to general acid catalysis of proton transfer. In the case of acetate buffer catalysis of bisulfite-mediated 5-iodocytosine dehalogenation, the values of k_{obsd} increase and then become invariant with increasing concentrations of acetate buffer. This result argues for an acetate buffer-dependent change in rate-determining step for the dehalogenation of 5-iodocytosine by bisulfite buffer. Based on the fact that sulfite addition to the 5-halouracils is discrete from proton transfer to form 5-halo-5,6-dihydrouracil 6-sulfonate and that dehalogenation of the conjugate acid of 5-iodo-5,6-dihydrouracil 6-sulfonate appears to be fast relative to its formation (e.g., it does not accumulate), it might be concluded that sulfite addition to the conjugate acid of 5-iodocytosine is separate from proton transfer to the resulting anion, thus explaining the change in rate-determining step as a function of acetate buffer. Preliminary evidence from our laboratory on buffer catalysis of sulfite elimination from 3-methyl-5,6-dihydrocytidine 6-sulfonate [37] tends to support this view; however, concerted general-acid-catalyzed addition of sulfite to the conjugate acid of the cytosines cannot, without further research, be eliminated as a mechanistic possibility.

DEHALOGENATION OF HALOPYRIMIDINES
BY SIMPLE THIOL COMPOUNDS

The reaction of bisulfite buffers and other nucleophiles with both substituted and unsubstituted pyrimidines is of considerable interest when one
considers the fact that these reagents are mutagenic and that the reactions can
be used to modify nucleic acid components [20]. Similar reactions of simple
thiol compounds, however, may be more relevant to enzymatic catalysis
because this functionality can be an integral component of the active site of
an enzyme. Indeed, thiol-promoted C-5 hydrogen–deuterium exchange reactions [38–40], which presumably proceed via reversible addition of thiol
anions and protons across the 5,6 double bond of various uracil derivatives
[Eq. (15)], have been valuable in terms of deducing the mechanism of action
of enzymes such as thymidylate synthetase, which also catalyzes hydrogen–
deuterium exchange at C-5 of deoxyuridylate [41] and forms a covalent
enzyme–5-fluoro-2-deoxy-5,6-dihydrouridylate adduct upon incubation with
5-fluoro-2-deoxyuridylate [42–48].

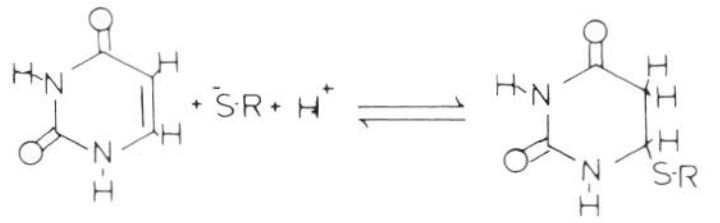

$$(15)$$

In the case of simple thiols reacting with pyrimidine bases, it has not been
possible to isolate and characterize thioether containing 5,6-dihydropyrimidine
adducts similar to 5,6-dihydrouracil 6-sulfonate, probably because the
equilibrium constants for thiol addition are so small that product isolation is
not possible. It has been possible, however, to isolate and characterize the
product resulting from the intramolecular attack of a thiol on C-6 of a uracil
ring system [49–51]. Equation (16) shows the formation of 5'-deoxy-5',6-*epi*-
thio-5,6-dihydro-2',3'-*o*-isopropylideneuridine resulting from the intramolecular attack of the 5'-thiol group on C-6 of the uracil ring system of
5'-deoxy-5'-thio-2',3'-*o*-isopropylideneuridine [49].

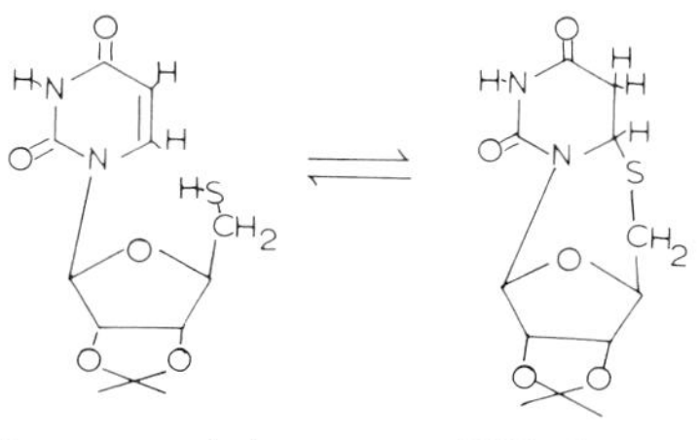

$$(16)$$

In early 1973, work from our laboratory [52] demonstrated the ability of
simple thiol compounds, such as cysteine and 2-mercaptoethanol, to dehalogenate 5-iodo- and 5-bromouracil under mild conditions of temperature and

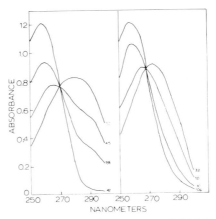

Fig. 7 Ultraviolet absorption spectra of 5-bromouracil (right) and 5-iodouracil (left) reacting with 0.20 M cysteine, 23°C, ionic strength 1.0 M, pH 7.72, after quenching by 3× dilution with 1.0 M HCl. Numbers next to curves refer to time of quenching in minutes [50].

pH. Figure 7 shows the ultraviolet absorption spectra obtained upon HCl quenching of reaction mixtures containing cysteine (pH 7.72) and either 5-iodo- or 5-bromouracil. Thin-layer chromatography and measurement of spectral ratios of the products showed that uracil was the final pyrimidine product of the reaction. On the basis of these observations, we proposed that the reactions proceeded via a 5-halo-6-cysteinyl-5,6-dihydrouracil intermediate in a manner similar to bisulfite-mediated 5-halopyrimidine dehalogenation. Later that same year, Wataya et al. [53] confirmed and extended our results by showing that both 5-bromo-2'-deoxyuridine and 5-bromouracil were dehalogenated by cysteine, yielding both 2'-deoxyuridine (uracil) and 5-cysteinyl-2'-deoxyuridine (5-cysteinyluracil) as final products. This product ratio varied with both pH and cysteine concentration. The 5-cysteinyl derivatives were favored at lower total cysteine concentrations and at higher pH values. These 5-substituted uracil products were not observed by Sedor and Sander [52], perhaps because they passed their reaction mixtures through a mixed-bed ion-exchange column to remove excess cysteine prior to pyrimidine product analysis, a procedure that would probably remove the highly charged 5-cysteinyluracil product.

On the basis of their results, Wataya et al. [53] proposed the reaction pathway shown in Eqs. (17)–(19) for the cysteine-promoted dehalogenation of 5-bromouracil derivatives. This reaction pathway involves the initial formation of 5-bromo-6-cysteinyl-5,6-dihydrouracil (**VIII**), followed by a cysteine thiol anion S_N2 displacement of bromide to yield 5,6-dicysteinyl-5,6-dihydrouracil (**IX**). This dicysteinyl intermediate (**IX**) was then predicted either to be attacked at the C-5 sulfur atom by another mole of cysteine thiol anion to yield uracil and cystine [Eq. (18)] or to be subject to the intramolecular general-base-catalyzed elimination of the cysteine at C-6 to yield 6-cysteinyl-uracil as a final product [Eq. (19)]. Szabo *et al.* [54] proposed a similar mechanism for the dehalogenation of 5-bromouracil promoted by sodium hydrogen sulfide. Since this reaction pathway suffers from some large steric and electrostatic problems associated with both the formation and subsequent reaction of the 5,6-dicysteinyl-5,6-dihydrouracil (**IX**), further work is required to test its general applicability to thiol-promoted 5-halouracil dehalogenation.

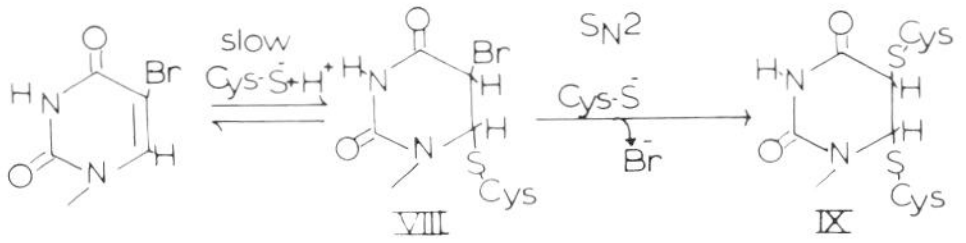

$$\text{(17)}$$

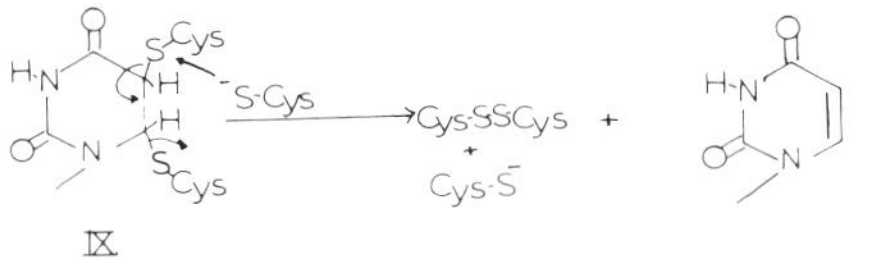

$$\text{(18)}$$

$$\underline{\text{OR}}$$

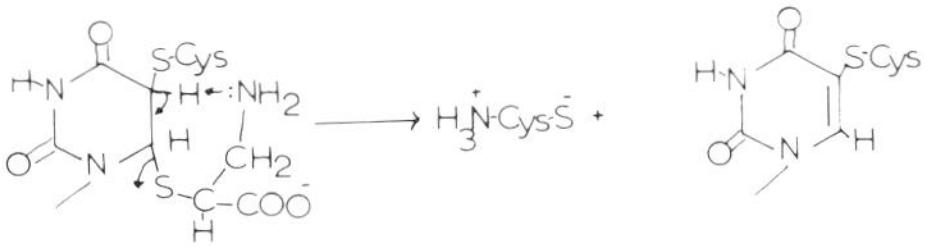

$$\text{(19)}$$

To further elucidate the reactions involved in thiol-promoted 5-halo-pyrimidine dehalogenation, the kinetics of both cysteine and 2-mercapto-ethanol reacting with 5-iodouracil were studied by Sedor *et al.* [55]. This work showed that the rate of the overall dehalogenation reaction depends on the concentrations of neutral 5-iodouracil and the species of cysteine ^-OOC—$(NH_3{}^+)$—CH—CH_2—S^- (^-S—Cys—$NH_3{}^+$), and 2-mercapto-ethanol (^-S—CH_2—CH_2—OH). The reaction with cysteine has a strict first-order dependence on ^-S—Cys—$NH_3{}^+$ concentration, and the kinetics are compatible with a rate law which assumes that the attack of ^-S—Cys—$NH_3{}^+$ on 5-iodouracil is rate determining, a conclusion that agrees with results from Hayatsu's laboratory [53]. The reaction of 2-mercaptoethanol with 5-iodouracil is subject to catalysis by general acids, such as imidazole · HCl and tris(hydroxymethyl)aminomethane · HCl, while the reaction of cysteine is not subject to catalysis by added buffers. Solvent hydrogen–deuterium isotope effects were measured for the addition of both thiol anions to 5-iodouracil. In the case of 2-mercaptoethanol, an isotope effect of 1.13 was observed, a value that is comparable to the previously discussed work on the general-acid-catalyzed addition of sulfite to both 5-fluoro- and 5-bromouracil. The results with cysteine and 5-iodouracil were strikingly different. This reaction, which was not catalyzed by external buffers, exhibited a solvent hydrogen–deuterium isotope effect of 4.10. An isotope effect of this magnitude has not previously been observed for the addition of a sulfur nucleophile to a pyrimidine base but is of exactly the same magnitude as isotope effects observed for the general-base-catalyzed elimination of sulfite from the sulfite–dihydropyrimidine adducts of 1,3-dimethyluracil [29] and 5-fluorouracil [31]. The magnitude of this solvent hydrogen–deuterium isotope effect and the fact that the addition of $^-$S—Cys—NH$_3^+$ to 5-iodouracil is not catalyzed by external buffers can be explained in terms of either a concerted [Eq. (20)] or a stepwise [Eq. (21)]

$$(20)$$

$$(21)$$

addition of cysteine to C-6 of the uracil ring with the protonated α-amino group acting as an intramolecular general acid catalyst. No real distinction can be made between these mechanisms; however, the magnitude of the isotope effect (4.1) does argue for proton transfer to a carbon atom. Furthermore, since other sulfur nucleophiles, such as sulfite and 2-mercaptoethanol, add to pyrimidines via stepwise, external general-acid-catalyzed mechanisms with isotope effects of about 1.1, it might be argued that the addition of cysteine, a molecule that also contains a potential general acid and exhibits an isotope effect of 4.1, might proceed via the concerted process shown in Eq. (20). As previously pointed out in the discussion on the bisulfite system, further research is required to completely explain these solvent isotope effects.

Kinetic analysis of the overall dehalogenation of the 5-halouracils with either cysteine, 2-mercaptoethanol, or other organic thiol compounds is limited by the fact that the rate of thiol anion and proton addition to the 5,6 double bond of the 5-halouracils controls the rate of reaction [53,55] and hence cannot yield much information about the subsequent dehalogenation of the proposed 5-halo-6-cysteinyl-5,6-dihydrouracil intermediate.

To avoid this difficulty, we have recently studied the cysteine-promoted dehalogenation of 5-bromo-6-methoxy-5,6-dihydrothymine [56]. Rork and Pitman [24] considered 5-bromo-6-methoxy-5,6-dihydrouracil a model for 5-bromo-5,6-dihydrouracil 6-sulfonate, the intermediate in the bisulfite-mediated dehalogenation of 5-bromouracil. Thus, by analogy, 5-bromo-6-methoxy-5,6-dihydrothymine can be considered a model for the intermediate,

5-bromo-6-cysteinyl-5,6-dihydrouracil, proposed to be involved in cysteine-promoted 5-bromouracil dehalogenation. The dihydrothymine rather than the dihydrouracil analog was selected for this purpose to avoid base-catalyzed 5,6-elimination of methanol as a side reaction which might occur at higher pH values. These studies illustrated the following about the cysteine-promoted dehalogenation of 5-bromo-6-methoxy-5,6-dihydrothymine (BrMDHT). First, thymine formation from BrMDHT is much faster than cysteine addition to either 5-iodo- or 5-bromouracil. Second, the reaction has a strict first-order dependence on cysteine thiol anion with a second-order rate constant equal to 1652 ± 77 M^{-1} min^{-1} at 25°C, $\mu = 1.0$ M. Third, the products of the reaction in an argon atmosphere are thymine, cystine, bromide, and cysteic acid. Fourth, thiol titration data indicated that, under conditions of excess cysteine, 2 moles of cysteine are consumed per mole of BrMDHT dehalogenated. In addition, about 88% of the added BrMDHT is converted to thymine, which argues for the formation of 6-methoxy-5,6-dihydrothymine as a product. Fifth, under conditions of excess BrMDHT relative to cysteine, the ratios of thymine produced to cysteine initially present are consistently above 0.50. Sixth, the rate of reaction is not sensitive to catalysis by external buffer systems.

Two considerably different reaction pathways have been proposed for the sulfur nucleophile-promoted dehalogenation of the 5-halouracils. Both involve the initial formation of a 5-halo-5,6-dihydropyrimidine adduct; however, vastly different mechanisms have been proposed for the subsequent

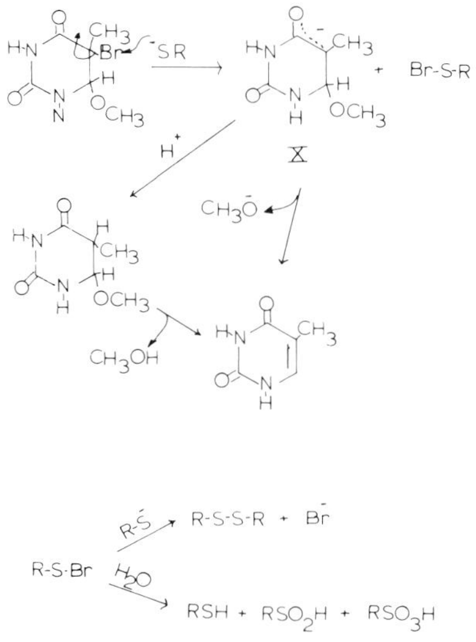

Scheme 6 E2 Hal.

dehalogenation of this 6-substituted 5-halo-5,6-dihydropyrimidine intermediate. Both of these mechanisms are illustrated for BrMDHT dehalogenation by cysteine. Scheme 6 shows the E2 Hal mechanism proposed to be
involved in the bisulfite buffer- and cysteine-mediated dehalogenation of the
5-halouracils [23–25,56], the thiophenoxide anion debromination of 1,3-
dibromocyclohexane [57], and, partially, the thiolphenol-promoted dehalogenation of 5-bromouracil and 5-bromoisocytosine [58]. Scheme 7 shows the
S_N2 mechanism proposed by Wataya *et al.* [53] for the cysteine-mediated
dehalogenation of 5-bromouracil derivatives. None of the data from our
laboratory [56] can eliminate the S_N2 pathway proposed by Wataya *et al.* [53]
and illustrated in Scheme 7 for BrMDHT dehalogenation as a mechanistic
possibility. The kinetic dependence of the reaction of cysteine thiol anion, the
observation that cystine is the major sulfur-containing product, the 2:1
relationship between cysteine and BrMDHT utilization under conditions of
excess thiol, and the fact that 88% of the BrMDHT is converted to thymine
can be accommodated by both the E2 Hal (Scheme 6) and the S_N2 (Scheme 7)
pathways. However, under conditions in which the reaction is run in limiting
cysteine relative to BrMDHT concentration, the S_N2 mechanism would
predict that the ratio of thymine produced relative to cysteine utilized could
not exceed 0.50, a restriction not placed on the E2 Hal mechanism because
the second mole of cysteine thiol anion required for cystine formation is
involved in the reaction with a sulfenyl bromide (Scheme 6) and not with
5-cysteinyl-6-methoxy-5,6-dihydrouracil (**XI**), an intermediate directly in-

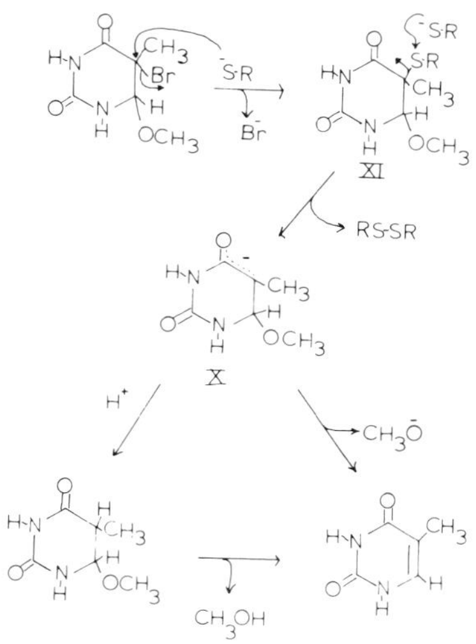

Scheme 7 S_N2.

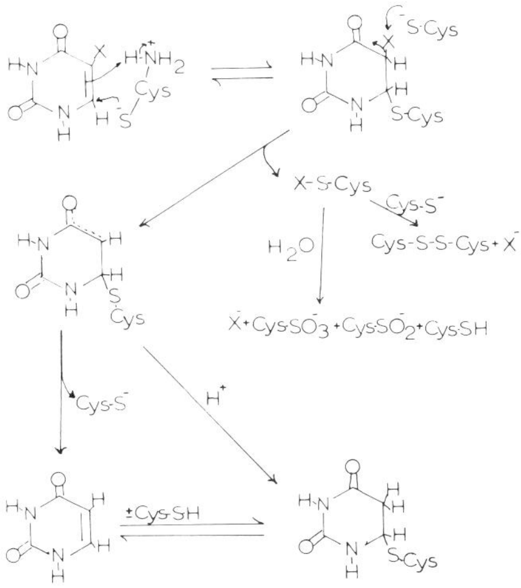

Scheme 8

volved in thymine formation (Scheme 7). In addition, only the E2 Hal mechanism (Scheme 6) can account for the formation of cysteic acid as a product via the reaction of the sulfenyl bromide with water. Consequently, the E2 Hal mechanism most be involved, at least in part, in the cysteine-promoted dehalogenation of BrMDHT.

Thus, in accord with the more firmly established mechanism for 5-halouracil dehalogenation by bisulfite, the E2 Hal mechanism for cysteine-promoted 5-halouracil dehalogenation must be considered. Both the E2 Hal mechanism proposed in Scheme 8 and the S_N2 process proposed by Wataya *et al.* [53] may be operative; however, based on the steric problems illustrated in Eqs. (17)–(19) and the data presented by Rork and Pitman [24], which argue against involvement of the S_N2 process in the sulfite-promoted dehalogenation of 5-halodihydropyrimidines that have a leaving group at C-6, the S_N2 process might be more restricted in its general applicability. More work is required to determine the importance of these two pathways using different 5-halo substituents and experimental conditions.

RELATIONSHIP BETWEEN SULFUR NUCLEOPHILE AND ENZYMATIC HALOPYRIMIDINE DEHALOGENATION

The postulated mechanism for the *in vivo* dehalogenation of the 5-halouracils [Eqs. (1) and (2)] cannot be considered completely valid because, under the same experimental conditions of temperature, pH, etc., we [59] have observed that the rate of uracil formation from 5-iodo-5,6-dihydrouracil is about four orders of magnitude too slow to account for the enzymatic release of $^{125}I^-$ from [5-^{125}I]uracil observed by Cooper and Greer [12]. Principally on the basis of reactions of bisulfite buffers and cysteine with 5-halouracils, we [55] consequently proposed a possible enzymatic mechanism that could accommodate the NADP requirement for enzymatically catalyzed iodide release from 5-iodouracil, observed by Cooper and Greer [12], without the involvement of the nonenzymatically catalyzed E2 elimination of HI [Eq. (2)] to yield uracil. This hypothetical enzymatic process (Scheme 9) involves an active site that contains, in addition to two thiol anions, a potential general acid.

The first step in this postulated enzymatically catalyzed mechanism involves the formation of a covalent enzyme–5-halo-5,6-dihydropyrimidine inter-mediate (**XII**) via a mechanism similar to the concerted addition of cysteine to 5-iodouracil. This intermediate (**XII**) is similar to the covalent intermediate involved in thymidylate synthetase catalysis. The second step then involves nucleophilic attack of the second active-site thiol anion on the halogen atom via the E2 Hal process, yielding uracil as a product and an enzyme–sulfenyl halide (**XIII**), which eliminates iodide ion, resulting in oxidized enzyme (**XIV**). Reduction of the oxidized enzyme might then require reduced pyridine nucleotides in the form of NADP to regenerate the active enzyme.

Recent work from Santi's laboratory [14] has given some support to the first part of this hypothetical process in that they have shown that thymidylate synthetase can catalyze the dehalogenation of both 5-bromo- and 5-iodo-2'-

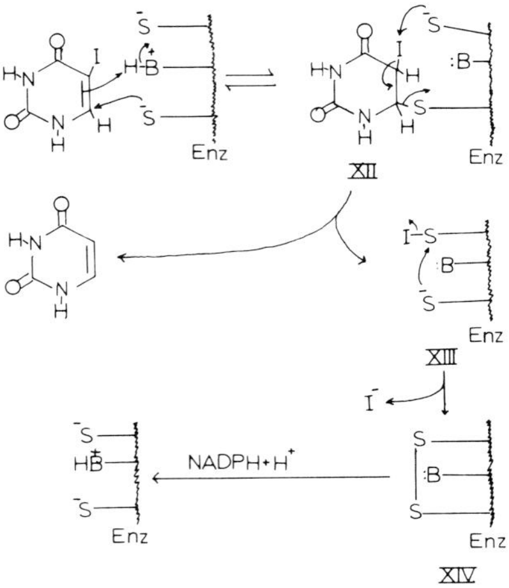

Scheme 9

deoxyuridylate via a process that presumably involves covalent adduct formation between the enzyme and the 5-halo-2'-deoxyuridylate (XdUMP). Subsequent steps involved in the dehalogenation of the thymidylate synthetase–XdUMP covalent adduct may not, however, be catalyzed by the enzyme since the overall dehalogenation reaction seems to require, in addition to the enzyme, dithiolthreitol, a dithiol compound that could easily account for dehalogenation of the covalent enzyme–XdUMP intermediate. Thus, the dehalogenation of XdUMP's by thymidylate synthetase, although an interesting reaction that should give further insight into the mechanism of action of the enzyme, may not be physiologically important in halopyrimidine dehalogenation at the nucleotide level. The clear-cut demonstration of thymidylate synthetase catalysis of XdUMP dehalogenation in the absence of added thiols would clarify this point.

The hypothetical mechanism for NADP-requiring enzymatic 5-halopyrimidine dehalogenation (Scheme 9) may have little bearing on the dehalogenation of halouracils at the level of the pyrimidine bases. Recent results from our laboratory [60] indicate that the enzymes responsible for in vivo uracil catabolism are also involved in this process. Using ^{125}I-labeled 5-iodo-5,6-dihydrouracil as substrate, it has been possible to demonstrate an enzyme activity present in soluble rat liver fractions which can catalyze the release of [125]I$^-$ as well as catalyze the hydrolysis of 5,6-dihydrouracil and 5,6-dihydrothymine. Partial purification of this enzyme, followed by incubation with 5-iodo-5,6-[2-^{14}C]dihydrouracil, yielded iodide and 2-amino-2-oxazoline-5-carboxylic acid (**XV**) as products. Thus, it can be concluded that one pathway for 5-halouracil dehalogenation involves NADP-linked, dihydrouracil dehydrogenase-catalyzed reduction of the 5,6 double bond of the halouracil [Eq. (1)], dihydropyrimidinase-catalyzed hydrolysis of the dihydropyrimidine ring system [Eq. (22)], and iodide elimination via intra-

$$\text{(22)}$$

molecular attack of the ureido oxygen atom on the carbon atom attached to the halogen substituent [Eq. (23)]. A similar reaction involving intramolecular

$$\text{(23)}$$

attack of the ureido nitrogen has been shown by Fox's laboratory [61,62] to be involved in the dehalogenation of 2',3'-o-isopropylidene-5-halouridine in strongly alkaline medium, and Hegarty and Bruice [63] have shown that intramolecular nucleophilic reactions involving ureido functions may proceed by both nitrogen and oxygen attack. It would be of considerable interest to see if the oxazoline (**XV**) shown in Eq. (23) is a urinary metabolite of *in vivo* 5-halouracil catabolism.

ACKNOWLEDGMENTS

I wish to thank Dr. F. A. Sedor and Mr. D. G. Jacobson for their comments during the preparation of this manuscript and Dr. H. Hayatsu for sending preprints of references 20 and 33 prior to publication. Special thanks to Ms. Nancy Scherzer, who typed the manuscript. This work was supported by Grant CA–12971 awarded by the National Cancer Institute, DHEW.

REFERENCES

1. J. Sugar and H. E. Kaufman, in "Selective Inhibitors of Viral Functions" (W. A. Carter, ed.), p. 295. CRC Press, Cleveland, Ohio, 1973.
2. W. H. Prusoff and B. Goz, *Fed. Proc., Fed. Am. Soc. Exp. Biol.* **32**, 1679 (1973).
3. D. Shugar, *FEBS Lett.* **40**, Suppl., S48 (1974).
4. G. M. Timmis and D. C. Williams, "Chemotherapy of Cancer." Butterworth, London, 1967.
5. S. S. Cohen, J. G. Flaks, H. D. Barner, M. R. Loeb, and J. Lichtenstein, *Proc. Natl. Acad. Sci. U.S.A.* **44**, 1004 (1958).
6. W. H. Prusoff, J. J. Jaffe, and H. Günther, *Biochem. Pharmacol.* **3**, 110 (1960).
7. H. B. Pahl, M. P. Gordon, and R. R. Ellison, *Arch. Biochem. Biophys.* **79**, 245 (1959).
8. J. P. Kriss, Y. Maruyama, L. A. Tung, S. B. Bond, and L. Revesz, *Cancer Res.* **23**, 260 (1963).
9. J. P. Kriss and L. Revesz, *Cancer Res.* **22**, 254 (1962).
10. E. G. Hampton and M. L. Eidinoff, *Cancer Res.* **21**, 345 (1961).
11. H. W. Barrett and R. A. West, *J. Am. Chem. Soc.* **78**, 1612 (1956).
12. G. M. Cooper and S. Greer, *Cancer Res.* **30**, 2937 (1970).
13. M. T. Dorsett, P. A. Morse, Jr., and G. A. Gentry, *Cancer Res.* **29**, 79 (1969).
14. Y. Wataya and D. V. Santi, *Biochem. Biophys. Res. Commun.* **67**, 818 (1975).
15. R. E. Notari, *J. Pharm. Sci.* **56**, 804 (1967).
16. R. Shapiro and R. S. Klein, *Biochemistry* **5**, 2358 (1966).
17. R. Shapiro, R. E. Servis, and M. Welcher, *J. Am. Chem. Soc.* **92**, 422 (1970).
18. H. Hayatsu, Y. Wataya, and K. Kai, *J. Am. Chem. Soc.* **92**, 724 (1970).
19. H. Hayatsu, Y. Wataya, K. Kai, and S. Iida, *Biochemistry* **9**, 2858 (1970).
20. H. Hayatsu, *Prog. Nucleic Acid Res. Mol. Biol.* **16**, 75 (1976).
21. E. G. Sander and C. L. Deyrup, *Arch. Biochem. Biophys.* **150**, 600 (1972).
22. J. Fourrey, *Bull. Soc. Chim. Fr.* p. 4580 (1972).
23. G. S. Rork and I. H. Pitman, *J. Am. Chem. Soc.* **97**, 5559 (1975).
24. G. S. Rork and I. H. Pitman, *J. Am. Chem. Soc.* **97**, 5566 (1975).
25. F. A. Sedor, D. G. Jacobson, and E. G. Sander, *J. Am. Chem. Soc.* **97**, 5572 (1975).
26. E. R. Garrett, T. Suzuki, and D. J. Weber, *J. Am. Chem. Soc.* **86**, 4460 (1964).
27. E. R. Garrett, P. B. Chemburkar, and T. Suzuki, *Chem. Pharm. Bull.* **13**, 1113 (1965).

28. F. A. Sedor and E. G. Sander, *Arch. Biochem. Biophys.* **161**, 632 (1974).
29. R. W. Erickson and E. G. Sander, *J. Am. Chem. Soc.* **94**, 2086 (1972).
30. G. S. Rork and I. H. Pitman, *J. Am. Chem. Soc.* **96**, 4654 (1974).
31. F. A. Sedor, D. G. Jacobson, and E. G. Sander, *Bioorg. Chem.* **3**, 221 (1974).
32. S. Y. Wang, *J. Org. Chem.* **24**, 11 (1959).
33. H. Hayatsu, T. Chikuma, and K. Negishi, *J. Org. Chem.* **40**, 3862 (1975).
34. D. G. Jacobson, F. A. Sedor, and E. G. Sander, *Bioorg. Chem.* **4**, 72 (1975).
35. M. Sono, Y. Wataya, and H. Hayatsu, *J. Am. Chem. Soc.* **95**, 4745 (1973).
36. R. Shapiro, V. DiFate, and M. Welcher, *J. Am. Chem. Soc.* **96**, 906 (1974).
37. D. G. Jacobson and E. G. Sander, unpublished observations.
38. S. R. Heller, *Biochem. Biophys. Res. Commun.* **32**, 998 (1968).
39. T. I. Kalman, *Biochemistry* **10**, 2567 (1971).
40. Y. Wataya, H. Hayatsu, and Y. Kawazoe, *J. Am. Chem. Soc.* **94**, 8927 (1972).
41. M. I. S. Lomax and G. R. Greenberg, *J. Biol. Chem.* **242**, 1302 (1967).
42. D. V. Santi and C. S. McHenry, *Proc. Natl. Acad. Sci. U.S.A.* **69**, 1855 (1972).
43. D. V. Santi, C. S. McHenry, and H. Sommer, *Biochemistry* **13**, 471 (1974).
44. D. V. Santi, C. S. McHenry, and E. R. Perriard, *Biochemistry* **13**, 467 (1974).
45. C. S. McHenry and D. V. Santi, *Biochem. Biophys. Res. Commun.* **57**, 204 (1974).
46. H. Sommer and D. V. Santi, *Biochem. Biophys. Res. Commun.* **57**, 689 (1974).
47. R. J. Langenbach, P. V. Danenberg, and C. Heidelberger, *Biochem. Biophys. Res. Commun.* **48**, 1565 (1972).
48. P. V. Danenberg, R. J. Langenbach, and C. Heidelberger, *Biochemistry* **13**, 926 (1974).
49. R. W. Chambers and V. Kurkov, *J. Am. Chem. Soc.* **85**, 2160 (1963).
50. E. J. Reist, A. Benitez, and L. Goodman, *J. Org. Chem.* **29**, 554 (1964).
51. D. M. Brown and C. M. Taylor, *J. Chem. Soc., Perkin Trans. 1* p. 2385 (1972).
52. F. A. Sedor and E. G. Sander, *Biochem. Biophys. Res. Commun.* **50**, 328 (1973).
53. Y. Wataya, K. Negishi, and H. Hayatsu, *Biochemistry* **12**, 3992 (1973).
54. L. Szabo, T. I. Kalman, and T. J. Bardos, *J. Org. Chem.* **35**, 1434 (1970).
55. F. A. Sedor, D. G. Jacobson, and E. G. Sander, *Biorg. Chem.* **3**, 154 (1974).
56. F. A. Sedor and E. G. Sander, *J. Am. Chem. Soc.* **98**, 2314 (1976).
57. E. C. F. Ko and A. J. Parker, *J. Am. Chem. Soc.* **90**, 6447 (1968).
58. B. Roth and G. H. Hitchings, *J. Org. Chem.* **26**, 2770 (1961).
59. E. G. Sander, E. Young, and F. A. Sedor, *Bioorg. Chem.* **5**, 231 (1976).
60. B. D. Kim, S. Keenan, J. Bodnar, and E. G. Sander, *J. Biol. Chem.* **251** 6909 (1976.)
61. B. A. Otter, E. A. Falco, and J. J. Fox, *J. Org. Chem.* **33**, 3593 (1968).
62. B. A. Otter, E. A. Falco, and J. J. Fox, *J. Org. Chem.* **34**, 1390 (1969).
63. A. F. Hegarty and T. C. Bruice, *J. Am. Chem. Soc.* **92**, 6575 (1970).

12

Vitamin D: Chemistry and Biochemistry of a New Hormonal System

H. K. Schnoes and H. F. DeLuca

INTRODUCTION

Research of the past decade has brought about a fundamental revision in our understanding of the biochemistry of the D vitamins, comparable in impact to the initial discovery of these antirachitic factors more than 50 years ago. The vitamin is now considered to be a prohormone that gives rise by a well-regulated metabolic process to a hormone with a specific function—the control of calcium and phosphate homeostasis. The seminal event initiating this conceptual evolution was the demonstration, in 1966, of biologically active vitamin D metabolites and the subsequent realization that metabolic conversion of the vitamin is a prerequisite for function. That function basically involves the regulation of calcium and phosphate metabolism to maintain blood levels of these ions within narrow limits (e.g., ca. 10 mg/100 ml for calcium), thereby ensuring their availability for diverse biochemical processes in which calcium (muscle contraction, nerve function, and blood clotting) or calcium and phosphate (bone and egg-shell formation) play an important role. Several major processes affect the calcium and phosphate balance of an organism: intestinal absorption from the diet, deposition and/or resorption of calcium phosphate from bone, transport and redistribution within the organism via the blood, and excretion via urine and feces. At least three of these processes—intestinal calcium and phosphate absorption and the liberation of calcium and phosphate from bone—are responsive to vitamin D.

The vitamin stimulates calcium transport across intestinal cells and it stimulates, in concert with another hormone, parathyroid hormone, the liberation of calcium from bone mineral. Phosphate is liberated with calcium from bone, and it represents the normal anion accompanying calcium transport in intestine. In addition, however, vitamin D also stimulates an intestinal phosphate transport system that is independent of the calcium transport mechanism. There is no convincing evidence that vitamin D participates directly in the calcification of bone. Its well-known antirachitic effect is rather thought to be a consequence of maintaining blood calcium and phosphate at the required levels for deposition of apatite mineral at bone nucleation sites. Although the D vitamins have been implicated in biochemical processes in other organs, specifically kidney and muscle, only their effect on intestine and bone has been subjected to detailed experimental scrutiny, and "activity"—at least in the context of this review—thus refers to the response elicited by a vitamin metabolite or analog in the intestinal and bone systems.

The demonstration that the activity of the D vitamins depends on prior metabolism and the characterization of active metabolites represent major results of recent research. Vitamin D_3 metabolism leads to $1\alpha,25$-dihydroxyvitamin D_3 [$1\alpha,25$-$(OH)_2D_3$] via 25-hydroxyvitamin D_3 (25-OH-D_3) as an intermediate, whereas vitamin D_2 yields $1,25$-$(OH)_2D_2$ by an entirely analogous pathway. Metabolite identification led directly to studies of the enzymology of their formation, and this work yielded another important advance: the realization that vitamin D is a key element of a finely tuned endocrine system involving production of the active hormonal form [$1\alpha,25$-$(OH)_2D_3$ or $1,25$-$(OH)_2D_2$] in one organ (the kidney), expression of activity in other tissues (intestine and bone), and feedback control of hormone synthesis by the "products" of its action (calcium and phosphate levels).

In contrast to vitamin D metabolism and its regulation, in which the main outlines, if not the nuances, are reasonably securely established, details of the mode of action of the hormone in target organs remain to be defined. Current attempts to rationalize vitamin action in terms of a steroid-hormone-like mechanism are promising but as yet in need of much experimental support; the concept does offer a useful working model, and the recent recognition of a specific intestinal receptor protein for $1\alpha,25$-$(OH)_2D_3$ may provide an important experimental entry into this complex problem area.

Progress in vitamin D biochemistry has stimulated, and has to some extent been made possible by, a renewed interest and activity in vitamin D_3 synthetic chemistry. The preparation of most of the known metabolites, of selected structural analogs, and of radiolabeled derivatives provided required material for detailed biochemical and exploratory clinical investigation. The definition of stereochemical details of metabolite structures, information on structure–

activity relationships, and successful clinical applications are some of the chief practical results of this fusion of chemical and biochemical efforts.

In this review, key aspects or recent developments in vitamin D research are considered. More comprehensive surveys of the voluminous research literature and critical discussions of specific topics may be consulted for exhaustive presentations of the body of experimental data that has shaped current concepts of vitamin D biochemistry [1–7] and chemistry [1,8–11].

VITAMIN D METABOLISM AND ITS REGULATION

Provitamins and Vitamins

Vitamin D is made available to an organism either through the diet or through photochemical synthesis in the skin from a steroidal 5,7-diene precursor (the provitamin). Vitamin D_3 (cholecalciferol) is an irradiation product of 7-dehydrocholesterol, while vitamin D_2 (ergocalciferol) derives from ergosterol. Since 7-dehydrocholesterol is a normal product of animal sterol metabolism, vitamin D_3 is not an essential dietary factor (i.e., a "vitamin" in the classic sense) given adequate exposure to sunlight, and the compound is perhaps described more precisely as a prohormone, produced in the skin, converted to its hormonal form in the kidney, and acting on specific target organs (bone and intestine). Vitamin D_2, on the other hand, originating from the fungal sterol ergosterol, is not a product of animal metabolism and should be regarded as a biologically active structural analog of the natural prohormone, derived as such from the diet.

The solution photochemistry of the provitamins has been the subject of detailed studies, principally by Velluz and Havinga and their co-workers [11,12]. As illustrated in Fig. 1, formation of vitamin D involves a two-step reaction. The first is a photochemical event leading to previtamin D, which is in thermal equilibrium with the corresponding vitamin D isomer. At equilibrium the pre-D and D isomers occur in a ration of ca. 1:4. Previtamin D, in addition, undergoes several photochemical isomerizations, among which the conversion to tachysterol is quantitatively the most important reaction. Ring closure of previtamin D to the two $\Delta^{5,7}$-isomers of Fig. 1 appears to occur only at wavelengths greater than 300 nm [13,14]. Prolonged irradiation (24–48 hr) leads to the accumulation of other products (the toxisterols and photofragments), several of which have been recently characterized [15–17].

The corresponding *in vivo* conversion of the provitamin to the vitamin has received no detailed attention, and, although there is persuasive evidence [1] for the production of vitamin D_3 by irradiation of skin, the biochemistry of

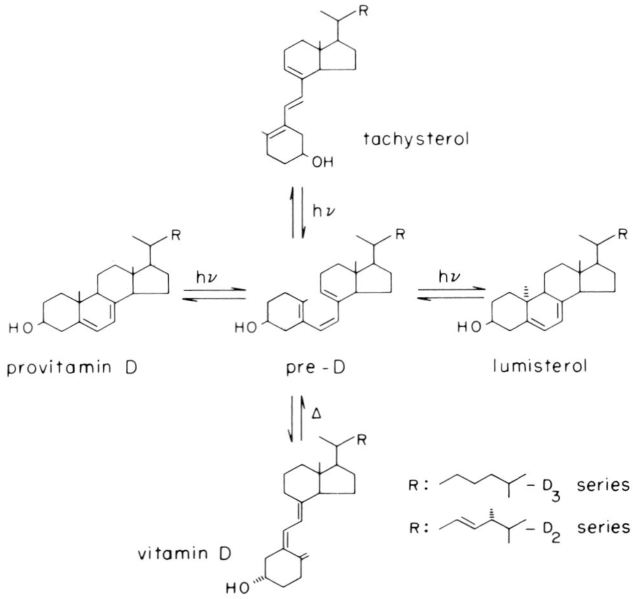

Fig. 1 Irradiation products of the provitamins D and formation of vitamins D_2 and D_3.

the reaction is not understood. The involvement of enzymes in this process, the existence and nature of intermediates, and the formation of triene isomers and/or overirradiation products are only a few of the topics that appear to be worthy of closer scrutiny.

Vitamin D_3 Metabolism

The main features of the conversion of vitamin D_3 to active metabolites are by now well defined. As summarized in Fig. 2, biosynthesis of currently-known metabolites involves an initial hydroxylation of vitamin D_3 to 25-hydroxyvitamin D_3. This central intermediate undergoes a second hydroxylation at C-1 to give $1\alpha,25$-dihydroxyvitamin D_3, the presumed tissue-active form of vitamin D_3. An alternative hydroxylation at C-24 converts 25-OH-D_3 to $(24R)$-24,25-dihydroxyvitamin D_3 [$(24R)$-24,25-$(OH)_2D_3$], and further metabolism of either $1\alpha,25$-$(OH)_2D_3$ or 24,25-$(OH)_2D_3$ yields $(24R)$-$1\alpha,24,25$-trihydroxyvitamin D_3 [$(24R)$-$1\alpha,24,25$-$(OH)_3D_3$]. Finally, 25-OH-D_3 is very likely also the precursor for a fifth metabolite, 25,26-dihydroxyvitamin D_3 [25,26-$(OH)_2D_3$]. Also indicated in Fig. 2 are the key physiological parameters whose effect on vitamin D metabolism is considered in the ensuing sections.

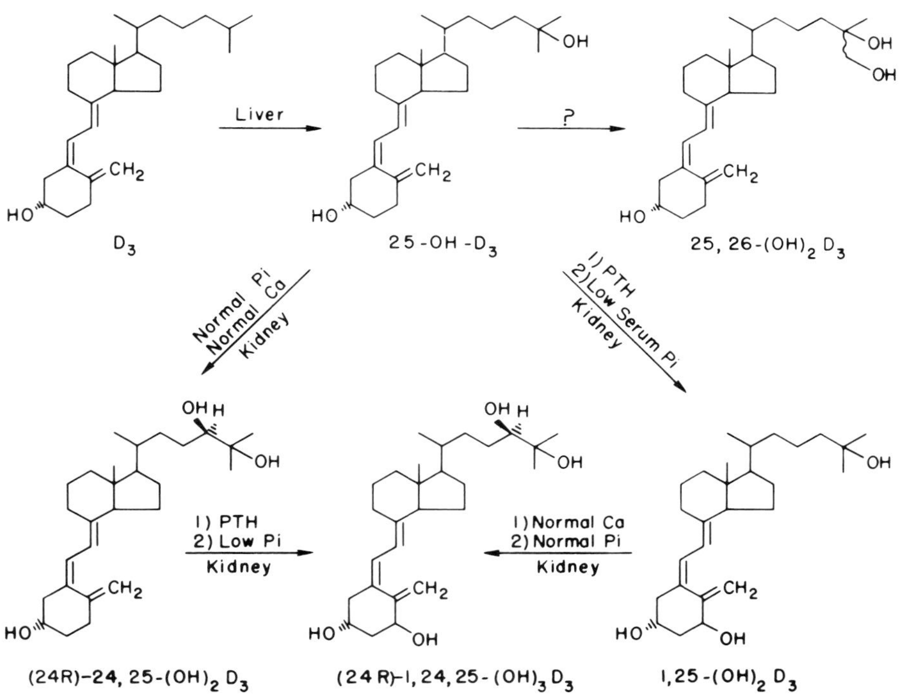

Fig. 2 Metabolism of vitamin D_3.

25-OH-D_3

Once tritiated vitamin D_3 of high specific activity became available [18] it could be demonstrated clearly that vitamin D_3 administered in physiological amounts is rapidly metabolized to more polar forms exhibiting biological activity (intestinal calcium transport stimulation, bone calcium liberation) equal to or greater than vitamin D_3 itself [19]. Isolation of the major circulating metabolite from pig plasma and characterization as 25-OH-D_3 (see Fig. 2) were accomplished shortly thereafter [20]. Liver is a major site for the conversion of D_3 to 25-OH-D_3 [21,22]. This is certainly true for the rat, in which hepatectomy either abolishes or markedly reduces the synthesis of 25-OH-D_3 [23]. In the chick, hydroxylase activity in the intestine and kidney has been detected [24], but it is doubtful at present whether conversion in these organs is quantitatively significant, under physiological conditions, since absorbed vitamin is concentrated by liver tissue and circulating levels are low [18,21,23,25].

Rat liver 25-hydroxylase requires two cellular components—a microsomal and a cytosolic fraction—for maximal hydroxylation activity [26]. *In vitro* conversions of D_3 are relatively low (7%) but adequate to account for the observed *in vivo* accumulation of the 25-OH-D_3 product. Only fragmentary evidence pertaining to the mechanism of 25-hydroxylation is available at present [6]. Hydroxylation in rat liver homogenates requires molecular oxygen, magnesium ions, and NADPH as the reductant, but a cytochrome *P*-450 does not seem to be involved in the reaction [22,24,26]. The microsomal 25-hydroxylase is suppressed, however, by prior administration of vitamin D_3 [6,26,27]. The hepatic level of 25-OH-D_3 appears to be the actual controlling parameter of hydroxylase activity, but the nature of this phenomenon— whether it represents simple product inhibition or another more subtle regulatory mechanism—is not understood. Administration of large amounts of D_3 overcomes 25-hydroxylase suppression and leads to increased circulating levels of 25-OH-D_3 [20,27–29]. Since rat liver contains in addition to the microsomal 25-hydroxylase, a mitochondrial system capable of hydroxylating, vitamin D_3 (as well as other steroids) at high substrate concentrations [27], it is possible to suggest that increased production of 25-OH-D_3 upon administration of pharmacological doses of D_3 might involve such a relatively nonspecific mitochondrial hydroxylase. Extrahepatic hydroxylation [24] is an alternative rationalization.

As indicated earlier, 25-OH-D_3 represents the major circulating metabolite of vitamin D_3 [2,6]. It exhibits considerably higher antirachitic potency than D_3, it acts more rapidly in inducing intestinal calcium absorption and bone mineral mobilization [30], and, unlike D_3, it is capable of calcium resorption from fetal bone cultures [31]. There is no evidence, however, that, under physiological circumstances, the compound plays a direct functional role in any target organ; it merely serves as an obligatory intermediate for the synthesis of other hydroxylated vitamin metabolites.

$(24R)$-24,25-$(OH)_2D_3$ AND 25,26-$(OH)_2D_3$

Two dihydroxy derivatives, isolated from the plasma of hogs dosed with tritiated vitamin D_3, were identified as 25,26-dihydroxyvitamin D_3 [32] and 24,25-dihydroxyvitamin D_3 [33], respectively. Very little is known about the former metabolite; site and mode of biosynthesis have not been investigated, although it appears reasonable to presume that the compound is indeed derived from 25-OH-D_3. It is, in any case, a minor metabolite that exhibits moderate intestinal calcium transport activity but is ineffective on bone [34]. Although a functional role is not excluded, it is, at this stage, perhaps more reasonable to assume that 25,26-$(OH)_2D_3$ represents a partially active, but nonfunctional, metabolite eventually destined for side-chain degradation and excretion.

The other dihydroxylated isomer, $(24R)$-24,25-$(OH)_2D_3$, is, next to 25-OH-D_3, the most prominent vitamin D_3 metabolite in animals maintained on a normal or high calcium and phosphate diet. Chemical synthesis of the (24R)- and (24S)-epimers (35) has established the (24R) configuration for the natural product [36]. The (24R)-isomer exhibits a biological activity pattern similar to that of 25-OH-D$_3$; it stimulates intestinal calcium transport, bone calcium mobilization, and the calcification of bone [36,37]. The synthetic (24S) analog, by contrast, is active only in intestine and but moderately so [36]. Hydroxylation of 25-OH-D$_3$ at C-24 is catalyzed by an enzyme system in kidney mitochondria [37,38]. In intact mitochondria, the reaction is supported by Krebs cycle intermediates, utilizes molecular oxygen, and is blocked by inhibitors of oxidative phosphorylation. However, the conversion is not sensitive to carbon monoxide and does not appear to involve cytochrome P-450. The 25-OH-D$_3$ 24-hydroxylase is subject to regulation by calcium and phosphate such that (24R)-24,25-(OH)$_2$D$_3$ is produced only at normal or high circulating levels of calcium and phosphate. Since this control mechanism is intimately related to the regulation of 1α,25-(OH)$_2$D$_3$ production, it is discussed more fully in connection with the latter metabolite.

None of these metabolites, although biologically active in normal animals, appears to have a functional role per se. They are active only after one other metabolic conversion—the introduction of an additional hydroxyl group at carbon 1 of the vitamin skeleton.

1α,25-(OH)$_2$D$_3$

From the point of view of function, C-1 hydroxylation of 25-OH-D$_3$ to yield 1α,25-(OH)$_2$D$_3$ is the crucial metabolic step. A polar metabolite in intestine exhibiting high potency was first detected by Haussler et al. [39]. When the existence of this substance and its very rapid action were confirmed in other laboratories, and after demonstration that 25-OH-D$_3$ was a precursor (see [2–6] for reviews), characterization efforts were initiated which culminated in 1971 in the identification of the substance (isolated in low yield, 2 μg, from intestines of chicks dosed with tritiated vitamin D$_3$) as a 1,25-dihydroxyvitamin D$_3$ derivative [40]. Lawson et al. [41] had earlier shown that the conversion of [1α-^{3}H]vitamin D$_3$ to this metabolite involved loss of 1α-^{3}H label, but proof of the α configuration for the C-1 hydroxy function depended on the results of subsequent stereospecific chemical syntheses [42–44]. Independently and simultaneously, Lawson et al. [45], taking advantage of their earlier findings [46] that kidney was the site of synthesis for the metabolite, isolated the compound from chick kidney homogenates incubated with 25-OH-D$_3$ and proposed the same structure. Kidney is the exclusive site of synthesis [46–48]. Nephrectomized animals cannot hydroxylate 25-OH-D$_3$ at C-1, while normal or sham-operated controls produce 1α,25-(OH)$_2$D$_3$ from this substrate [47].

The biological activity of $1\alpha,25\text{-}(OH)_2D_3$ has been the subject of exhaustive studies. A considerable body of data discussed in greater detail in other reviews [2,4–6] points very convincingly to this metabolite as the active form of the vitamin in target tissues. For our purposes, a summary of the results may suffice. (1) $1\alpha,25\text{-}(OH)_2D_3$ is the most potent and most rapidly acting vitamin D derivative in promoting calcium transport in intestine and calcium liberation from bone [49–53]. (2) It is effective in mobilizing phosphate. Although this aspect of vitamin D activity has received much less attention [6], recent work [54,55] has established the $1\alpha,25\text{-}(OH)_2D_3$ promotes intestinal phosphate transport by a mechanism independent of the calcium transport process. (3) It is fully active in nephrectomized animals, where its precursor, $25\text{-}OH\text{-}D_3$ (as well as any other derivative lacking a C-1 hydroxy group), is ineffective [51–54]. (4) It is by far the most effective compound in promoting calcium resorption from fetal bone in culture, exceeding $25\text{-}OH\text{-}D_3$ by approximately three orders of magnitude [56,57]. (5) It localizes predominantly in target tissues, such as intestine [58,59]. (6) It exhibits very high affinity for a cytosol–chromatin receptor complex of chick intestine [60,61] which has the properties (specificity, saturability, temperature dependence of the cytosol → chromatin transfer) characteristic of target-tissue receptors of steroid hormones.

Although these data clearly define the functional significance of $1\alpha,25\text{-}(OH)_2D_3$, the possibility that the compound undergoes further structural modification in the target tissues before initiating a response (or in the process of doing so) is not totally excluded. Earlier studies failed to detect other metabolites in intestine [6], but recent work [62] using $1\alpha,25\text{-}(OH)_2D_3$ of very high specific activity has in fact shown that the compound is further metabolized even before the intestine responds; at present, the functional relevance of this observation is obscure.

Hydroxylation of $25\text{-}OH\text{-}D_3$ at C-1 is catalyzed by a mitochondrial enzyme system [46,63]. A series of detailed studies [6,63–67] has identified the 1α-hydroxylase of chick kidney mitochondria as a three-component system including a flavoprotein, renal ferredoxin, and cytochrome *P*-450, which, with NADPH as the electron donor, effects reduction of molecular oxygen [64] and hydroxylation of the substrate. It is interesting that, although vitamin D_3 itself is not hydroxylated at C-1, other 25-hydroxyvitamin derivatives, such as 25-hydroxyvitamin D_2 and $(24R)$- and $(24S)\text{-}24,25\text{-}(OH)_2D_3$, do serve as substrates for the 1α-hydroxylase.

$(24R)\text{-}1\alpha,24,25\text{-}(OH)_3D_3$

From *in vitro* incubations of $(24R)\text{-}24,25\text{-}(OH)_2D_3$ with chick kidney mitochondria a more polar product could be isolated and identified as

$1,24,25\text{-}(OH)_3D_3$ [68]. Since this compound has now been shown to arise biosynthetically both by C-1 hydroxylation of $(24R)\text{-}24,25\text{-}(OH)_2D_3$ [69] and by C-24 hydroxylation of $1\alpha,25\text{-}(OH)_2D_3$ [70], it is defined stereochemically as $(24R)\text{-}1\alpha,24,25\text{-}trihydroxyvitamin D_3$. Synthesis of $(24R)\text{-}1\alpha,24,25\text{-}(OH)_3D_3$ from $1\alpha,25\text{-}(OH)_2D_3$ may represent the most important route *in vivo* since $1\alpha,25\text{-}(OH)_2D_3$, as is discussed later, does stimulate or induce the 24-hydroxylase of chick kidney mitochondria.

$(24R)\text{-}1\alpha,24,25\text{-}(OH)_3D_3$ represents, of course, the "active form" of $(24R)\text{-}24,25\text{-}(OH)_2D_3$; i.e., the latter is inactive unless converted to the 1-hydroxylated derivative. This fact is readily demonstrated in nephrectomized animals, where $(24R)\text{-}24,25\text{-}(OH)_2D_3$ is ineffective, while $(24R)\text{-}1\alpha,24,25\text{-}(OH)_3D_3$ exhibits full activity [69]. In normal and nephrectomized animals $(24R)\text{-}1\alpha,24,25\text{-}(OH)_3D_3$ stimulates all vitamin D-responsive systems (with the possible exception of intestinal phosphate transport), although the compound exhibits lower activity (ca. 60%) than $1\alpha,25\text{-}(OH)_2D_3$ [68,69]. It is not possible at present to define a functional role for this metabolite, and it is likely that formation of 24-hydroxylated derivatives represents the first step of a degradative route.

SUMMARY

We might conclude this section with a brief recapitulation. Metabolic conversion of vitamin D_3 to its biologically active hormonal form involves the sequence $D_3 \rightarrow 25\text{-OH-}D_3 \rightarrow 1\alpha,25\text{-}(OH)_2D_3$. Hydroxylation of 25-OH-D_3 and $1\alpha,25\text{-}(OH)_2D_3$ at C-24, in view of its regulation (see below) by various physiological parameters, is clearly a significant event *in vivo*, but the resulting derivatives [$(24R)\text{-}24,25\text{-}(OH)_2D_3$ and $(24R)\text{-}1\alpha,24,25\text{-}(OH)_3D_3$, respectively] are more likely to represent precursors to eventual excretory products rather than tissue-active hormones. Whether C-26 hydroxylation is a step of some physiological relevance or merely a reaction triggered by pharmacological amounts of administered vitamin remains to be established.

Vitamin D_2 Metabolism

Although vitamin D_2 metabolites have been investigated much less extensively, the characterization of some key transformation products establishes clearly that vitamin D_2 metabolism to its tissue-active form involves an analogous hydroxylation sequence. Thus, hydroxylation in the liver yields initially 25-hydroxyvitamin D_2 (25-OH-D_2) [71], and this intermediate can serve as a substrate for 1-hydroxylation by a chick kidney mitochondrial system to yield $1,25\text{-}(OH)_2D_2$ [72]. The stereochemistry at carbon 1, although not established, is presumably 1α. In addition, small amounts of two other metabolites have been isolated, which on the basis of preliminary spectral data are believed to be 24,25-(OH)$_2$D$_2$ and 24-OH-D$_2$ (G. Jones, H. K. Schnoes, and H. F. DeLuca, unpublished).

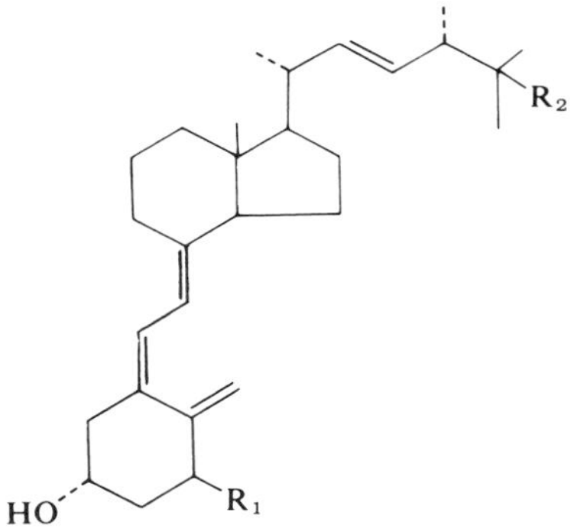

Vitamin D$_2$: R$_1$ = R$_2$ = H
25-OH-D$_2$: R$_1$ = H, R$_2$ = OH
1,25-(OH)$_2$D$_2$: R$_1$ = R$_2$ = OH

It is well known that vitamin D$_2$ is about ten times less potent than vitamin D$_3$ in birds and New World monkeys, whereas in all other mammals the compounds show equal activity. The same is true for vitamin D$_2$ metabolites. In the rat, 25-OH-D$_2$ and 1,25-(OH)$_2$D$_2$ are as effective as the corresponding D$_3$ analogs, but in the chick they exhibit only about 10–20% of the activity of the D$_3$ metabolites [71,73]. Thus, the discrimination of chicks against D$_2$ is not due to impaired metabolism, and it can indeed be shown [74] that both chick liver 25-hydroxylase and chick kidney 1α-hydroxylase convert the appropriate D$_2$ substrate as effectively as the D$_3$ relative. Similarly, there seems to be no discrimination against the active form, 1,25-(OH)$_2$D$_2$, in the target tissue. The cytoplasmic 1α,25-(OH)$_2$D$_3$-binding protein in chick intestine appears to have equal affinity for the 1,25-(OH)$_2$D$_2$ analog [75]. These results and the demonstration that injection of radioactive D$_2$ into chicks yields only low levels of circulatory 25-OH-D$_2$ and low intestinal levels of 1,25-(OH)$_2$D$_2$ [74] suggest that the low potency of D$_2$ in birds may be a consequence of an as yet unknown metabolic pathway that leads to the rapid degradation and/or excretion of D$_2$ and its metabolites. It is thought that the C-24 substituent may be the key structural parameter that diverts D$_2$ compounds into a degradative route, a hypothesis that is suggested by the observation that other C-24-substituted vitamins, i.e., (24R)-24,25-(OH)$_2$D$_3$ and (24R)-1α,24,25-(OH)$_3$D$_3$, also exhibit very low activity in the chick and can be shown to be rapidly eliminated from circulation [76].

The preceding discussion points to a comparative study of the excretory metabolites of the D vitamins and their characterization as a topic of considerable interest. Little progress has been made in this area to date [1,6], but current research aided by powerful new chromatographic and physico-chemical techniques may fill this void in the foreseeable future.

Regulation of the Renal Hydroxylases

Both renal hydroxylases, the 25-OH-D$_3$ 1α-hydroxylase and the 25-OH-D$_3$ (24R)-hydroxylase, are subject to an intricate regulatory mechanism that controls the production of 1α,25-(OH)$_2$D$_3$ and (24R)-24,25-(OH)$_2$D$_3$. In animals maintained on a vitamin D source, a strict inverse relationship exists between serum calcium level and 1α,25-(OH)$_2$D$_3$ production. Hypo-calcemia stimulates 1α,25-(OH)$_2$D$_3$ synthesis [77–79]; normal or high calcium levels suppress it while stimulating the synthesis of (24R)-24,25-(OH)$_2$D$_3$ [77]. The same effect can be demonstrated with mitochondrial preparations. Kidney mitochondria isolated from chicks maintained on a low calcium diet produce 1α,25-(OH)$_2$D$_3$ exclusively; mitochondria from chicks on a high calcium diet synthesize (24R)-24,25-(OH)$_2$D$_3$ [80].

It should be stressed that this regulation by calcium occurs only in animals maintained on vitamin D. Vitamin D-deficient animals produce 1α,25-(OH)$_2$-D$_3$ regardless of calcium levels [77]. The regulatory effect of calcium on 1α,25-(OH)$_2$D$_3$ synthesis is mediated by parathyroid hormone (PTH), which is secreted from the parathyroid glands under hypocalcemic conditions. Thus, thyroparathyroidectomy of rats abolishes their ability to produce 1α,25-(OH)$_2$-D$_3$, even under hypocalcemic conditions, and leads to the accumulation of (24R)-24,25-(OH)$_2$D$_3$ [81]. Administration of PTH to these animals restores 1α,25-(OH)$_2$D$_3$ synthesis and suppresses the production of (24R)-24,25-(OH)$_2$D$_3$ [81]. Studies of kidney 1α-hydroxylase in chicks [82] and direct measurements of 1α,25-(OH)$_2$D$_3$ levels in the rat [79] have confirmed the stimulating effect of PTH on 1α-hydroxylase activity and its essential regulatory role in 1α,25-(OH)$_2$D$_3$ production.

Blood phosphate levels also affect 1α,25-(OH)$_2$D$_3$ synthesis. The mechanism of regulation by phosphate is obscure, but it is clear that PTH is not involved. Tanaka and DeLuca [83] have reported that phosphate deprivation results in a marked enhancement of 1α,25-(OH)$_2$D$_3$ production both in normal and in thyroparathyroidectomized rats. Again 1α,25-(OH)$_2$D$_3$ and (24R)-24,25-(OH)$_2$D$_3$ synthesis exhibit the familiar inverse relationship: Low serum phosphorus stimulates production of 1α,25-(OH)$_2$D$_3$, and normal or high levels lead to the accumulation of the (24R)-24,25-dihydroxy isomer. A more recent investigation [79] of the effect of serum phosphate on circulatory 1α,25-(OH)$_2$D$_3$ levels has confirmed these observations.

As mentioned earlier, the regulation of 1α-hydroxylase activity by PTH and phosphate depends on the presence of some form of vitamin D [77].

Recent results [84–86] suggest that this requirement may be a consequence of the fact that $1\alpha,25\text{-}(OH)_2D_3$ itself has a regulatory role. It can be shown, for example, that administration of $1\alpha,25\text{-}(OH)_2D_3$ to vitamin D-deficient rats causes a marked suppression of the 1α-hydroxylase with a concomitant stimulation of the 24-hydroxylase [85]. It is thought that $1\alpha,25\text{-}(OH)_2D_3$ may in fact induce the 24-hydroxylase, but details of this complex regulatory scheme are as yet obscure.

The Vitamin D Endocrine System

The influence of the various parameters mentioned in the preceding section (calcium, PTH, phosphate) on production of $1\alpha,25\text{-}(OH)_2D_3$ can be interpreted [3,55,87] in terms of an intricate control mechanism in which the interplay of several factors accomplishes the *independent* regulation of calcium and phosphate supply. The current model, shown in Fig. 3, is an intellectually satisfying synthesis of available data but may, of course, require some revision and refinement as new results become available.

The scheme envisions two signals—low serum calcium and low serum phosphate levels—as the primary controlling parameters that trigger the appropriate corrective mechanism. Thus, hypocalcemia causes the release of PTH, which in turn stimulates production of $1\alpha,25\text{-}(OH)_2D_3$ in the kidney. This hormone then acts on intestine to promote the transport of calcium and phosphate and on bone, where, in conjunction with PTH, it stimulates the mobilization of calcium and phosphate from the bone matrix. These effects of the hormone would tend to increase serum levels of both calcium and phosphate, but PTH inhibits reabsorption of phosphate in renal tubules and leads to excretion of phosphate in the urine. The result, therefore, is a rise of

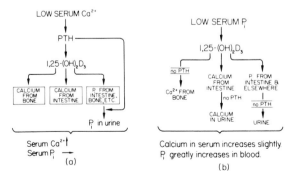

Fig. 3 (a) Schematic representation of the regulation of $1\alpha,25\text{-}(OH)_2D_3$ synthesis by serum calcium. (b) Schematic representation of the regulation of $1\alpha,25\text{-}(OH)_2D_3$ synthesis by serum phosphate.

serum calcium while phosphate levels remain constant. Increasing serum calcium suppresses PTH secretion and consequently $1\alpha,25\text{-(OH)}_2\text{D}_3$ synthesis.

Low serum phosphate stimulates $1\alpha,25\text{-(OH)}_2\text{D}_3$ production without PTH secretion. The metabolite cannot liberate bone calcium and phosphate since PTH is not secreted while both calcium and phosphate transport in intestine is stimulated. The absence of PTH also causes renal retention of phosphate but increased excretion of calcium. The net effect of the hypophosphatemic stimulus is thus the selective increase of serum phosphate, which upon reaching normal levels inhibits $1\alpha,25\text{-(OH)}_2\text{D}_3$ synthesis.

Current results give no more than a rough outline of this new endocrine system. Exact molecular events underlying observed phenomena are poorly understood. In particular, it is not possible at present to describe the molecular mechanism by which PTH and phosphate control $1\alpha,25\text{-(OH)}_2\text{D}_3$ and $(24R)\text{-}24,25\text{-(OH)}_2\text{D}_3$ synthesis, and factors not yet recognized or not as yet sufficiently explored may prove to be important elements of the regulatory scheme (see discussion in reference [6]). One cannot even be certain whether it is the 1α- or the 24-hydroxylase or both enzymes that are subject to regulation. Available data are not inconsistent with the view that the 24-hydroxylase is a major control point [85]. The known regulatory effects of $1\alpha,25\text{-(OH)}_2\text{D}_3$, PTH, and phosphate on $1\alpha,25\text{-(OH)}_2\text{D}_3$ and $(24R)\text{-}24,25\text{-}$ $\text{(OH)}_2\text{D}_3$ synthesis might be rationalized, for example, by postulating induction of the 24-hydroxylase by $1\alpha,25\text{-(OH)}_2\text{D}_3$, stimulation of enzyme activity by phosphate, and its inhibition by PTH. If one assumes that 24-hydroxylation represents the first step of a degradative metabolism, then the above scheme would imply that modulation of vitamin D degradation is the primary mechanism for gross regulation of $1\alpha,25\text{-(OH)}_2\text{D}_3$ levels. Only additional experimentation can transform such speculative guesses into valid concepts.

VITAMIN D CHEMISTRY: STRUCTURE
AND ACTIVITY

Aside from the intellectual challenge, a number of practical considerations have stimulated recent synthetic efforts in the vitamin D area. One important factor is the extremely limited availability of metabolites from natural sources. The plasma level of 25-OH-D_3, for example, ranges from 15 to 30 ng/ml for normal individuals [88], and for $1\alpha,25\text{-(OH)}_2\text{D}_3$ values of ca. 50 pg/ml have been reported [89]. Although the potency of $1\alpha,25\text{-(OH)}_2\text{D}_3$ is such that only extremely small doses are required in clinical applications (e.g., < 1 μg/day), synthetic material was clearly essential for any extensive biomedical experimentation. Confirmation of gross structures as well as definition of stereo-

TABLE 1

Synthetic Vitamin D Metabolites and Analogs

Vitamin D_3 analogs	25-OH-D_3 and analogs	1α-Hydroxylated compounds
4α-OH-D_3 [150]	25-OH-D_3 [92–94,105]	1α,25-(OH)$_2D_3$ [42–44,105]
2α-OH-D_3 [130]	24-Nor-25-OH-D_3 [157]	1α-OH-D_3 [91,114,115,117,128]
5,6-trans-D_3 [148]	27-Nor-25-OH-D_3 [155,156]	1α-OH-D_2 [120]
3-epi-D_3 [160]	26,27-Bisnor-25-OH-D_3 [155,156]	1α-OH-PC [122]
5,6-trans-3-epi-D_3 [160]	20,25-(OH)$_2D_3$ [156]	3-Deoxy-1α-OH-D_3 [116,146,147]
20-OH-PC [155]	27-Nor-20,25-(OH)$_2D_3$ [156]	3-Deoxy-1α,25-(OH)$_2D_3$ [138]
(24R)-24-OH-D_3 [109,110]	24,25-(OH)$_2D_3$ [95,98]	1α-OH-D_3 3-methyl ether [139]
(24S)-24-OH-D_3 [109,110]	(24R)-24,25-(OH)$_2D_3$ [35,105]	1α-OH-3-epi-D_3 [123,124]
DHT$_3$ [145]	(24S)-24,25-(OH)$_2D_3$ [35]	24-Nor-1α,25-(OH)$_2D_3$ [139]
DHT$_2$ [145]	25,26-(OH)$_2D_3$ [34,96]	(24R)-1α,24,25-(OH)$_3D_3$. [110,119]
D_4 [152,153]	5,6-trans-25-OH-D_3 [148]	(24S)-1α,24,25-(OH)$_3D_3$ [110,119]
(22S)-22-OH-D_4 [154]		(24R)-1α,24-(OH)$_2D_3$ [110,121]
		(24S)-1α,24-(OH)$_2D_3$ [110,121]

chemical details was another necessary task, and the intriguing question, finally, of structure–activity relationships in the vitamin D series called for the systematic elaboration of diverse structural types. Tangible results of recent synthetic work are summarized in Table 1 and Fig. 4 (see also Fig. 2 for metabolite structures) and include all known vitamin D_3 metabolites and a spectrum of structural analogs. The practical consequence has been the complete definition of metabolite structure [C-25 stereochemistry of 25,26-(OH)$_2D_3$ excepted], initial but very promising clinical and agricultural applications [90], and some basic information on the effect of specific structural parameters on expression of biological activity.

Recent Synthetic Approaches

With the exception of the total synthesis approach of Lythgoe's group [91] all syntheses of metabolites and analogs have adopted the same basic scheme:

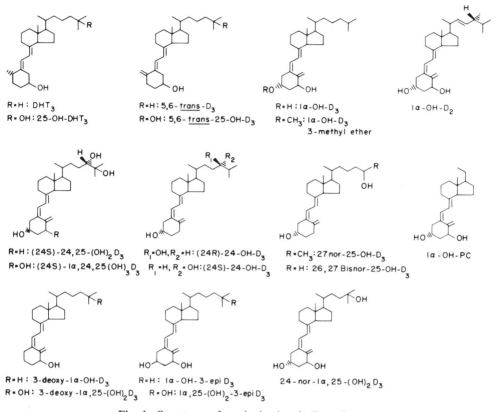

Fig. 4 Structures of synthetic vitamin D analogs.

the preparation of a suitably substituted provitamin derivative (the steroidal 5,7-diene) and photochemical conversion of the latter to the vitamin skeleton. In almost all cases the required diene is prepared by allylic bromination and dehydrobromination of the appropriate Δ^5-sterol derivative, an adequate although not highly efficient method since the desired 5,7-diene is accompanied by considerable amounts of the $\Delta^{4,6}$-isomer, necessitating careful chromatographic purification of the target product. Vitamin D metabolite and analog synthesis via modification of a sterol then involves two general tasks: construction of the hydroxylated side chain and/or introduction of oxygen into ring A.

In many of the published syntheses the former objective has been achieved by direct modification of available intermediates featuring suitably functionalized side chains. Reactions (1)–(3) summarize typical conversions yielding 25-hydroxy-, 24,25-dihydroxy-, and 25,26-dihydroxycholesterol derivatives from homocholenic acid ester [reaction (1)], 27-nor-25-ketocholesterol [reaction (2)], or demosterol [reaction (3)]. Direct photooxygenation of the cholesterol side chain [100] provides another reasonably efficient route to 25-hydroxy derivatives.

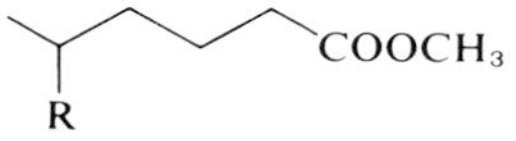

$$\text{R} \longrightarrow \text{COOCH}_3 \longrightarrow \text{25-OH- [42,92]} \tag{1}$$

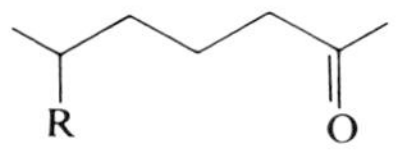

$$\longrightarrow \begin{array}{l} \text{25-OH- [93,94]; 24,25-(OH)}_2\text{- [95];} \\ \text{25,26-(OH)}_2\text{- [34,96]} \end{array} \tag{2}$$

$$\longrightarrow \begin{array}{l} \text{25-OH- [97]; 24,25-(OH)}_2\text{- [35,97,98];} \\ \text{25,26-(OH)}_2\text{- [99]} \end{array} \tag{3}$$

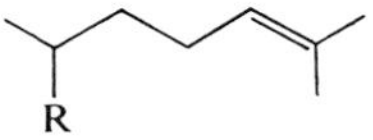

R = Δ^5 or $\Delta^{5,7}$-sterol nucleus.

Starting materials for reactions (1)–(3) are, unfortunately, not readily available from natural or commercial sources, although they can be prepared by degradation of the appropriate natural products. For example, homocholenic acid ester [reaction (1)] is accessible via homologation of cholenic acid [92], which together with the 25-keto derivative [reaction (2)] can be obtained as minor products of cholesterol degradation, and the conversion of fucosterol, a product of marine algae, to desmosterol [101] is a practical route to the latter if the former can be procured.

In general, however, resynthesis of the desired side chains onto readily available C_{19}, C_{21}, and C_{22} steroids represents the more practical, adaptable, and efficient approach and has been explored, therefore, in a number of studies. A recent preparation [102] of desmosterol, for example, utilized a 23-aldehyde derivative as an intermediate for side chain elongation via Grignard condensation [reaction (4)]. The required aldehyde is derived from the commercially available C_{22} acid (3β-acetoxybisnorcholenic acid) by

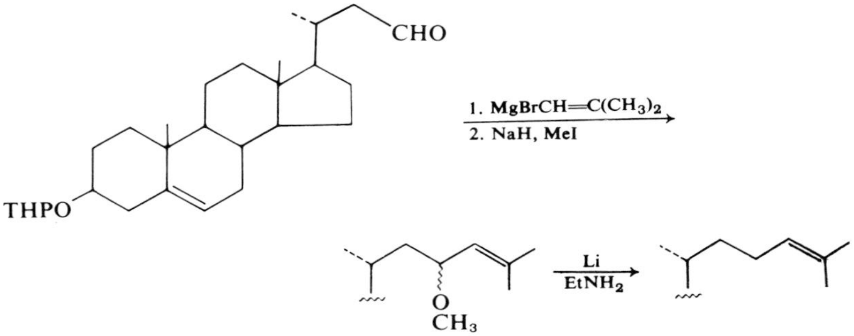

$$\frac{\text{1. MgBrCH=C(CH}_3)_2}{\text{2. NaH, MeI}} \tag{4}$$

$$\frac{\text{Li}}{\text{EtNH}_2}$$

Arndst–Eistert homologation followed by reduction of the 23-ester to the aldehyde and protection of the 3β-alcohol function as the tetrahydropyranyl ether (THP). An alternative approach [102] involves the direct introduction of an allyl unit by the coupling reaction between a 22-iodo steroid and π-(dimethylallyl)nickel bromide [reaction (5a)]. The iodo compound is available

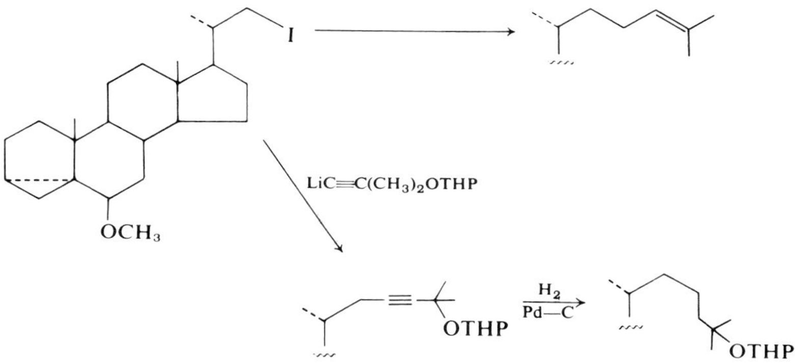

(5a)

LiC≡C(CH$_3$)$_2$OTHP

(5b)

from 3β-acetoxybisnorcholenic acid in a straightforward five-step sequence [102], or, more directly, from stigmasterol [103] in six steps [i.e. stigmasterol → i-ether → 22-al → 22-ol → 22-tosyloxy → 22-iodo]. Reaction of the same 22-iodo derivative with the appropriate lithium acetylide [reaction (5b)] represents an efficient and direct route to the 25-hydroxylated side chain [103].

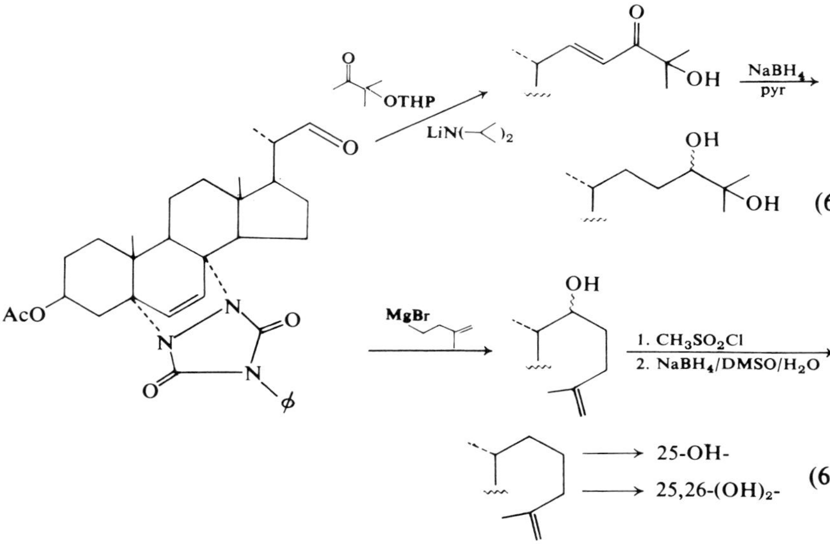

(6a)

(6b)

25-OH-
25,26-(OH)$_2$-

This approach, which gives 25-hydroxycholesterol in ca. 30% overall yield from stigmasterol, has the further advantage of providing, via modification of the 23,24-yne intermediate, a convenient and stereoselective route to both (24R)- and (24S)-24,25-dihydroxycholesterol [104], convertible to the respective vitamin D metabolite [105] and its 24-epimer by standard methods.

The 22-aldehyde obtainable from ergosterol by Diels–Alder adduct formation (to protect the 5,7-diene) and ozonolysis has been used as nucleus for the synthesis of several vitamin D metabolites [106]. After elaboration of the desired side chain as illustrated by reactions (6a) and (6b), the 5,7-diene is regenerated by reductive removal of the triazoline protecting group.

25-Hydroxycholesterol has also been prepared [107] in 25–28% yield from pregnenolone acetate by a four-step sequence [reaction (7)] and in 42% yield from androstenolone acetate in ten steps [108] as summarised in reaction (8). Thus, reaction of pregnenolone acetate with vinylmagnesium chloride gives an allylic alcohol that serves as the substrate for side-chain extension using diketene with collidine as catalyst in refluxing decalin. The resulting mixture of *cis*- and *trans*-ketoolefins can be reduced directly (without protection of the Δ^5 unsaturation) to a mixture of the saturated (20R)- and (20S)-25-ketones, from which the desired (20R)-isomer is obtained by fractional crystallization. A Grignard reaction then yields 25-hydroxycholesterol.

In reaction (8), a two-carbon side chain is added to 3β-acetoxy-5-androsten-

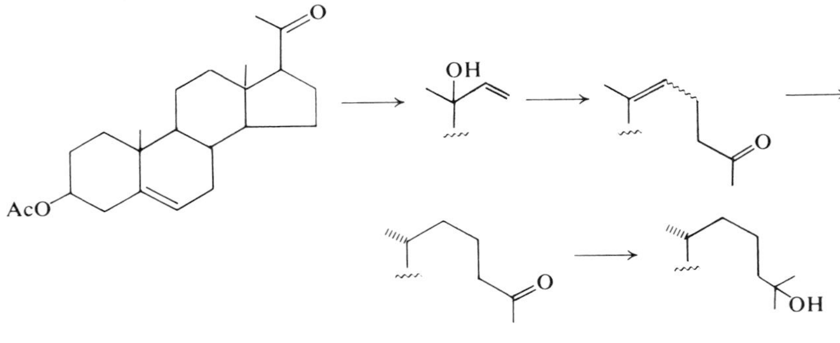

$$(7)$$

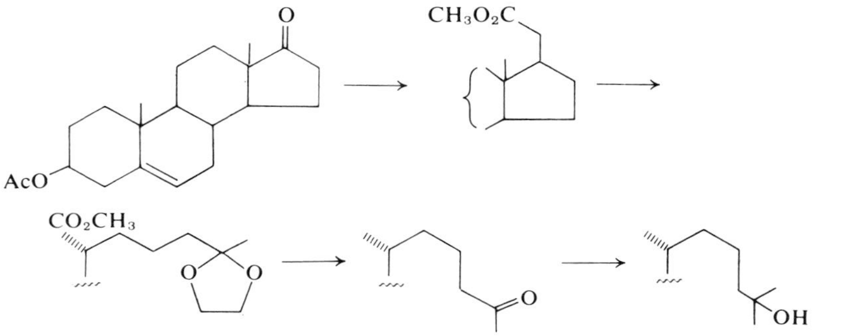

$$(8)$$

17-one by Reformatsky reaction, dehydration, and selective reduction. Side-chain extension is accomplished by alkylation of the resulting ester (after conversion of the 3β-Ac to the THP ether) with the ethylene ketal derivative of 5-bromo-2-pentanone and diisopropyllithium amide as base. Alkylation appears to be highly stereospecific, giving the correct isomer exclusively. Reduction of the 21-ester to methyl (via the 21-ol and 21-tosylate) and removal of the ketal lead eventually to 25-hydroxycholesterol after a final Grignard reaction.

With the exception of the Hoffmann–La Roche synthesis [104] all reported routes to the 24,25- and 25,26-dihydroxy sterols yield epimeric mixtures. The C-24 epimers of the 24,25-diol prepared by Ikekawa's group [97] have been separated, however, and absolute configurations have been assigned [35,36] by a modification of Horeau's method. The C-24 stereochemistry of the 24-OH-D_3 epimers was established similarly [109,110].

Introduction of oxygen at C-1 of the sterol has been accomplished by a number of routes. The synthesis of 1α,25-dihydroxycholesterol by Narwid *et al.* [44] may serve to illustrate one typical scheme [reaction (9)] involving initial protection of the Δ^5 unsaturation of the sterol starting material (25-hydroxycholesterol) followed by ring A modification and regeneration of the 5,6 double bond. The sequence gives the 1α-hydroxylated product in ca. 18% yield.

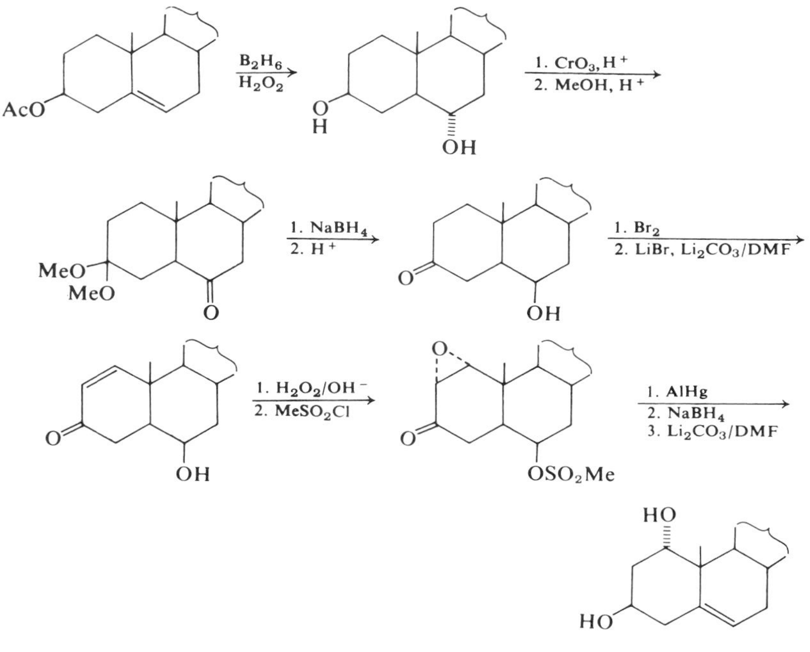

$$(9)$$

Reported preparations of 1α-hydroxycholesterol [111,112], 1α,25-dihydroxy-cholesterol [113] and 1α-OH-D₃ [114,115], 1α,25-(OH)₂D₃ [42], and 3-deoxy-1α-OH-D₃ [116] are based on essentially analogous synthetic concepts.

An alternative approach [117] to the 1α-hydroxy-5-cholestene system involves reduction (Li/liquid NH₃/THF), of a 1α,2α-epoxy-4,6-dien-3-one intermediate, in turn generated from cholesterol (or 25-hydroxycholesterol [43]) by dehydrogenation with dichlorodicyanoquinone (DDQ) and epoxidation [118]; 1α-hydroxycholesterol is obtained in ca. 25% overall yield [reaction (10)]. The reaction has since been used for the small-scale prepara-

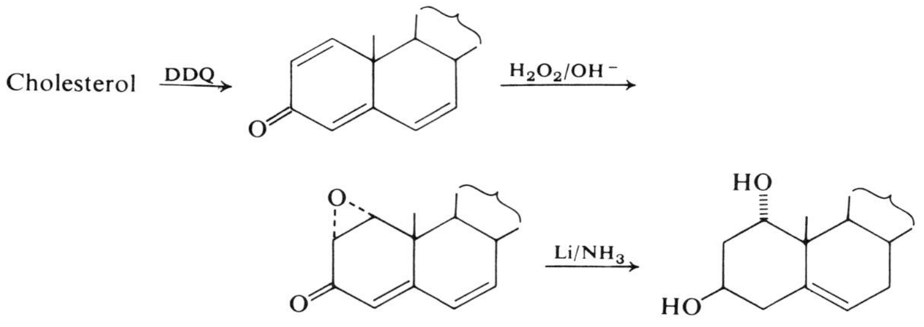

$$\text{Cholesterol} \xrightarrow{\text{DDQ}} \qquad \xrightarrow{\text{H}_2\text{O}_2/\text{OH}^-} \qquad \xrightarrow{\text{Li/NH}_3} \qquad (10)$$

tion of a number of vitamin metabolites, analogs, or the steroid precursors, including 1α,25-(OH)₂D₃ [43], 1α,24,25-(OH)₃D₃ [119], 1α-OH-D₃ [117], 1α-OH-D₂ [120], 1α,24-(OH)₂D₃ [121], 1α-hydroxypregnacalciferol [122], 1α-OH-3-*epi*-D₃ [123,124], 1α,25-dihydroxycholesterol [125], 1α-hydroxy-7-dehydrocholesterol [126], and 1α-hydroxy-5-cholestene [127].

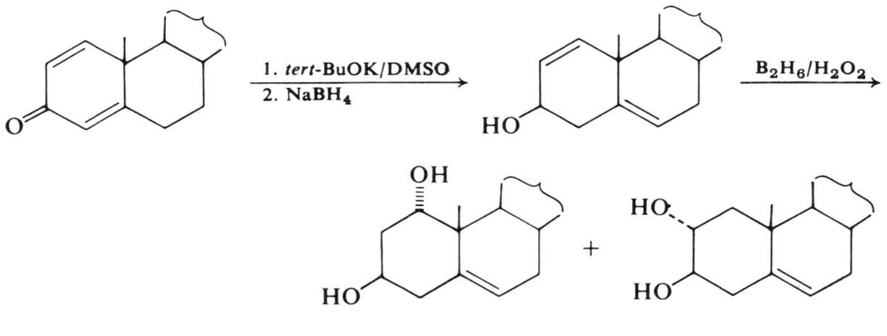

$$\xrightarrow[\text{2. NaBH}_4]{\text{1. } tert\text{-BuOK/DMSO}} \qquad \xrightarrow{\text{B}_2\text{H}_6/\text{H}_2\text{O}_2} \qquad + \qquad (11)$$

Deconjugation of a 1,4-dien-3-one intermediate to the 1,5-diene followed by ketone reduction and hydroboration is a third approach to the 1α,3β-dihydroxy-Δ⁵-sterol skeleton [128,129] [reaction (11)]. The procedure yields both 1α-hydroxy- and 2α-hydroxycholesterol, and the latter intermediate has also been converted to the corresponding 2α-OH-D₃ analog [130]. Application of this method to 1,4,6-cholestatrien-3-one leads directly to the 1α-hydroxylated provitamin skeleton [131].

Lythgoe's group has made major contributions to the total synthesis of the D vitamins. Their general synthetic concept involves the independent preparation of the ring A and ring C/D portions of the molecule and their coupling as a late step in the overall scheme. A partial illustration of their recent synthesis of 1α-OH-D_3 [91] may serve as an example [reaction (12), R = C_8H_{17}]. The sequence shown yields 1α-OH-D_3 in ca. 20% yield based on chloroketone starting material.

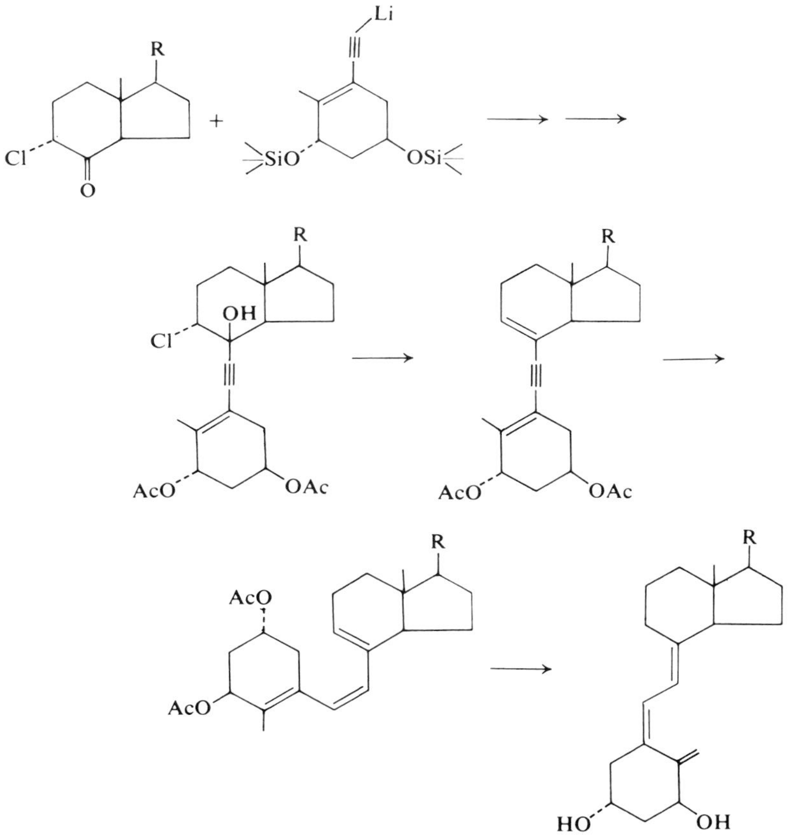

(12)

The development of efficient synthetic routes to vitamin D metabolites and analogs and the availability of suitable intermediates have also been exploited for the preparation of several radiolabeled derivatives, including [1,2-^{3}H]-vitamin D_3 [18], [3α-^{3}H]vitamin D_2 [72], [26,27-^{3}H]25-OH-D_3 [132], [1,2-^{3}H]dihydrotachysterol [133], [2-^{3}H]- and [6-^{3}H]1α-OH-D_3 [134,135], and [23,24-^{3}H]25-OH-D_3 (S. Yamada, H. K. Schnoes, H. F. DeLuca, unpublished).

Structure–Activity Relationships

One contribution of the synthetic work has been the definition, at least in rough outline, of the relative importance of specific structural units as determinants of biological activity. The preparation of structural variants has made possible a crude assessment of the functional significance of the hydroxy groups in $1\alpha,25\text{-}(OH)_2D_3$, of side-chain and triene chromophore structure, and of stereochemical factors. These assessments are based on *in vivo* and *in vitro* activity comparisons between vitamin metabolites and analogs. *In vivo* assays include intestinal calcium transport, bone calcium and phosphate mobilization, and antirachitic activity, whereas *in vitro* comparisons depend on measurements of calcium resorption from fetal rat bone in culture and of relative binding affinity for the cytosol–chromatin receptor system from chick intestine. A third *in vitro* system—measurement of calcium-binding protein synthesis in response to vitamin treatment in cultured embryonic chick intestine—has been used less extensively. Although available results are as yet too fragmentary to allow detailed cross-comparisons, they do suffice to identify major relationships and trends.

THE HYDROXY GROUPS OF $1\alpha,25\text{-}(OH)_2D_3$

It can be stated at the outset that all three hydroxy groups of $1\alpha,25\text{-}(OH)_2D_3$ are required for full activity. Modification of any of them reduces the biological potency of the resulting product (as measured by both *in vivo* and *in vitro* assays). The hydroxy functions differ, however, in their relative effect on activity. Modification at C-1 or C-25 affects potency very dramatically; modification at C-3 affects potency only moderately. Thus, there can be no doubt that under physiological circumstances biological activity depends on the presence of 1α-hydroxy function. The activity patterns of the natural metabolites provide clear support for this conclusion. (a) In physiological doses, $25\text{-}OH\text{-}D_3$ and $24,25\text{-}(OH)_2D_3$ elicit no response *in vivo* when 1α-hydroxylation is prevented by nephrectomy, whereas $1,25\text{-}(OH)_2D_3$ and $1,24,25\text{-}(OH)_3D_3$ are fully active [37,51–54,68]. (b) Structural analogs possessing a 1α-hydroxy group are highly potent *in vivo*; $1\alpha\text{-}OH\text{-}D_3$ [115,136], $1\alpha\text{-}OH\text{-}D_2$ [120], 3-deoxy-$1\alpha\text{-}OH\text{-}D_3$ [116,137], and 3-deoxy-$1\alpha\text{-}25\text{-}(OH)_2D_3$ [138] are pertinent examples (see Fig. 4 for structures). (c) Even more drastically modified compounds are active if a 1α-hydroxy group is present; analogs such as 5,6-*trans*-D_3, dihydrotachysterol$_3$ (DHT$_3$), and their 25-hydroxy derivatives, in which rotation of ring A has transformed the original 3β-OH into a pseudo 1α-hydroxy group, illustrate this point [139]. (d) Hydroxylation at C-1 always increases the affinity of an analog for the intestinal receptor protein of $1\alpha,25\text{-}(OH)_2D_3$ and increases its activity in the *in vitro* bone resorption assay (see below). This is not to say that C-1 deoxy compounds are totally inactive. At extremely high doses, for example, 25-OH-D_3 does elicit a biological response *in vivo*, even when C-1 hydroxylation is prevented by nephrectomy [6]. A 1-deoxy compound thus can activate the biological machinery but represents a poor structural analog of the physiologically active material.

The necessity for a 25-hydroxy group appears equally clear, in spite of the fact that 25-deoxy analogs, such as 1α-OH-D_3, 1α-OH-D_2, and 3-deoxy-1α-OH-D_3, exhibit very pronounced *in vivo* activity [115,116,120,136]. Extensive *in vivo* comparisons of 1α-OH-D_3 vs. 1α,25-$(OH)_2D_3$ indicate, for example, that in the rat 1α-OH-D_3 is about half as active as the natural hormone [140], and the compounds have been assessed as equally potent in the chick [136]. In the case of 1α-OH-D_3 it can be shown, however, that activity almost certainly depends on prior metabolism to 1α,25-$(OH)_2D_3$ [134,135,141,142]. The conversion of tritiated 1α-OH-D_3 to 1α,25-$(OH)_2D_3$ in both chicks [142] and rats [134,141] has been established, and it was shown that the appearance of product precedes onset of biological activity by at least 2 hr. Although this result does not totally preclude a direct action of 1α-OH-D_3 on intestine or bone, careful monitoring of the time course of accumulation of tritiated 1α,25-$(OH)_2D_3$ in the tissues vs. the onset of a biological response after a pulse dose of tritiated 1α-OH-D_3 leaves little doubt that 1α,25-$(OH)_2D_3$ is the active species [141,142]. The same conclusion can be drawn from receptor binding experiments [143]. It is of some interest that 25-hydroxylation of 1α-OH-D_3 occurs in the liver of rats [141] and in liver and intestine of chicks [142]. The ability of one of the target organs in the chick to hydroxylate 1α-OH-D_3 (and thus ensure efficient utilization of this compound) may explain the observation that the 1α-OH-D_3 and 1α,25-$(OH)_2D_3$ appear to have roughly equal potency in the chick [136]. On the basis of these results, expression of activity by other 25-deoxy analogs (e.g., DHT_3, 1α-OH-D_2) should then also require prior side-chain hydroxylation. This appears to be the case; metabolism of 5,6-*trans*-vitamin D_3 yields 25-OH-5,6-*trans*-D_3 [144], and DHT_3 [145] is converted to 25-OH-DHT_3 [133]. Furthermore, one might speculate that analogs that cannot be hydroxylated at C-25 should be inactive. Results of biological assays of 1α-hydroxypregnacalciferol (1α-OH-PC [122]) and 24-nor-1α,25-$(OH)_2D_3$ [139] (see Fig. 4), although admittedly not perfect models for testing this question, are in accord with this prediction: neither compound elicits a response *in vivo* [122,139].

Assays of a series of analogs suggest that the 3β-OH function has a relatively modest effect on activity. For example, 3-deoxy-1α,25-$(OH)_2D_3$ [138] is a highly potent analog, and two related compounds, 3-deoxy-1α-OH-D_3 [116,137,146] and 1α-OH-D_3 3-methyl ether [139], likewise elicit a pronounced response, as measured by intestinal calcium transport, bone calcium and bone phosphate mobilization, and antirachitic activity [116,137]. Direct comparison of 3-deoxy-1α-OH-D$_3$ in the rat shows the former analog to be about 20–50 times less active *in vivo*, depending on the specific process assayed [147]. These observations and the activity of compounds such as DHT$_3$, 25-OH-DHT$_3$, 5,6-*trans*-D$_3$, and 25-OH-5,6-*trans*-D$_3$ [139,148,149], which in a biological sense also represent "3-deoxy" derivatives, suggest that among the hydroxyl groups of 1α,25-(OH)$_2$D$_3$ the C-3 substituent is least important for biological function. This assumes that hydroxylation at C-3 does not occur *in vivo*. The nearly identical time course of the response elicited by 3-deoxy-1α-OH-D$_3$ and 1α-OH-D$_3$ in bone [147] and the fact that 3-deoxy-1α-OH-D$_3$ and 1α-OH-D$_3$ 3-methyl ether exhibit very similar activity patterns [139] suggest that this is a reasonable assumption. Data from *in vitro* assays of receptor binding and bone resorption corroborate this conclusion (see below).

Other ring A modifications, for example, introduction of hydroxyl into the 4α [150] and 2α positions [130] of vitamin D$_3$, lead to inactive derivatives, perhaps because these analogs cannot be metabolized to the required 1α,25-dihydroxy compounds.

SIDE-CHAIN STRUCTURE

The foregoing results suggest that proper side-chain length, to allow placement of a hydroxy group at the normal C-25 position, is of prime consequence. The contrast in potency between 1α-OH-PC and 1α-OH-D$_3$ underscores this point [122]. The system appears to be quite specific for a 25-OH group, since even 24-nor-1α,25-(OH)$_2$D$_3$ (see Fig. 4) does not elicit a response *in vivo* [139]. If a 25-OH group is present (or if it can be introduced metabolically), modification of the hydrocarbon skeleton may have no or only moderate influence on activity. The equal potency of vitamins D$_3$ and D$_2$ (as well as their metabolites) and the comparable activities of the corresponding dihydrotachysterol pair, DHT$_3$ and DHT$_2$ [145,151], are cases in point, although the degree of toleration is species dependent. We mentioned earlier that in birds vitamin D$_2$ and its metabolites possess only one-tenth the potency of the corresponding D$_3$ relatives. Biological activity data on 22,23-dihydrovitamin D$_2$ (vitamin D$_4$) adhere to this pattern [152,153], but (20S)-22-hydroxyvitamin D$_4$ [(20S)-22-OH-D$_4$] is reported to exhibit no antirachitic activity [154].

Activity comparisons between side-chain-modified 25-OH-D$_3$ derivatives show similar trends. Removal of methyl substituents (e.g., 27-nor-25-OH-D$_3$ or the 26,27-bisnor compound [155,156]) markedly reduces but does not abolish activity of the analogs [155], but alteration of side-chain length (e.g., 24-nor-25-OH-D$_3$ [157] or 20-hydroxypregnacalciferol [155]) leads to inactive compounds [155,157]. Any activity of these compounds depends, of course, on *in vivo* conversion to their respective 1α-hydroxy analogs [149], and impaired metabolism, therefore, may be an important cause of diminished activity [155].

Introduction of additional hydroxy functions always reduces activity somewhat, but the effect can be accentuated by stereochemical factors (see next section). For example, the potency of $(24R)$-$1\alpha,24,25$-$(OH)_3D_3$ is estimated to be about 60% that of $1\alpha,25$-$(OH)_2D_3$, and $(24R)$-$24,25$-$(OH)_2D_3$ is also somewhat less active than 25-OH-D_3 in rats [68,69]; in the chick, as already discussed, any substitution at C-24 results in a marked reduction of potency (e.g., ca. 90%). Furthermore, slight alteration of hydroxy substitution pattern has no deleterious effect: $(24R)$-24-OH-D_3 exhibits an activity pattern comparable to that of 25-OH-D_3 and $(24R)$-$24,25$-$(OH)_2D_3$, perhaps because it undergoes metabolism to the latter [109,158].

STEREOCHEMICAL EFFECTS

It would be surprising if stereochemical differences between analogs were not reflected by activity profiles. The few examples available to test the question do show expected effects. Deviation of hydroxy stereochemistry from the natural always reduces activity, although the magnitude of the effect depends on the specific vitamin D-dependent process examined. Thus, $(24R)$-$24,25$-$(OH)_2D_3$ (the natural metabolite) stimulates bone calcification, intestinal calcium transport, and bone calcium mobilization, but its 24-epimer, $(24S)$-$24,25$-$(OH)_2D_3$, exhibits no antirachitic activity and has no effect on bone calcium liberation but does exhibit activity approaching that of the 24-epimer in the intestinal transport assay [36]. The same activity pattern is observed [158] for the pair $(24R)$-24-OH-D_3 and $(24S)$-24-OH-D_3. Since none of these compounds shows activity in nephrectomized animals it is clear that the active species in all cases are the corresponding 1-hydroxylated derivatives, and one would thus expect that the recently prepared, but not yet thoroughly tested, epimeric pairs $(24R)$- and $(24S)$-$1\alpha,24$-$(OH)_2D_3$ [121] and $(24R)$- and $(24S)$-$1\alpha,24,25$-$(OH)_3D_3$ [119] exhibit rather similar activity profiles.

There are as yet insufficient data to accurately assess the effect of ring A hydroxy stereochemistry on activity. Preliminary results on $1\alpha,25$-$(OH)_2$-3-epi-D_3 [159] show that inversion of stereochemistry at C-3 reduces potency very dramatically, much more so in fact than the lack of the C-3 substituent. Other C-3 epimers, including 3-epi-D_3 [160], its 5,6-trans isomer [160], and 1α-OH-3-epi-D_3 [123,124], have been prepared, but appropriate biological data are not available.

Modification of the triene system, in particular the change of the 5,6 double bond form the natural 5,6-cis to 5,6-trans geometry, reduces potency. Thus, analogs such as 5,6-trans-D_3, 25-OH-5,6-trans-D_3, DHT_3, and 25-OH-DHT_3 stimulate intestinal calcium transport and bone calcium resorption, although relatively high doses (2.5–25 μg, i.e., 10–100 times greater than those effective for the potent analogs) are required to elicit a pronounced response [139]. These compounds are, of course, relatively poor models for specifically assessing the influence of triene structure, since they all lack a hydroxy function of the " C-3 " position, and their activity therefore reflects the effect of both structural changes. More relevant models, such as $1\alpha,25$-$(OH)_2$-5,6-*trans*-D_3 have not been tested, and directly comparable *in vivo* data on appropriate analog pairs such as 5,6-*trans*-D_3 vs. 3-deoxy-1α-OH-D_3, or 25-OH-5,6-*trans*-D_3 vs. 3-deoxy-$1\alpha,25$-$(OH)_2D_3$ are not available. *In vivo* assays do indicate, however, that the *cis*-triene analog is the more effective member of each pair, and *in vitro* measurements (see below) tend to corroborate this conclusion.

ACTIVITY IN *in vitro* SYSTEMS

Interpretation of *in vivo* assay results of analogs is complicated by factors that may be unrelated to the mechanism of action of the compound. The intrinsic activity of a given analog may be masked *in vivo* by its inability to reach its site of action. Impaired transport to the organ or into the appropriate cells (because of low affinity for required transport proteins [2,7]), susceptibility to degradative metabolism, and excretion are but some of the factors that can influence profoundly the target-tissue concentration of the material. Conversely, an inactive or relatively poorly active species may undergo metabolism to a highly potent derivative; the conversion of 1α-OH-D_3 to $1\alpha,25$-$(OH)_2D_3$ is a pertinent example. In different animal species these effects may be accentuated. The discrimination of birds against the D_2 side chain, apparently because of a rapid degradative metabolism, is a telling example. *In vivo* activity then provides a measure of the overall efficacy of a structural analog as a substitute for the natural material, but it may be difficult to distinguish between structural parameters effecting activity per se from those affecting other processes. *In vitro* assays circumvent some of these complications and can yield fairly clear-cut comparative data on structure–activity relationships if it is remembered that the biological relevance of results obtained for *in vitro* (i.e., artificial) systems depends ultimately on the degree to which they can be corroborated by tests in the intact animal. We will consider here data from *in vitro* assays that have been used quite extensively: (a) calcium resorption from fetal rat bone in tissue culture and (b) the relative binding affinity of vitamin metabolites and analogs for a cytosol–chromatin receptor complex from chick intestine. Data from both assays are in good agreement in spite of the fact that they relate to different organs from different species and measure two very distinct phenomena—a complex and multistep biological process (Ca resorption) in one case and a strictly physical–chemical interaction (vitamin–receptor binding) in the other.

In the bone resorption assay, $1\alpha,25$-$(OH)_2D_3$ is the most potent compound [56,57,161,162], exhibiting a pronounced response in the concentration range of 10^{-11}–10^{-10} M. Removal of either the C-25 or the C-1 hydroxy group (i.e., 25-OH-D_3 and 1α-OH-D_3) has roughly the same effect, reducing activity by two to three orders of magnitude [31,161]. Analogs such as 25-OH-DHT_3 [163]

and 5,6-*trans*-25-OH-D_3 [161], which lack a "C-3" hydroxy function (and feature a modified triene system), exhibit about the same potency as 1α-OH-D_3 and 25-OH-D_3. The degree to which C-24 hydroxy substitution diminishes potency depends markedly on stereochemistry. Thus (24R)-1α,24,25-(OH)$_3D_3$ is about 10 times less effective than 1α,25-(OH)$_2D_3$, but the unnatural (24S)-isomer appears to be about 1000 times less potent [162]. A similar relative order of activity is observed for the group 25-OH-D_3, (24R)-24,25-(OH)$_2D_3$, and (24S)-24,25-(OH)$_2D_3$ [164]. These results are, of course, in accord with *in vivo* studies [36,158] mentioned earlier, showing (24R) compounds to possess somewhat reduced potency but the (24S) analog to have no effect on calcium mobilization from bone. Also in agreement with *in vivo* results is the observation that 1,25-(OH)$_2D_2$ is about as active as 1α,25-(OH)$_2D_3$ *in vitro* [162] and that other side-chain modifications drastically diminish potency; 24-nor-25-OH-D_3 is essentially inactive [162], and 1α-hydroxypregnacalciferol, which does not elicit a response *in vivo*, is active *in vitro* only at high (10^{-6} M) concentrations [122]. None of the monohydroxylated compounds, such as D_3, DHT$_3$, 5,6-*trans*-D_3, and 3-deoxy-1α-OH-D_3, which are active *in vivo* (because of metabolism to 25-hydroxy or 1α,25-dihydroxy derivatives) are effective in the *in vitro* system, even at the highest feasible concentrations (10^{-5}–10^{-4} M).

The chick intestinal receptor binding assay is an even "simpler" *in vitro* system, measuring the affinity of a vitamin ligand for its macromolecular receptor [60,89,165]. In practice, the combined cytosol–chromatin receptor system (in which vitamin binding to chromatin via the cytoplasmic receptor is determined) has been studied most extensively, but a procedure using the cytoplasmic receptor alone gives entirely analogous results [166].

Affinity for the receptor is again determined primarily by the hydroxy functions of the vitamin D molecule. The natural hormone, 1α,25-(OH)$_2D_3$, is the most effective ligand known ($K_D \sim 10^{-9}$ M), while its analogs form a sequence of relative affinity rather similar to the activity progression observed for the *in vitro* bone assay. Removal of the C-3 hydroxy group [3-deoxy-1α,25-(OH)$_2D_3$] reduces binding by a factor of 8 [167], and removal of either the C-1 or C-25 hydroxyl diminishes affinity by a factor of 500–1000 [165,167]. To a first approximation, at least, the effect of hydroxy groups on binding is strictly additive; 3-deoxy-1α-OH-D_3, lacking both the C-3 and C-25 hydroxy functions, is 5000 times, and vitamin D_3 itself is more than 10,000 times, less effective than 1α,25-(OH)$_2D_3$ in competing for the receptor sites. Addition of a 24-hydroxy group reduces affinity for the receptor; 1α,24,25-(OH)$_3D_3$ binds about 10 times less well than 1α,25-(OH)$_2D_3$ (J. A. Eisman and H. F. DeLuca, unpublished), and the same relationship is observed between 24,25-(OH)$_2D_3$ and 25-OH-D_3 [167]. In fact, introduction of a C-24 hydroxyl reduces binding affinity by about the same factor as removal of the C-3 function (compare

3-deoxy-1α,25-(OH)$_2$D$_3$ vs. 1α,24,25-(OH)$_3$D$_3$). Of special interest is the observation that C-24 stereochemistry has no effect, with both (R) and (S) analogs binding equally well to the receptor [167], a finding that contrasts bone and intestinal specificities (see above) and is in complete accord with *in vivo* activities of these derivatives. Alteration of triene structure has a relatively modest effect; 25-OH-DHT$_3$ and 5,6-*trans*-25-OH-D$_3$ bind about 10 and 75 times less effectively than 3-deoxy-1α,25-(OH)$_2$D$_3$ [167]. Competitive binding experiments thus establish the following order of vitamin–receptor affinity: 1α,25-(OH)$_2$D$_3$ $\approx$ 1α,25-(OH)$_2$D$_2$ > 3-deoxy-1α,25-(OH)$_2$D$_3$ $\approx$ (24R)-1α,24,25-(OH)$_3$D$_3$ $\approx$ (24S)-1α,24,25-(OH)$_3$D$_3$ > 25-OH-DHT$_3$ > 5,6-*trans*-25-OH-D$_3$ > 1α-OH-D$_3$ $\approx$ 25-OH-D$_3$ > 3-deoxy-1α-OH-D$_3$ $\approx$ (24R)-24,25-(OH)$_2$-D$_3$ $\approx$ (24S)-24,25-(OH)$_2$D$_3$ $\gg$ D$_3$. The bone resorption assay suggests roughly the same sequence with two general exceptions. Stereochemistry at C-24 markedly affects potency, and the C-3 hydroxy group appears to be relatively more important for expression of activity, a specificity that might be exploitable for the design of organ-selective analogs.

MECHANISM OF ACTION OF VITAMIN D

How does 1α,25-(OH)$_2$D$_3$ stimulate intestinal calcium and phosphate transport and calcium resorption from bone? To this most basic question we will respond with only a brief review of recent results, chiefly because a precise answer is not yet possible in any case and secondarily because existing discussions [2,4–6] of the biochemistry and physiology of these processes already provide much more comprehensive introductions than could be attempted in this space.

Calcium transport in intestine is an active process against both an electrical and concentration gradient. There appears to be general agreement that vitamin D [or more specifically 1α,25-(OH)$_2$D$_3$] brings about an alteration of the brush border region (the microvilli membranes projecting into the intestine lumen) of intestinal mucosal cells to facilitate and promote calcium translocation into the cell. What needs to be defined, then, is the sequence of biochemical events triggered by 1α,25-(OH)$_2$D$_3$ to initiate this process. A popular concept at present and one that, because of its close analogy to the mechanism of action of other steroid hormones, has considerable appeal describes vitamin D action in terms of several discreet chemical events, including (a) initial noncovalent binding of 1α,25-(OH)$_2$D$_3$ to a specific receptor protein in the cytoplasm of intestinal cells, to allow (b) entry of this protein–ligand complex into the nucleus and interaction with specific sites (receptors) on the nuclear chromatin, causing (c) activation and transcription of DNA sequences into messenger RNA, which in turn codes for (d) a specific protein or proteins required for the calcium transport process. The hypothesis is attractive, both because it is reasonable *a priori* and because for each of its components some experimental evidence can be marshalled.

The scheme requires that administered vitamin becomes eventually associated with the nucleus, and experiments on the subcellular distribution of radioactive $1\alpha,25\text{-}(OH)_2D_3$ accumulating in intestine have shown the nuclear fraction to contain most of the radioactivity [168,169]. Association with the chromatin was suggested by the isolation of chromatin fractions containing bound $1\alpha,25\text{-}(OH)_2D_3$, but the highly impure state of such preparations made it impossible to infer a specific chromatin–vitamin interaction. More recent work on cytoplasmic–nuclear receptors of $1\alpha,25\text{-}(OH)_2D_3$ has placed the concept of a nuclear action of the vitamin on somewhat firmer footing. The presence (in chick intestine) of a cytoplasmic receptor protein specific for $1\alpha,25\text{-}(OH)_2D_3$ appears to be reasonably well established [60,61,165–167]. This cytosol receptor–vitamin complex in turn can be shown to associate with isolated crude chromatin from mucosal cells. The cytoplasmic–nuclear receptor system exhibits properties analogous to those established for target-tissue receptors of other steroid hormones. The receptor machinery is specific to the target tissue intestine; binding of vitamin to chromatin material requires prior association with the cytoplasmic receptor protein; binding sites are selective for $1\alpha,25\text{-}(OH)_2D_3$ and finite (i.e., saturable); transfer of the cytoplasmic receptor–vitamin complex to chromatin binding sites is temperature dependent [60,61,170]. All of these results pertain to chick intestine, and for that tissue the demonstration of target receptors for $1\alpha,25\text{-}(OH)_2D_3$ appears to have a sound basis. In the case of rat intestine, unfortunately, attempts to demonstrate a similar $1\alpha,25\text{-}(OH)_2D_3$-specific binding protein have not been successful. Rat intestine, in fact, contains a binding protein, but one that has higher affinity for $25\text{-}OH\text{-}D_3$ than for $1\alpha,25\text{-}(OH)_2D_3$, and since this protein can also be demonstrated in several nontarget tissues, a role as $1\alpha,25\text{-}(OH)_2D_3$ receptor can hardly be assumed [171]. Whether failure to detect a specific binding protein in rat intestinal cytosol reflects experimental difficulties (i.e., extreme lability of the protein) or more fundamental differences between the chick and rat system remains to be established.

A lag phase of several hours between administration of $1\alpha,25\text{-}(OH)_2D_3$ and intestinal calcium transport response suggests protein synthesis as an intervening process. For chick intestine, *in vivo* stimulation of RNA synthesis after administration of $1\alpha,25\text{-}(OH)_2D_3$ has been reported [172], and data have been presented to show that inhibition of protein synthesis by actinomycin D prevents expression of activity of $1\alpha,25\text{-}(OH)_2D_3$ both *in vivo* [173] and *in vitro* [174]. Actinomycin D pretreatment, however, appears to have no effect on $1\alpha,25\text{-}(OH)_2D_3$ activity in rat intestine (although it does block expression of activity in rat bone) [6]. Nevertheless, even in the rat, the administration of large doses of vitamin D_3 to vitamin D-deficient animals increases target-tissue chromatin template capacity [175], and for chick intestine Zerwekh *et al.* [176] have recently shown that $1\alpha,25\text{-}(OH)_2D_3$ increases nuclear template activity for RNA synthesis both *in vivo* and *in vitro*. To demonstrate this nuclear response *in vitro*, incubation with intestinal cytosol (i.e., the presence of the cytoplasmic receptor) is essential, and, in agreement with the relative binding affinities of 25-OH-D_3 and $1\alpha,25\text{-}(OH)_2D_3$ for the cytoplasmic–chromatin receptor system discussed earlier, it was shown that 25-OH-D_3 at 500-fold higher concentration could substitute for the natural hormone as activator of the nuclear template mechanism.

Intestine of chicks, rats, and other species contains a calcium binding protein that is synthesized *in vivo* in response to vitamin D treatment [177]. Vitamin D-dependent synthesis of this protein has also been demonstrated in organ cultures of embryonic chick intestine, where again $1\alpha,25\text{-}(OH)_2D_3$ is by far the most potent inducer [174], and in cell-free preparations of chick intestinal polysomes [178]. Evidence of this kind points to calcium binding protein as a strong candidate for the ultimate product of $1\alpha,25\text{-}(OH)_2D_3$ action and as an important element of the calcium transport machinery. Its synthesis is clearly induced by $1\alpha,25\text{-}(OH)_2D_3$, it binds calcium ion (four ions per protein), its appearance in intestine correlates approximately·with calcium transport activity, and its apparent location on the surface of intestinal epithelial cells would be in accord with a role as chelator of dietary calcium and mediator of its translocation.

The relationship between calcium binding protein and calcium transport is, however, not as tight and clear-cut as might be desired (see discussion in reference [6]), suggesting that, if calcium binding protein is a key component of the calcium transporting machinery, it may not be the only one. Another candidate may be a brush border protein recently described by Moruichi and DeLuca [179] that appears to be modified by $1\alpha,25\text{-}(OH)_2D_3$ treatment and has other properties (affinity for calcium, time course of appearance in relation to transport, alkaline phosphatase activity) in accord with a role in the calcium translocation mechanism.

Movement of calcium across the epithelial cell and transport across its basal membrane are other outstanding issues. Translocation via a mitochondrial shuttle and eventual transport across the basal membrane via a sodium–calcium exchange reaction are favorite models, more fully explored in other reviews [2,6].

Mechanistic details on the effect of $1\alpha,25\text{-}(OH)_2D_3$ on intestinal phosphate transport are completely lacking, and little more can be said about the effect of the hormone on bone. As already mentioned, there is no clear demonstration of a direct action of the vitamin in the calcification of bone, but its involvement in the calcium mobilization process is well established. Vitamin activity in bone requires the presence of parathyroid hormone and, since expression of activity is blocked by actinomycin D [180] the presumption that the hormone acts via induction of synthesis of specific proteins is warranted. Nothing, however, is known about the overall process [6].

It appears, then, that vitamin D function in intestine and bone involves, at least in part, an induction process resulting in the synthesis of specific proteins. The steroid-hormone-like mechanism of action proposed for the vitamin in chick intestine must currently serve as the most useful working model, both on the strength of available evidence and for lack of an alternative equally consonant with experiment.

ACKNOWLEDGMENTS

Some of the original research summarized here was supported by Grant AM-14881 and Contract 72-2226 from the National Institutes of Health, by contract E (11-1)-1668 from the U.S. Energy Research and Development Administration, and by funds from the Wisconsin Alumni Research Foundation.

REFERENCES

1. W. H. Sebrell, Jr. and R. H. Harris, eds., "The Vitamins," 2nd ed., Vol. 3, Chapter 7, pp. 155–297. Academic Press, New York, 1971.
2. J. L. Omdahl and H. F. DeLuca, *Physiol. Rev.* **53**, 327 (1973).
3. H. F. DeLuca, *Fed. Proc., Fed. Am. Soc. Exp. Biol.* **33**, 2211 (1974).
4. R. S. Harris, P. L. Munson, E. Diczfalusy, and J. Glover, eds., "Vitamins and Hormones," Vol. 32, pp. 277–428. Academic Press, New York, 1974.
5. A. W. Norman and H. L. Henry, *Recent Prog. Horm. Res.* **30**, 431 (1974).
6. H. F. DeLuca and H. K. Schnoes, *Annu. Rev. Biochem.* **46**, 631 (1976).
7. L. V. Avioli and J. G. Haddad, *Metab., Clin. Exp.* **22**, 507 (1973).
8. L. F. Fieser and M. Fieser, "Steroids," Chapter 4, pp. 90–168. Van Nostrand-Reinhold, Princeton, New Jersey, 1959.
9. H. H. Inhoffen and K. Irmscher, *Fortschr. Chem. Organ. Naturst.* **17**, 70 (1959).
10. C. J. W. Brooks, *in* "Rodd's Chemistry of Carbon Compounds" (S. Coffey, ed.), 2nd ed., Vol. 2, Part D, pp. 134–148. Elsevier, Amsterdam, 1970.
11. E. Havinga, *Experientia* **29**, 1181 (1973).
12. L. Velluz, G. Amiard, and B. Goffinet, *Bull. Soc. Chim. Fr.* **22**, 1341 (1955).
13. K. Pfoertner and J. P. Weber, *Helv. Chim. Acta* **55**, 921 (1972).
14. K. Pfoertner, *Helv. Chim. Acta* **55**, 937 (1972).
15. F. Boosma, H. J. C. Jacobs, E. Havinga, and A. van der Gen, *Tetrahedron Lett.* p. 427 (1975).
16. A. G. M. Barrett, D. H. R. Barton, M. H. Pendlebury, R. A. Russell, and D. A. Widdowson, *J. Chem. Soc., Chem. Commun.* p. 101 (1975).
17. A. G. M. Barrett, D. H. R. Barton, R. A. Russell, and D. A. Widdowson, *J. Chem. Soc., Chem. Commun.* p. 102 (1975).
18. P. F. Neville and H. F. DeLuca, *Biochemistry* **5**, 2201 (1966).

19. J. Lund and H. F. DeLuca, *J. Lipid Res.* **7**, 739 (1966).
20. J. W. Blunt, H. F. DeLuca, and H. K. Schnoes, *Biochemistry* **7**, 3317 (1968).
21. G. Ponchon and H. F. DeLuca, *J. Clin. Invest.* **48**, 1273 (1969).
22. M. Horsting and H. F. DeLuca, *Biochem. Biophys. Res. Commun.* **36**, 251 (1969).
23. E. B. Olson, Jr., J. C. Knutson, M. H. Bhattacharyya, and H. F. DeLuca, *J. Clin. Invest.* **57**, 1213 (1976).
24. G. Tucker, III, R. E. Gagnon, and M. R. Haussler, *Arch. Biochem. Biophys.* **155**, 47 (1973).
25. A. W. Norman and H. F. DeLuca, *Biochemistry* **2**, 1160 (1963).
26. M. Bhattacharyya and H. F. DeLuca, *Arch. Biochem. Biophys.* **160**, 58 (1974).
27. M. H. Bhattacharyya and H. F. DeLuca, *J. Biol. Chem.* **248**, 2974 (1973).
28. A. W. Norman, *Am. J. Med.* **57**, 21 (1974).
29. J. G. Haddad and T. C. B. Stamp, *Am. J. Med.* **57**, 57 (1974).
30. J. W. Blunt, Y. Tanaka, and H. F. DeLuca, *Proc. Natl. Acad. Sci. U.S.A.* **61**, 1503 (1968).
31. C. L. Trummel, L. G. Raisz, J. W. Blunt, and H. F. DeLuca, *Science* **163**, 1450 (1969).
32. T. Suda, H. F. DeLuca, H. K. Schnoes, G. Ponchon, Y. Tanaka, and M. F. Holick, *Biochemistry* **9**, 2917 (1970).
33. M. F. Holick, H. K. Schnoes, H. F. DeLuca, R. W. Gray, I. T. Boyle, and T. Suda, *Biochemistry* **11**, 4251 (1972).
34. H.-Y. Lam, H. K. Schnoes, and H. F. DeLuca, *Steroids* **25**, 247 (1975).
35. M. Seki, N. Koizumi, M. Morisaki, and N. Ikekawa, *Tetrahedron Lett.* p. 15, (1975).
36. Y. Tanaka, H. F. DeLuca, N. Ikekawa, M. Morisaki, and N. Koizumi, *Arch. Biochem. Biophys.* **170**, 620 (1975).
37. I. T. Boyle, J. L. Omdahl, R. W. Gray, and H. F. DeLuca, *J. Biol. Chem.* **248**, 4174 (1973).
38. J. C. Knutson and H. F. DeLuca, *Biochemistry* **13**, 1543 (1974).
39. M. R. Haussler, J. F. Myrtle, and A. W. Norman, *J. Biol. Chem.* **243**, 4055 (1968).
40. M. F. Holick, H. K. Schnoes, and H. F. DeLuca, *Proc. Natl. Acad. Sci. U.S.A.* **68**, 803 (1971); M. F. Holick, H. K. Schnoes, H. F. DeLuca, T. Suda, and R. H. Cousins, *Biochemistry* **10**, 2799 (1971).
41. D. E. M. Lawson, P. W. Wilson, and E. Kodicek, *Biochem. J.* **115**, 269 (1969).
42. E. J. Semmler, M. F. Holick, H. K. Schnoes, and H. F. DeLuca, *Tetrahedron Lett.* p. 4147 (1972).
43. D. H. R. Barton, R. H. Hesse, M. M. Pechet, and E. Rizzardo, *J. Chem. Soc., Chem. Commun.* p. 203 (1974).
44. T. A. Narwid, J. F. Blount, J. A. Iacobelli, and M. R. Uskoković, *Helv. Chim. Acta* **47**, 781 (1974).
45. D. E. M. Lawson, D. R. Fraser, E. Kodicek, H. R. Morris, and D. H. Williams, *Nature (London)* **230**, 228 (1971).
46. D. R. Fraser and E. Kodicek, *Nature (London)* **228**, 764 (1970).
47. R. Gray, I. Boyle, and H. F. DeLuca, *Science* **172**, 1232 (1971).
48. A. W. Norman, R. J. Midgett, F. Myrtle, and H. G. Nowicke, *Biochem. Biophys. Res. Commun.* **42**, 1082 (1971).
49. M. R. Haussler, D. W. Boyce, E. T. Littledike, and H. Rasmussen, *Proc. Natl. Acad. Sci. U.S.A.* **68**, 177 (1971).
50. Y. Tanaka, H. Frank, and H. F. DeLuca, *Endocrinology* **92**, 411 (1973).
51. I. T. Boyle, L. Miravet, R. W. Gray, M. F. Holick, and H. F. DeLuca, *Endocrinology* **90**, 605 (1972).

52. R. G. Wong, A. W. Norman, C. R. Reddy, and J. W. Coburn, *J. Clin. Invest.* **52**, 1287 (1972).
53. M. F. Holick, M. Garabedian, and H. F. DeLuca, *Science* **176**, 1146 (1972).
54. T. C. Chen, L. Castillo, M. Korycka-Dahl, and H. F. DeLuca, *J. Nutr.* **104**, 1056 (1974).
55. H. F. DeLuca, Y. Tanaka, and L. Castillo, *Proc. Int. Parathyroid Horm. Conf., 5th, 1974*, Excerpta Med. Found. Int. Congr. Ser. No. 346, p. 305 (1975).
56. L. G. Raisz, C. L. Trummel, M. F. Holick, and H. F. DeLuca, *Science* **175**, 768 (1972).
57. J. J. Reynolds, H. F. Holick, and H. F. DeLuca, *Calcif. Tissue Res.* **12**, 295 (1973).
58. J. F. Myrtle, M. R. Haussler, and A. W. Norman, *J. Biol. Chem.* **245**, 1190 (1970).
59. D. E. M. Lawson and P. W. Wilson, *Biochem. J.* **144**, 573 (1974).
60. P. F. Brumbaugh and M. R. Haussler, *J. Biol. Chem.* **249**, 1251 (1974).
61. H. C. Tsai and A. W. Norman, *J. Biol. Chem.* **248**, 5967 (1973).
62. R. Kumar and H. F. DeLuca, *Biochem. Biophys. Res. Commun.* **69**, 197 (1976); R. Kumar, D. Harnden, and H. F. DeLuca, *Biochemistry* **15**, 2420 (1976); D. Harnden, R. Kumar, M. F. Holick, and H. F. DeLuca, *Science* **193**, 493 (1976).
63. R. W. Gray, J. L. Omdahl, J. G. Ghazarian, and H. F. DeLuca, *J. Biol. Chem.* **247**, 7528 (1972).
64. J. G. Ghazarian, H. K. Schnoes, and H. F. DeLuca, *Biochemistry* **12**, 2555 (1973).
65. J. G. Ghazarian and H. F. DeLuca, *Arch. Biochem. Biophys.* **160**, 63 (1974).
66. H. L. Henry and A. W. Norman, *J. Biol. Chem.* **249**, 7529 (1974).
67. J. G. Ghazarian, C. R. Jefcoate, J. C. Knutson, W. H. Orme-Johnson, and H. F. DeLuca, *J. Biol. Chem.* **249**, 3026 (1974).
68. M. F. Holick, A. Kleiner-Bosaller, H. K. Schnoes, P. M. Kasten, I. T. Boyle, and H. F. DeLuca, *J. Biol. Chem.* **248**, 6691 (1973).
69. I. T. Boyle, J. L. Omdahl, R. W. Gray, and H. F. DeLuca, *J. Biol. Chem.* **248**, 4174 (1973).
70. A. Kleiner-Bossaller and H. F. DeLuca, *Biochim. Biophys. Acta* **338**, 489 (1974).
71. T. Suda, H. F. DeLuca, H. K. Schnoes, and J. W. Blunt, *Biochemistry* **8**, 3515 (1969).
72. G. Jones, H. K. Schnoes, and H. F. DeLuca, *Biochemistry* **14**, 1250 (1975).
73. G. Jones, L. A. Baxter, and H. F. DeLuca, *Biochemistry* **15**, 713 (1976).
74. G. Jones, H. K. Schnoes, and H. F. DeLuca, *J. Biol. Chem.* **251**, 29 (1976).
75. M. R. Haussler, *Clin. Endocrinol. (Oxford)* **5**, 1515 (1976).
76. M. F. Holick, L. A. Baxter, P. K. Schraufvogel, T. E. Tavela, and H. F. DeLuca, *J. Biol. Chem.* **251**, 397 (1976).
77. I. T. Boyle, R. W. Gray, and H. F. DeLuca, *Proc. Natl. Acad. Sci. U.S.A.* **68**, 2131 (1971).
78. H. L. Henry, R. J. Midgett, and A. W. Norman, *J. Biol. Chem.* **249**, 7584 (1974).
79. M. R. Hughes, P. F. Brumbaugh, M. R. Haussler, J. E. Wergedahl, and D. J. Baylink, *Science* **190**, 578 (1975).
80. J. L. Omdahl, R. W. Gray, I. T. Boyle, J. Knutson, and H. F. DeLuca, *Nature (London), New Biol.* **237**, 63 (1972).
81. M. Garabedian, M. F. Holick, H. F. DeLuca, and I. T. Boyle, *Proc. Natl. Acad. Sci. U.S.A.* **69**, 1673 (1972).
82. D. R. Fraser and E. Kodicek, *Nature (London), New Biol.* **241**, 163 (1973).
83. Y. Tanaka and H. F. DeLuca, *Arch. Biochem. Biophys.* **154**, 566 (1973).
84. Y. Tanaka and H. F. DeLuca, *Science* **183**, 1198 (1974).

85. Y. Tanaka, R. S. Lorenc, and H. F. DeLuca, *Arch. Biochem. Biophys.* **171**, 521 (1975).

86. R. G. Larkins, S. J. MacAuley, and I. MacIntyre, *Nature (London)* **252**, 412 (1974).

87. H. F. DeLuca, *N. Engl. J. Med.* **289**, 359 (1973).

88. S. Edelstein, *Vitam. Horm. (N.Y.)* **32**, 407 (1974).

89. P. F. Brumbaugh, D. H. Haussler, K. M. Bursac, and M. R. Haussler, *Biochemistry* **13**, 4091 (1975).

90. M. F. Holick and H. F. DeLuca, *Annu. Rev. Med.* **25**, 349 (1974).

91. R. G. Harrison, B. Lythgoe, and P. W. Wright, *J. Chem. Soc., Perkin Trans. 1* p. 2654 (1974).

92. J. A. Campbell, D. M. Squires, and J. C. Babcock, *Steroids* **13**, 567 (1969).

93. S. J. Halkes and N. P. van Vliet, *Recl. Trav. Chim. Pays-Bas* **88**, 1080 (1969).

94. J. W. Blunt and H. F. DeLuca, *Biochemistry* **8**, 671 (1969).

95. H.-Y. Lam, H. K. Schnoes, H. F. DeLuca, and T. C. Chen, *Biochemistry* **12**, 4851 (1973).

96. J. Redel, P. A. Bell, N. Bazely, Y. Calando, F. Delbarre, and E. Kodicek, *Steroids* **24**, 463 (1974).

97. M. Morisaki, J. Rubio-Lightbourn, and N. Ikekawa, *Chem. Pharm. Bull.* **21**, 457 (1973).

98. J. Redel, P. Bell, F. Delbarre, and E. Kodicek, *C.R. Hebd. Seances Acad. Sci., Ser. D* **278**, 529 (1974).

99. M. Seki, J. Rubio-Lightbourn, M. Morisaki, and N. Ikekawa, *Chem. Pharm. Bull.* **21**, 2783 (1973).

100. A. Rotman and Y. Mazur, *J. Chem. Soc., Chem. Commun.* p. 15 (1974).

101. N. Ikekawa, M. Morisaki, H. Ohtaka, and Y. Chiyoda, *Chem. Commun.* p. 1498 (1971).

102. S. K. Dasgupta, D. R. Crump, and M. Gut, *J. Org. Chem.* **39**, 1658 (1974).

103. J. J. Partridge, S. Faber, and M. R. Uskoković, *Helv. Chim. Acta* **57**, 764 (1974).

104. J. J. Partridge, V. Toome, and M. R. Uskoković, *J. Am. Chem. Soc.* **98**, 3739 (1976).

105. M. R. Uskoković, E. Baggiolini, A. Mahgoul, T. Narwid, and J. J. Partridge, *in* "Vitamin D and Problems Related to Uremic Bone Disease" (A. W. Norman *et al.*, eds.), pp. 270–283. de Gruyter, Berlin, 1975.

106. S. C. Eyley and D. H. Williams, *J. Chem. Soc., Perkin Trans. 1* p. 727 (1976); 731 (1976).

107. T. A. Narwid, K. E. Cooney, and M. R. Uskoković, *Helv. Chim. Acta* **57**, 771 (1974).

108. J. Wicha and K. Bal, *J. Chem. Soc., Chem. Commun.* p. 968 (1975).

109. N. Ikekawa, M. Morisaki, N. Koizumi, M. Sawamura, Y. Tanaka, and H. F. DeLuca, *Biochem. Biophys. Res. Commun.* **62**, 485 (1975).

110. N. Koizumi, M. Morisaki, N. Ikekawa, A. Suzaki, and T. Takeshita, *Tetrahedron Lett.* p. 2203 (1975).

111. B. Pelc and E. Kodicek, *J. Chem. Soc. C* p. 1624 (1970).

112. M. Morisaki, K. Bannai, and N. Ikekawa, *Chem. Pharm. Bull.* **21**, 1853 (1973).

113. J. Rubio-Lightbourn, M. Morisaki, and N. Ikekawa, *Chem. Pharm. Bull.* **21**, 1854 (1973).

114. A. Fürst, L. Labler, W. Meier, and K. H. Pfoertner, *Helv. Chim. Acta* **56**, 1108 (1973).

115. M. F. Holick, E. J. Semmler, H. K. Schnoes, and H. F. DeLuca, *Science* **180**, 190 (1973).

116. H.-Y. Lam, B. L. Onisko, H. K. Schnoes, and H. F. DeLuca, *Biochem. Biophys. Res. Commun.* **59**, 845 (1974).
117. D. H. R. Barton, R. H. Hesse, M. M. Pechet, and E. Rizzardo, *J. Am. Chem. Soc.* **95**, 2748 (1973).
118. B. Pelc and E. Kodicek, *J. Chem. Soc. C* p. 1568 (1971).
119. N. Ikekawa, M. Morisaki, N. Koizumi, Y. Kato, and T. Takeshita, *Chem. Pharm. Bull.* **23**, 695 (1975).
120. H.-Y. Lam, H. K. Schnoes, and H. F. DeLuca, *Science* **186**, 1038 (1974).
121. M. Morisaki, N. Koizumi, N. Ikekawa, T. Takeshita, and S. Ishimoto, *J. Chem. Soc., Perkin Trans. 1* p. 1421 (1975).
122. H.-Y. Lam, H. K. Schnoes, H. F. DeLuca, L. Reeve, and P. H. Stern, *Steroids* **26**, 422 (1975).
123. M. Sheves, E. Berman, D. Freeman, and Y. Mazur, *J. Chem. Soc., Chem. Commun.* p. 643 (1975).
124. W. Okamura and M. R. Pirio, *Tetrahedron Lett.* p. 4317 (1975).
125. M. Morisaki, J. Rubio-Lightbourn, N. Ikekawa, and T. Takeshita, *Chem. Pharm. Bull.* **21**, 2568 (1973).
126. D. Freeman, A. Acher, and Y. Mazur, *Tetrahedron Lett.* **4**, 261 (1975).
127. M. N. Mitra, A. W. Norman, and W. H. Okamura, *J. Org. Chem.* **39**, 2931 (1974).
128. C. Kaneko, S. Yamada, A. Sugimoto, Y. Eguchi, M. Ishikawa, T. Suda, M. Suzuki, S. Kakuta, and S. Sasaki, *Steroids* **23**, 75 (1973).
129. C. Kaneko, S. Yamada, A. Sugimoto, M. Ishikawa, S. Sasaki, and T. Suda, *Tetrahedron Lett.* p. 2339 (1973); *Chem. Pharm. Bull.* **22**, 2101 (1974).
130. C. Kaneko, S. Yamada, A. Sugimoto, M. Ishikawa, T. Suda, M. Suzuki, and S. Sasaki, *J. Chem. Soc., Perkin Trans. 1* p. 1104 (1975).
131. C. Kaneko, A. Sugimoto, Y. Eguchi, S. Yamada, M. Ishikawa, S. Sasaki, and T. Suda, *Tetrahedron* **30**, 2701 (1974).
132. T. Suda, H. F. DeLuca, and R. B. Hallick, *Anal. Biochem.* **43**, 139 (1971).
133. R. B. Hallick and H. F. DeLuca, *J. Biol. Chem.* **246**, 5733 (1971).
134. K. Fukushima, Y. Suzuki, Y. Tohira, I. Matsunaga, K. Ochi, H. Nagano, Y. Nishii, and T. Suda, *Biochem. Biophys. Res. Commun.* **66**, 632 (1975).
135. M. F. Holick, S. A. Holick, T. Tavela, B. Gallagher, H. K. Schnoes, and H. F. DeLuca, *Science* **190**, 576 (1975).
136. M. R. Haussler, J. E. Zerwekh, R. H. Hesse, E. Rizzardo, and M. M. Pechet, *Proc. Natl. Acad. Sci. U.S.A.* **70**, 2248 (1973).
137. A. W. Norman, M. N. Mitra, W. H. Okamura, and R. M. Wing, *Science* **188**, 1013 (1975).
138. W. H. Okamura, M. N. Mitra, D. A. Procsal, and A. W. Norman, *Biochem. Biophys. Res. Commun.* **65**, 24 (1975).
139. H. K. Schnoes and H. F. DeLuca, *Vitam. Horm.* (*N.Y.*) **32**, 385 (1974).
140. M. F. Holick, P. Kasten-Schraufrogel, T. Tavela, and H. F. DeLuca, *Arch. Biochem. Biophys.* **166**, 63 (1976).
141. M. F. Holick, T. Tavela, S. A. Holick, H. K. Schnoes, H. F. DeLuca, and B. M. Gallagher, *J. Biol. Chem.* **251**, 1020 (1976).
142. S. A. Holick, M. F. Holick, T. E. Tavela, H. K. Schnoes, and H. F. DeLuca, *J. Biol. Chem.* **251**, 1025 (1976).
143. J. E. Zerwekh, P. F. Brumbaugh, D. H. Haussler, D. J. Cork, and M. R. Haussler, *Biochemistry* **13**, 4097 (1974).
144. D. E. M. Lawson and T. A. Bell, *Biochem. J.* **142**, 37 (1974).

145. P. Westerhof and J. A. Keverling-Buisman, *Recl. Trav. Chim. Pays-Bas* **75**, 453 (1956); **75**, 1243 (1956); **76**, 679 (1957).
146. W. H. Okamura, M. N. Mitra, R. M. Wing, and A. W. Norman, *Biochem. Biophys. Res. Commun.* **60**, 179 (1974).
147. B. L. Onisko, H. Y. Lam, L. Reeve, H. K. Schnoes, and H. F. DeLuca, *Bioorg. Chem.* **6**, 203 (1977).
148. M. F. Holick, M. Garabedian, and H. F. DeLuca, *Biochemistry* **11**, 2715 (1972).
149. T. Suda, R. B. Hallick, H. F. DeLuca, and H. K. Schnoes, *Biochemistry* **9**, 1651 (1970).
150. B. Pelc, *J. Chem. Soc., Perkin Trans. 1* p. 1436 (1974).
151. H. B. Bosman and P. S. Chen, Jr., *J. Nutr.* **90**, 141 (1966).
152. A. Windaus and G. Trautmann, *Hoppe-Seylers Z. Physiol. Chem.* **247**, 185 (1937).
153. H. F. DeLuca, M. Weller, J. W. Blunt, and P. F. Neville, *Arch. Biochem. Biophys.* **124**, 122 (1968).
154. D. R. Crump, D. H. Williams, and B. Pelc, *J. Chem. Soc., Perkin Trans. 1* p. 2731 (1973).
155. F. Holick, M. Garabedian, H. K. Schnoes, and H. F. DeLuca, *J. Biol. Chem.* **250**, 226 (1975).
156. J. S. Bontekoe, A. Wignall, M. P. Rappoldt, and J. R. Roborgh, *Int. Z. Vitaminforsch.* **40**, 589 (1970).
157. R. L. Johnson, W. H. Okamura, and A. W. Norman, *Biochem. Biophys. Res. Commun.* **67**, 797 (1975).
158. Y. Tanaka, H. Frank, H. F. DeLuca, N. Koizumi, and N. Ikekawa, *Biochemistry* **14**, 3293 (1975).
159. M. F. Holick and H. F. DeLuca, *Steroid Biochem. Pharmacol.* **4**, 111–155 (1974).
160. D. J. Aberhart, J. Y.-R. Chu, and A. C.-T. Hsu, *J. Org. Chem.* **41**, 1067 (1976).
161. P. H. Stern, C. L. Trummel, H. K. Schnoes, and H. F. DeLuca, *Endocrinology* **97**, 1552 (1975).
162. P. H. Stern, T. Mavreas, C. L. Trummel, H. K. Schnoes, and H. F. DeLuca, *Mol. Pharmacol.* **12**, 879 (1976).
163. C. L. Trummel, L. G. Raisz, R. B. Hallick, and H. F. DeLuca, *Biochem. Biophys. Res. Commun.* **44**, 1096 (1971).
164. P. H. Stern, H. F. DeLuca, and N. Ikekawa, *Biochem. Biophys. Res. Commun.* **67**, 965 (1975).
165. P. F. Brumbaugh and M. R. Haussler, *Life Sci.* **13**, 1737 (1973).
166. J. A. Eisman, B. E. Kream, H. A. Hamstra, and H. F. DeLuca, *Arch. Biochem. Biophys.* **176**, 235 (1976).
167. D. A. Procsal, W. H. Okamura, and A. W. Norman, *J. Biol. Chem.* **250**, 8382 (1975).
168. M. R. Haussler, J. F. Myrtle, and A. W. Norman, *J. Biol. Chem.* **243**, 4055 (1968).
169. T. C. Chen, J. C. Weber, and H. F. DeLuca, *J. Biol. Chem.* **245**, 3776 (1970).
170. P. F. Brumbaugh and M. R. Haussler, *J. Biol. Chem.* **249**, 1258 (1974); **250**, 1588 (1975).
171. J. G. Haddad, T. J. Hahn, and S. F. Birge, *Biochim. Biophys. Acta* **329**, 93 (1973).
172. H. C. Tsai and A. W. Norman, *Biochem. Biophys. Res. Commun.* **54**, 622 (1973).
173. H. C. Tsai, R. J. Midgett, and A. W. Norman, *Arch. Biochem. Biophys.* **157**, 339 (1973).

174. R. A. Corradino, *Nature (London)* **243**, 41 (1973).
175. R. B. Hallick and H. F. DeLuca, *Proc. Natl. Acad. Sci. U.S.A.* **63**, 528 (1969).
176. J. E. Zerwekh, T. J. Lindell, and M. R. Haussler, *J. Biol. Chem.* **251**, 2388 (1976).
177. R. H. Wasserman and R. A. Corradino, *Annu. Rev. Biochem.* **40**, 501 (1971).
178. J. S. Emtage, D. E. M. Lawson, and E. Kodicek, *Nature (London)* **246**, 100 (1973).
179. S. Moriuchi and H. F. DeLuca, *Arch. Biochem. Biophys.* **174**, 367 (1976).
180. Y. Tanaka and H. F. DeLuca, *Arch. Biochem. Biophys.* **146**, 574 (1971).

Recent Structural Investigations of Diterpenes

James D. White and Percy S. Manchand

INTRODUCTION

The search for new natural products in the plant kingdom, an activity that seemed just a decade ago to be destined for the status of a chemical hobby, has recently gained new vigor. The resurgence of interest in this area can be traced in large part to advances in biological evaluation of plant extracts, which now permit continuous monitoring of fractionated material for pharmacological properties and which have led directly to the isolation of numerous plant substances of pronounced physiological activity. A second factor that has lately come to the fore in facilitating and, we believe, stimulating isolation work is the contribution of the X-ray crystallographer, who, through considerable innovation, has refined his technique so that complex structures can be solved rapidly and by direct methods. The assumption of structural responsibility by the X-ray crystallographer has, of course, denied the organic chemist the satisfaction that comes from the effort of fitting the pieces of a structure puzzle into place. On the other hand, freedom from the labor of a rigorous, chemical structure proof has allowed the organic chemist to pursue a broader range of objectives, in which the relationship among compounds becomes as important as the compounds themselves.

Diterpenes have been prominent among the host of new compounds discovered in plants in recent years. The classic summary of skeletal types compiled by Simonsen and Barton [1] has been extended and elaborated with novel structural variants in several excellent monographs and reviews [2–2c]. This chapter describes work carried out primarily in the authors' laboratories during the past several years, which had as its purpose the isolation of major constituents from certain plants indigenous to the West Indies and to which local lore attributed certain physiological effects [3].

PREFURANOID DITERPENES (LABDANE GROUP)

Primitive medical treatment of cancer has involved several species of the widespread family Labiatae, including *Leonotis nepetaefolia* (L.) R. Br. [4]. Extracts of *L. nepetaefolia* showed activity in the National Cancer Institute's Carcinosarcoma 256 (intramuscular) screening system which varied with collection site [5]. A sample of dried leaves of *L. nepetaefolia* collected in Trinidad contained a diterpene, ca. 0.2% by weight, which was subsequently shown to be **1** (nepetaefolin) [6]. Lesser amounts of the related furanoid diterpenes nepetaefuran (**2**) and nepetaefuranol (**3**) were also found, as well as the marrubiin derivative **4** (leonotin) [7].

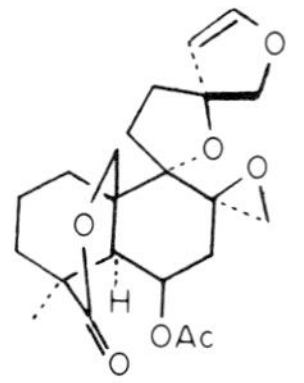

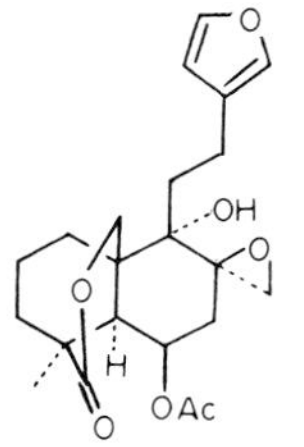

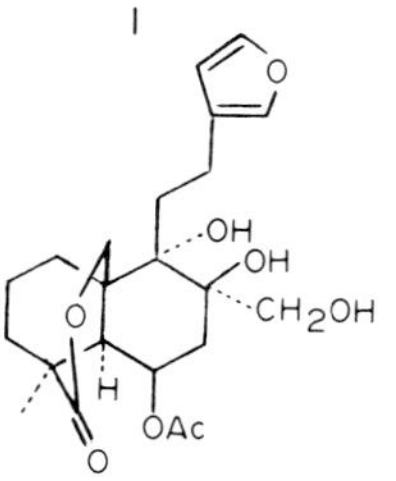

The structures, including stereochemistry, of these four substances were deduced entirely from chemical arguments, except for the configuration of the spirodihydrofuran carbon in **1** [8]. Assignments were subsequently confirmed in every detail by means of a single-crystal X-ray analysis [5].

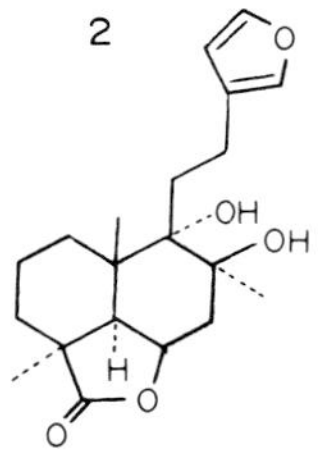

The novel spirodihydrofuran moiety of nepetaefolin was apparent at the outset from the characteristic vinyl proton chemical shifts and also from an exceedingly facile elimination of **1** to the 3-alkylfuran system of **2**. The development of a structural hypothesis for **1, 2**, and **3** hinged on three key reactions. First, saponification of **2** was found to give, not the product of straightforward hydrolysis of an acetate, but the primary alcohol **5**, in which a new oxirane had been formed. The formation of **5** implied that participation by the axial hydroxyl group at C-9 had occurred in displacement of the spiroepoxide in **2**, and this yielded information about ring B, which was

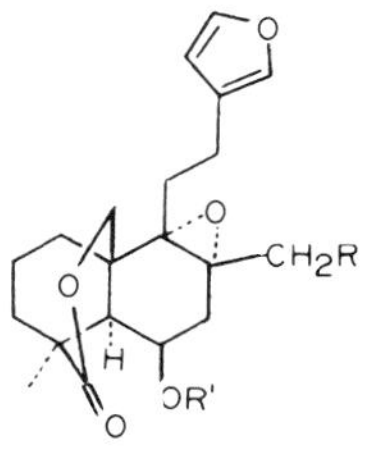

5, R = OH, R' = H
6, R = Cl, R' = Ac

corroborated by the finding that nepetaefuranol (**3**), upon reaction with phosphorous oxychloride, gave **6**. Second, oxidation of nepetaefuranol with sodium metaperiodate, followed by an elimination of acetic acid on alumina, afforded α,β-unsaturated ketone **7**. Oxidative cleavage of the terminal 8,17-glycol of **3**, in preference to the (*trans*)8,9-glycol, is expected on steric grounds, and the facile elimination from the resulting β-acetoxyketone is fully consistent with an axial disposition of the ester. This reaction, in context with other transformations of **2** and **3**, served to complete the placement of functionality in one carbocyclic ring of the diterpene skeleton. The third reaction, reduction of either **2** or **3** with lithium aluminum hydride, exposed the δ-lactone bridge (which could not be verified by hydrolysis). The reduction of **3** yielded a compound (**8**) containing three primary alcohol functions, all of which showed geminal but no vicinal proton coupling. Reduction of **2** likewise produced **9**, in which hydrogenolysis of the epoxide had given rise to a new methyl group ($\delta 1.24$, singlet).

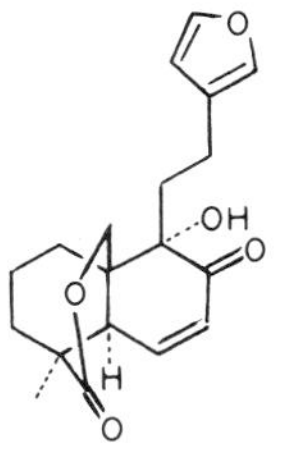

7

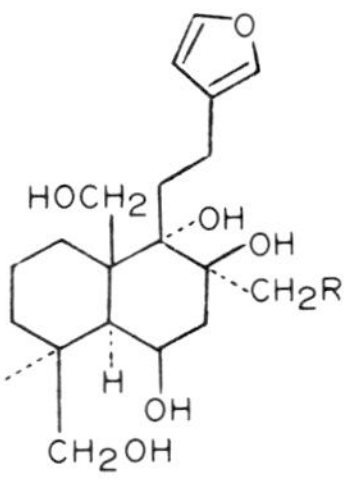

8, R = OH
9, R = H

With certain assumptions it was possible, on the basis of these chemical results, to postulate structures **1, 2,** and **3** for nepetaefolin, nepetaefuran, and nepetaefuranol, respectively, and in order to confirm these assignments a correlation with a substance of established structure and configuration was sought. This correlation is shown in Scheme 1. The reduction product **9** from nepetaefuran underwent a reaction with *p*-toluenesulfonyl chloride to give the two primary tosylates. Upon treatment with lithium aluminum hydride, the mixture of tosylates gave an array of products that included, in addition to

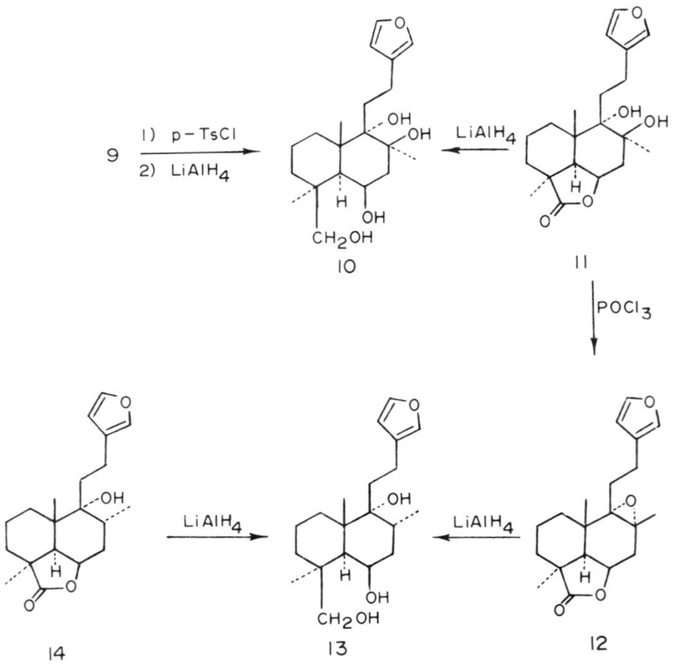

Scheme 1

cyclic ethers, a substance identical with leonotol (10). Leonotol had been previously prepared by reduction of leonotin (8β-hydroxymarrubiin, 11), a diterpene isolated from *L. nepetaefolia* by ourselves [7] and by Rivett, independently [9]. The structure of 11 was firmly established by interrelation with marrubiin [14] via dehydration to the epoxide 12 followed by reduction to marrubenol (13). The stereochemistry of marrubiin, for many years a matter of uncertainty, has now been rigorously proved by two independent structural investigations [10] and also confirmed by total synthesis [11]. Since the absolute configuration of marrubiin is known to be that represented by enantiomer 14 [12], the corresponding centers in 1, 2, and 3 are correctly designated in the absolute sense.

Completion of a structure proof for nepetaefolin (1) required the configuration of the spirocarbon, information that was not readily accessible by chemical methods. Consequently, a crystal structure determination was undertaken [5] which not only confirmed our structural assignments in all respects, but also allowed complete stereochemical designation of the seven chiral centers as (S)-C-4, (S)-C-5, (R)-C-6, (S)-C-8, (S)-C-9, (S)-C-10, and (R)-C-13. The X-ray analysis of 1 revealed some additional features of the structure which are distinctly unusual and which can be attributed to the very large degree of steric crowding in this molecule. Thus, the C-13—O-18 bond is exceptionally long (1.471 Å) for an ether, an observation that explains the facile cleavage that this weakened linkage undergoes in the rearrangement of 1 to 2. Steric effects are probably also responsible for the unusual reactivity associated with functional groups at C-8 and C-9, where the C—C internuclear distance (1.534 Å) is again abnormally large.

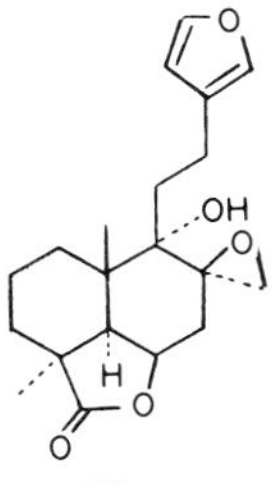

15

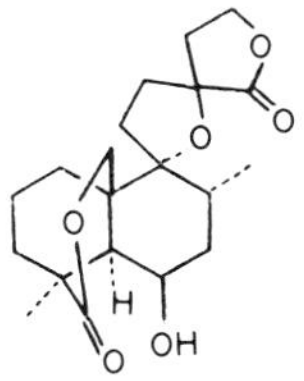

16

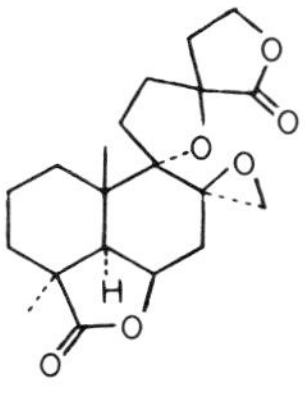

17

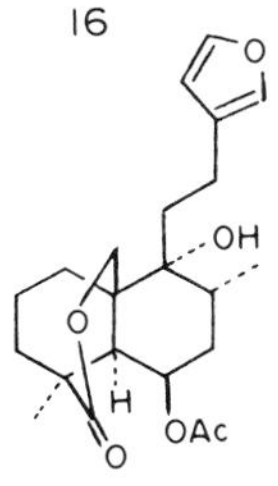

18

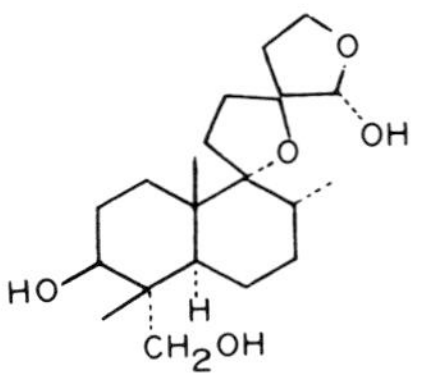

Recently, other variants of the nepetaefuran and leonotin families have emerged, including 15-methoxynepetaefolin [13], leonotinin (**15**), nepetaefolinol (**16**), and the epoxide **17**, all from *L. nepetaefolia* [14] and dubiin (**18**) from *L. dubia* [15]. Other members of the 9,13-epoxylabdane ("grindelane") group include lasiocoryin (**19**) from *Lasiocorys capensis* L. [16], lagochilin (**20**) from *Lagochilus inebrians* [17], and the acetal **21** from *Marrubium vulgare* L., "white horehound" [18]. Although some of these substances may turn out to be "prefuranoid" in a broad biogenetic context, the true members of the prefuranoid group are confined to **1**, premarrubiin (**22**) from *Marrubium vulgare* L. [19], presolidagenone (**23**) from *Solidago canadensis* L. [20], and prerotundifuran (**24**) from the seeds of *Vitex rotundifolia* [21,21a]. The latter three, as expected, undergo extremely facile conversions to the corresponding furanoids, in parallel with the conversion of **1** to **2** [8].

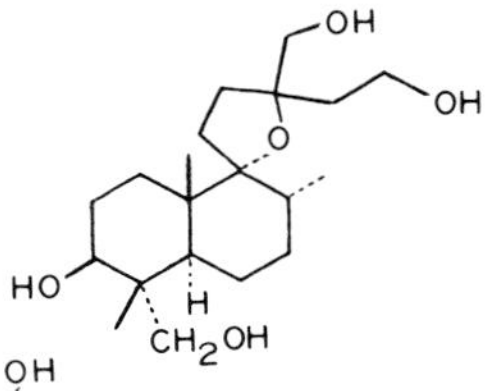

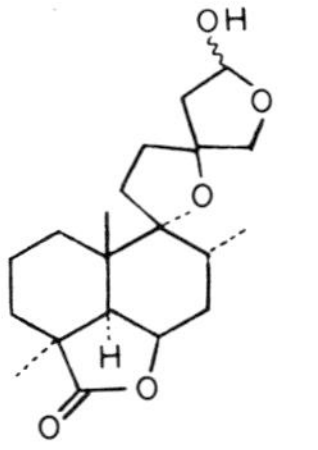

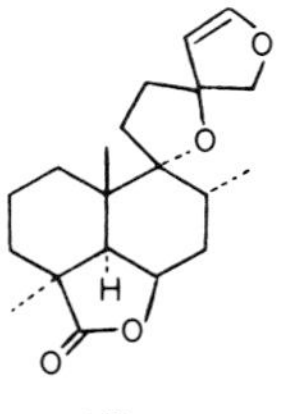

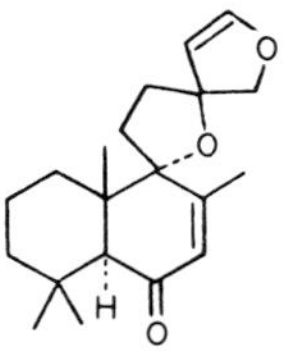

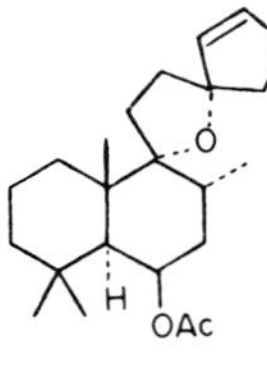

The question as to whether the furanoid diterpenes **2** and **14** are artifacts generated from the prefuranoid systems during extraction of plant material or isolation seems to have been answered in the affirmative by the experimental evidence available. The argument is particularly convincing in the instance of marrubiin (**14**), the structure of which has given rise to over 50 publications but which is totally absent in extracts of horehound, its "traditional" source, when extraction is carried out with cold acetone [8,19]. Rotundifuran, the furanoid derived from **24**, has also been shown to be an artifact [21]. Although minor amounts of nepetaefuran (**2**) were present in our extracts of *Leonotis nepetaefolia* under all extraction conditions, nepetaefolin (**1**) is clearly the dominant diterpene in this species, and it is entirely possible that the furan isolated was the result of an elimination reaction occurring during the process of storage or shipment of plant material.

The progression of oxidation steps that must take place in the side chain of the labdane system during biosynthesis of the prefuranoid diterpenes remains a matter for speculation. The proposal by McCrindle (Scheme 2) [18] that the 9,13-oxy bridge is formed via intramolecular Michael addition of a 9α-hydroxyl group to an α,β-unsaturated aldehyde, and that subsequently a derived hemiacetal undergoes elimination to the dihydrofuran, has considerable merit. This scheme readily accommodates structures such as **21**, 15-methoxynepetaefolin [13], and **20** as an aberrant. However, an equally plausible hypothesis for the origin of the prefuranoids, which became more attractive with the discovery that the configuration (*R*) at C-13 in both nepetaefolin [5] (**1**) and prerotundifuran [21a] (**24**) is identical with that of the corresponding carbon in the common labdanoids manool and sclareol, hinges on formation of the 9,13-oxy bridge via allylic oxidation of a manool type of precursor (Scheme 3). A close chemical analogy for this step has recently been described in a study of the oxidation of labdanes with a mixture of selenium dioxide and hydrogen peroxide [22]. It was found not only that a 9,13-ether linkage is formed in this reaction, but that the exocyclic olefin at C-8 is converted to a

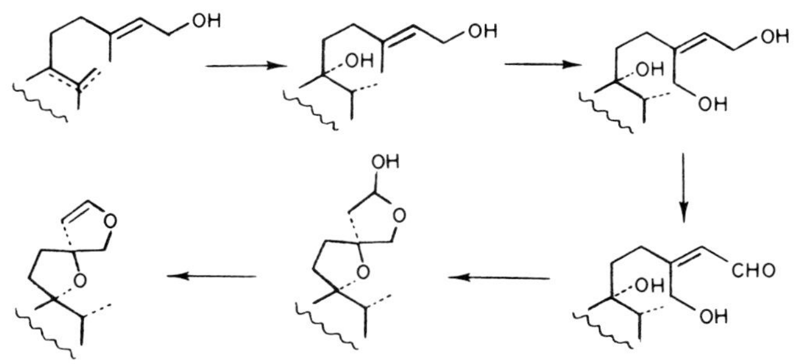

Scheme 2

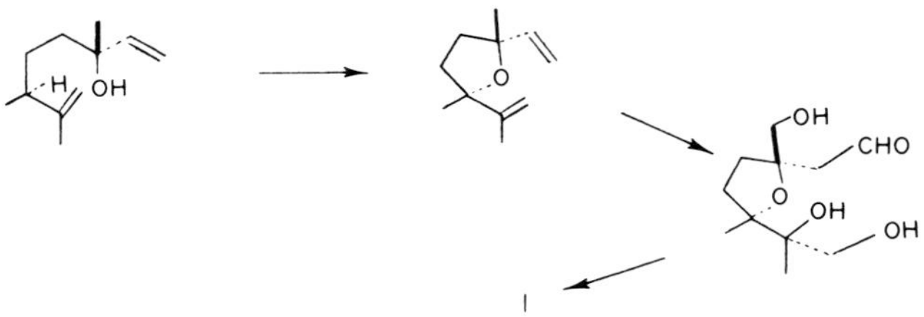

Scheme 3

8,17-diol in the functional and configurational sense observed in nepetae-furanol (**3**). Although mechanistic details of this remarkable biomimetic process are lacking, it is evident that the chemical oxidation parallels the biological one in several respects, including the preservation of configuration at C-9. Scheme 3 is also less restrictive than McCrindle's proposal in terms of permissible oxidation levels at the methyl and vinyl groups, which provide the dihydrofuran in **1**, and can easily be adapted to structures such as **17** and **19**. It is interesing to note that the C-9-hydroxylated *Laurencia* diterpene concinndiol (**25**) can also be derived along these lines.

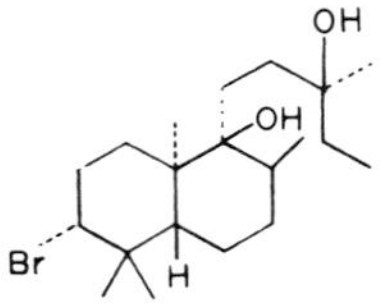

25

CLERODANE DITERPENES

The normal labdanoid skeleton, found in nepetaefolin (**1**) and related diterpenes discussed in the foregoing section, is modified biogenetically in a variety of ways. One of the more common modifications, represented by

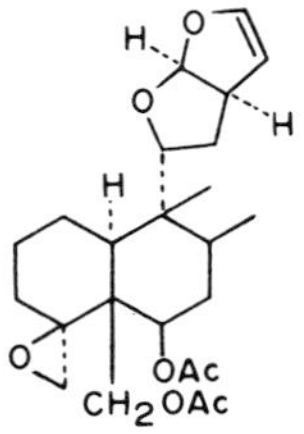

26

clerodin (26) [23], possesses a skeleton in which two methyl transpositions, one from C-4 to C-5 and one from C-10 to C-9, have taken place to give rise to what may be designated the "clerodane" family [24]. However, a larger than usual number of deviants are found within this family, both configurational (e.g., *cis*-fused types [25]) and skeletal (e.g., 28), so that classification and hence biogenetic rationale become somewhat arbitrary.

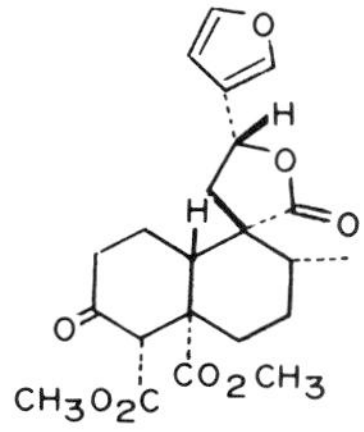

27 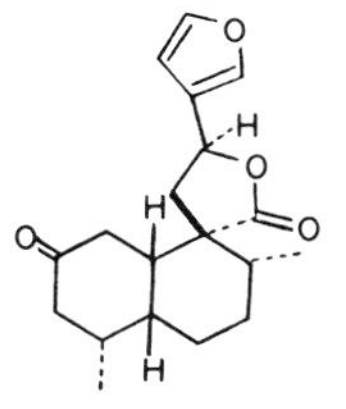   28

A recent examination of two Jamaican *Croton* species led to the isolation of corylifuran (27) from *C. corylifolius* [26] and crotonin (28) from *C. lucidus* [27]. Corylifuran contains the unusual (for terpenoids) β-ketoester function in addition to a second ester and γ-lactone grouping and yet, surprisingly, is recovered unchanged from both hot 2 N sulfuric acid and hot methanolic potassium hydroxide. The formation of an enolate anion of 27 in dilute alkali is indicated by an absorption maximum at 283 nm (ε, 9000), but the compound gives a negative response to ferric chloride. Reduction of corylifuran with sodium borohydride gave alcohol 29, corresponding to attack by hydride at C-3 from the more open β side; further reduction with this reagent then afforded lactone alcohol 30, which thereby established the relationship between functionalities at C-3 and C-5.

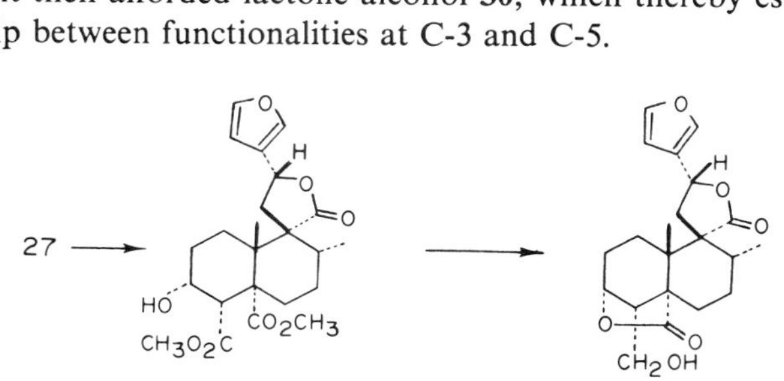

29   30

The full structure and relative configuration of corylifuran (27) were revealed by an X-ray crystallographic analysis [26]. Absolute configuration was determined by application of the octant rule to the optical rotatory dispersion curve of 27 and is seen to be the reverse of that found in clerodin.

Crotonin was first isolated several years ago and, on the basis of chemical studies, was shown to be a norditerpene of the rearranged labdane type. Structural and stereochemical work was recently completed by means of an X-ray analysis [28] and this, together with optical rotatory dispersion and circular dichroism data, established the absolute stereostructure represented by **28**. Thus, crotonin and corylifuran are both members of the set of clerodanes which includes, *inter alia*, hautriwaic acid (**31**) [29] and teucrin (**32**) [30] and which is enantiomeric with clerodin itself.

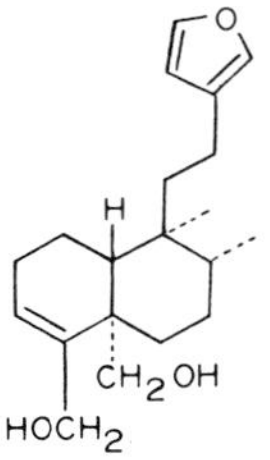

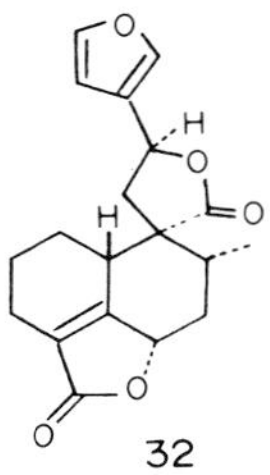

The origin of **27** and related diterpenes in this set is readily understood in terms of the backbone (friedo) rearrangement [2a] of an *ent*-labdane precursor (**33**, Scheme 4). A succession of alternate hydride and methyl 1,2 shifts toward C-8 probably terminates in loss of a proton from a hydroxyl

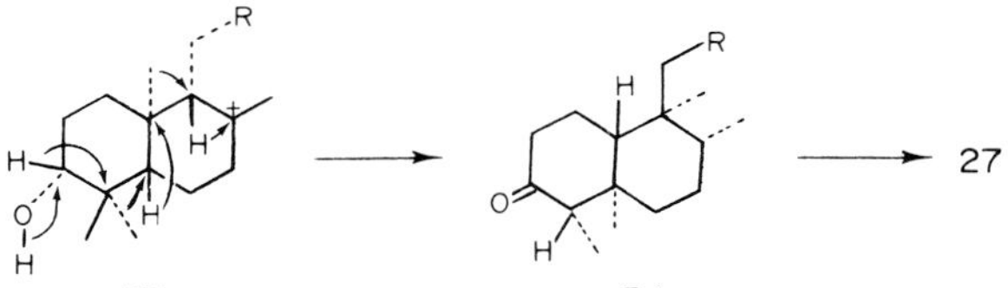

Scheme 4

group to yield the keto function at C-3, as in **34**. Subsequent modifications are largely oxidative, although the first chloroditerpene, gutierolide (**35**) [31], was recently found to be a member of the clerodane series.

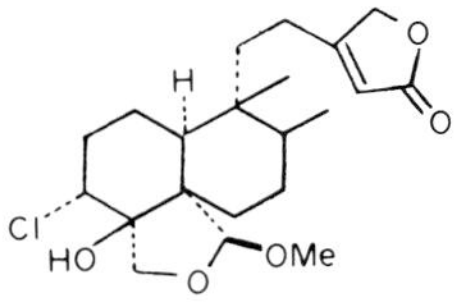

35

ABIETANE DITERPENES

The abietane group, typified by abietic acid (36) and its congeners, forms a large component of the family of tricyclic diterpenoids. Recently, a number of interesting representations of this class have emerged which, like the labdane and clerodane systems discussed in the foregoing sections, are extensively oxygenated. This oxygenation seems to be concentrated most heavily in ring

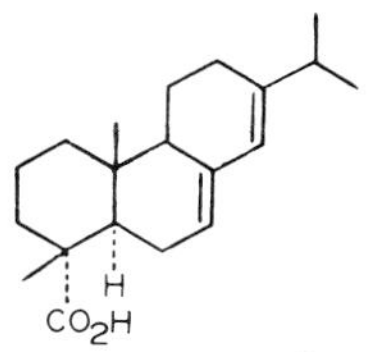

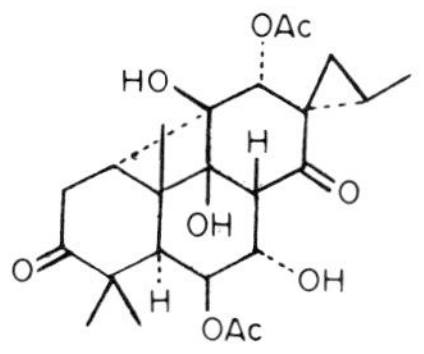

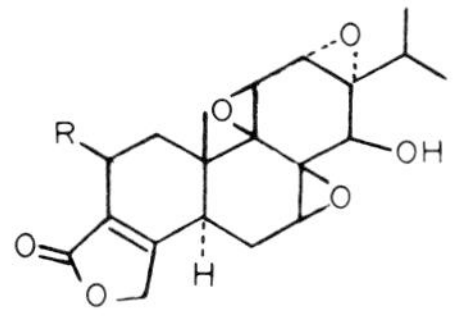

C of the abietane nucleus, although the A and B rings are not immune. Examples include cyclobutatusin (37) from *Coleus barbatus* [32], the highly active antileukemic principles triptolide (38) and triptdiolide (39) from *Tripterygium wilfordii* [33], coleon H (40) from *Coleus somaliensis* [34], and jolkinolide B (41) from *Euphorbia jolkini* [35].

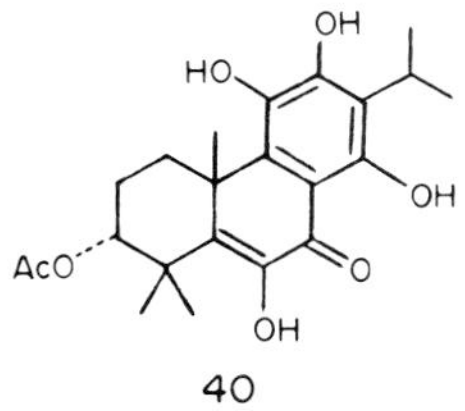

Our own isolation studies involving *Hyptis suaveolens* resulted in suaveolic acid (42) [36], an abietic acid derivative which, like diterpenes 37–41, is oxygenated at C-14. More recent work with *Stemodia maritima* has led to the isolation of another structural variant, stemolide (43) [37], closely related to triptolide (38). The structure and absolute configuration of stemolide were established by an X-ray crystallographic analysis which leaves no doubt that, while the structural resemblance between 43 and 38 is strong, the configuration of the 12,13-epoxide group in stemolide is the reverse of that in triptolide.

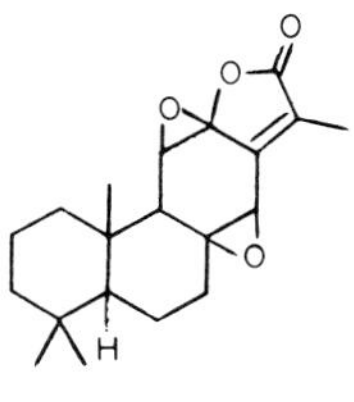

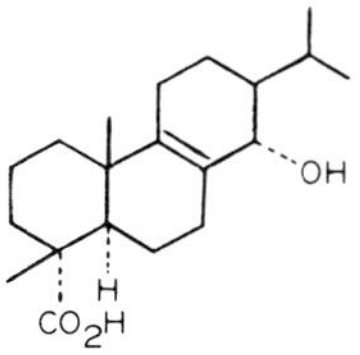

42 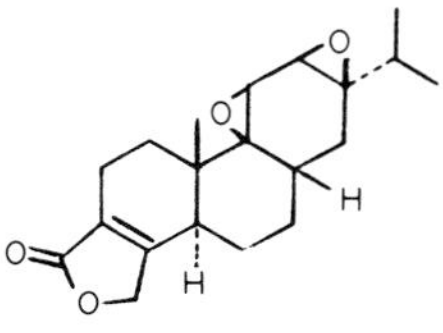   43

It is interesting that stemolide has shown no antitumor activity, suggesting that, if nucleophilic attack by thiol groups of proteins at the 9,11-epoxide of **38** is involved in tumor inhibition, then assistance in epoxide opening by the 14β-hydroxyl group may be a key factor, as proposed by Kupchan [38]. The migration of C-18 from C-4 to C-3, which gives rise to the abeo skeleton from the normal isoprenoid precursor, has been observed thus far only in stemolide, triptolide, triptdiolide, and coleons E and F [39] and, curiously, in three of these terminates as the carbonyl group of a 3,4-fused, α,β-unsaturated γ-lactone.

The *cis*-1,3-diepoxide moiety found in stemolide has proved to be a quite common functionality in natural products. It is present in crotepoxide (**44**) [40] and the antibiotic LL-Z1220 (**45**) [41] among others and could conceivably be formed via rearrangement of an *endo*-peroxide, as has been demonstrated for several analogous cases [42], including an *endo*-peroxide intermediate recently utilized in a total synthesis of **44** [43]. The partial

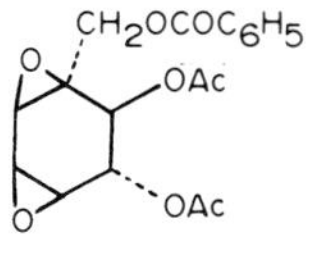

44 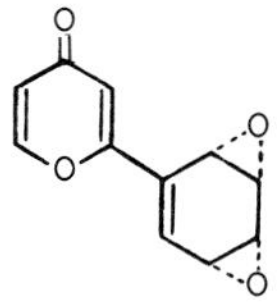   45

biogenesis of **43** put forward in Scheme 5 therefore posits **50**, an unknown member of the abietadiene group, as an intermediate. Oxygenation of **50**, which recent results have shown can be either an enzymatic or photosensitized process *in vivo* [44], would then yield **51**. The derivation of **50** assumes a precursor (**46**) having 9β-H configuration, and hence a *trans,syn* backbone, which undergoes a 1,2-hydride shift to give the observed 8β stereochemistry of **47**. This configuration has been characteristic of all diterpenes isolated from *S. maritima* so far; its implications are discussed more fully in the following section. A proton elimination from **47** and a conventional pimarane to abietane methyl shift in **48** lead to **49** and then the requisite diene **50**.

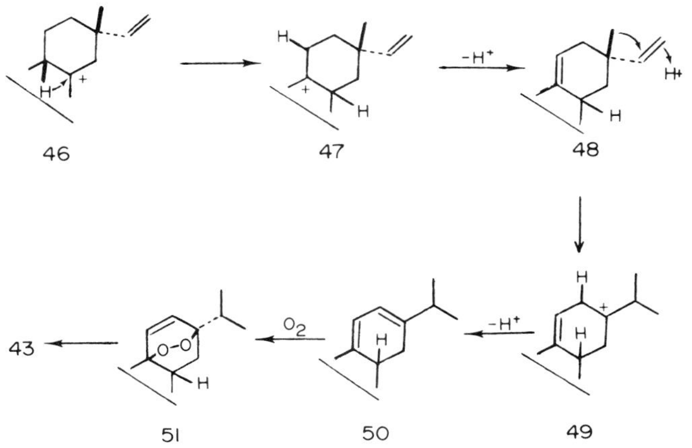

Scheme 5

STEMODANE, STEMARANE, AND APHIDICOLANE DITERPENES

Examination of the littoral plant *Stemodia maritima* L. (Scrophulariaceae) [45] has led to the isolation of a new family of tetracyclic diterpenes based on the stemodane skeleton [46]. The first members, $2\alpha,13\alpha$-stemodanediol (stemodin, **52**) and the corresponding ketone (stemodinone, **53**), have recently been joined by the $3\beta,13\alpha$-diol (maritimol, **54**) and the $13\alpha,18$-diol (stemodinol, **55**) [47]. The sequence of transformations through which the set of four natural stemodanes have been related is shown in Scheme 6.

The hydroxystemodane **56**, which is also present in *S. maritima*, is the focal point at which the degradation paths of **52**, **54**, and **55** intersect and, although it was not possible to unambiguously assign a carbon skeleton to these diterpenes on the basis of available chemical evidence, the dehydration product **57**, from **56**, was unusually informative. In particular, the vinyl proton of **57** and also that of dehydration product **58** from stemodinone (**53**) was a multiplet, indicating coupling of this hydrogen to a vicinal methylene group. This feature ruled out structures of the kaurene–phyllocladene type and, in concert with other spectral data establishing the tetracyclic nature of the skeleton, was suggestive of a novel bicyclo[3.2.1]octane nucleus.

A single-crystal X-ray analysis of stemodinone (**53**) [46] confirmed the structure and relative configuration shown, and hence, by correlation, the stereostructures **52**, **54**, and **55** are determined. Each of the six-membered rings in stemodinone was found in the chair conformation, with the five-membered ring in a normal envelope. Stemodinone displays a positive Cotton

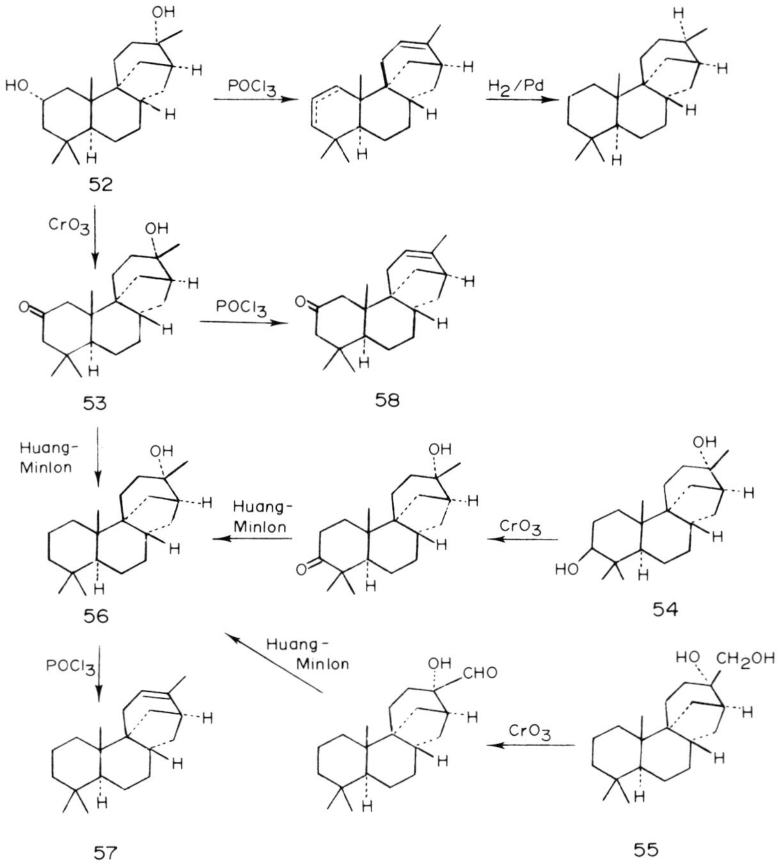

Scheme 6

effect ($[\alpha]_{308} = 2460$) in the optical rotatory dispersion spectrum, and the application of the octant rule indicates that the natural enantiomer has 5α-H and 10β-CH$_3$ configuration. Stemodinone and the other stemodanes are therefore correctly represented in the absolute sense by the enantiomers shown in Scheme 6.

Almost simultaneously with the discovery of the stemodanes, an antibiotic, aphidicolin, was isolated from the fungus *Cephalosporium aphidicola* Petch by Hesp's group at the Imperial Chemical Industries Laboratories in England [48]. The structure of aphidicolin, an antimitotic and antiviral metabolite, was established by X-ray analysis as **59** (absolute configuration shown) [49]. Thus, aphidicolin contains the same gross skeleton as the stemodanes but

differs in configuration at carbons 9, 13, and 14 [50]. It is particularly interesting to note that the aphidicolane and stemodane skeletons possess the same
(S) configuration at C-8. The possibility that a false solution had emerged
from the X-ray data in one of the two structures and that the stemodanes and
aphidicolin had the same stereochemistry was eliminated by a comparison
of the hydrocarbon **60** (mp, 101–105°C; [α]$_D$, +11.2°) from **59** with **57**
(mp, 52–53°C; [α]$_D$, +36.6°) from stemodin. Whether the stereochemical
variations represented in the stemodin and aphidicolin structures reflect
taxonomic origin will not be clear until a more thorough examination of
natural sources, particularly microbial ones, has been conducted. The
relationship between **52** and **59** is reminiscent of that existing between
kaurene and phyllocladene, in which there is indeed a taxonomic differentiation of the two skeletal types.

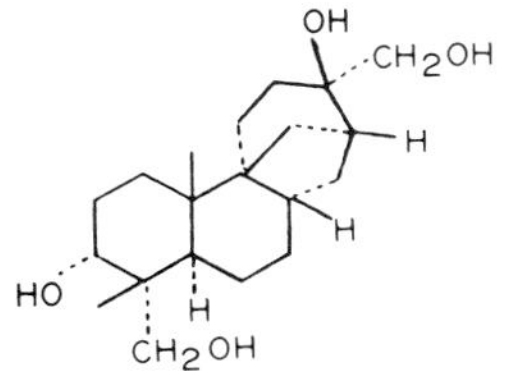

59 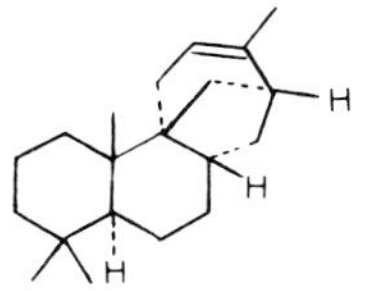   60

Recently, a further tetracyclic variant of these structures came to light with
the isolation of stemarin (**61**) from *S. maritima* [51]. The structure of stemarin,
deduced from an X-ray study, reveals that, like stemodin and aphidicolin, it
contains a bicyclo[3.2.1]octane moiety. However, in the case of **61**, the decalin
(AB) system is fused to the bicyclic framework between the bridgehead and
three-carbon bridge rather than toward the two-carbon bridge as in **52** and
59. The 8β-H configuration is again found in **61**, and the configuration at C-9
is that of the stemodane series.

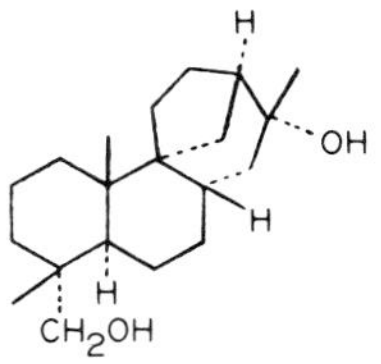

61

The intriguing structural relationship between the stemodane stemarane,
and aphidicolane skeletons suggests that these substances are derived via a
biogenetic pathway which diverges after the configuration at C-8 is fixed

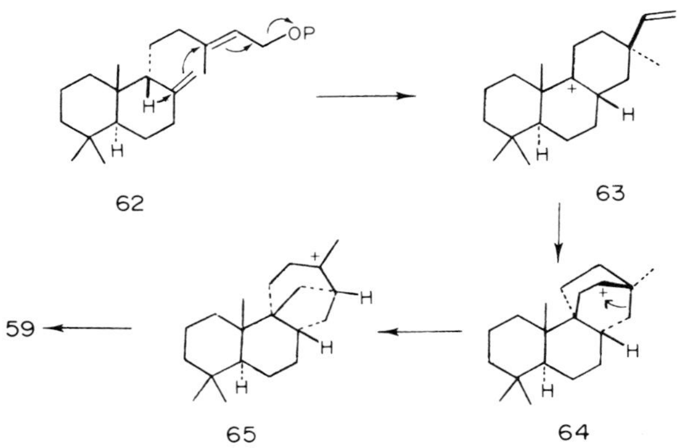

Scheme 7

but before the structural elements of the CD rings are in place. Biosynthetic studies by Bu'Lock have shown that incorporation of $(4R)$-$[2$-^{14}C,$^{3}H]$ mevalonate into aphidicolin (**59**) takes place with retention of three of the four possible tritium labels [52]. The missing label was lost from the 3α position in the parent aphidicolane during the hydroxylation (with retention) process, so that the tritium originally located at C-6 in the geranylgeraniol precursor is still present in aphidicolin. It was not possible to define the position of this label in the metabolite unambiguously, but the reasonable assumption was made that it had moved to C-8. Since the hydrogen at C-8 in all of the stemodane diterpenes as well as stemarin and aphidicolin is β, it is logical to infer that the precursor common to these systems is the 9-*epi*-labdanoid **62**. It has been suggested that such a configuration could arise from a chair–boat folding of the geranylgeraniol precursor [49].

The complete pathway leading to aphidicolin, as envisaged by Bu'Lock [52], is shown in Scheme 7. Cyclization of the 1,6-diene moiety of **62**, with a concomitant hydride shift from C-9 to C-8, leads to cation **63**. The C-13 configuration in this cation corresponds to the pimarane series, and subsequent bridging by the vinyl group to C-9 must therefore take place across the β face to give **64**. A Wagner–Meerwein rearrangement, involving migration of C-14 to the secondary cation, leads to tertiary ion **65** and then the aphidicolin structure.

As pointed out by Bu'Lock [52] and Hesp [49], a sequence analogous to that of Scheme 7 accounts for the stemodane diterpenes if the 13-*epi* intermediate **66** (isopimarane configuration) is invoked. Actually, it is not necessary to postulate separate tricyclic intermediates such as **63** and **66** since the stemodane and aphidicolane skeletons, as well as that of stemarin, can be reached via a set of transformations (shown in Scheme 8) which emanate

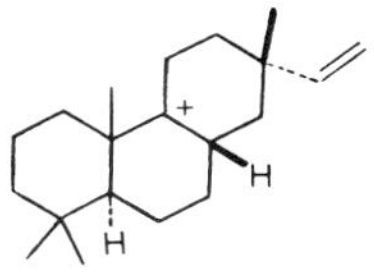

66

from a single tricyclic precursor, e.g., **67**. In this view, the three cyclopropane structures **69**, **70**, and **72**, which give rise by Markovnikov protonation in each case to the stemodane, stemarane, and aphidicolane skeletons, respectively, originate from **71**.

The reasoning that underlies this biogenetic scheme is clearly adumbrated by Wenkert [53] and Coates [54] in their rationale for the occurrence of kaurane, atiserane, and beyerane families (Scheme 9) and for which the discovery of the trachylobane skeleton (**73**) was a particularly cogent event [55]. The divergence between Scheme 8 and the Wenkert–Coates hypothesis arises as a consequence of bridging of the C-13 vinyl group of **67** at C-9

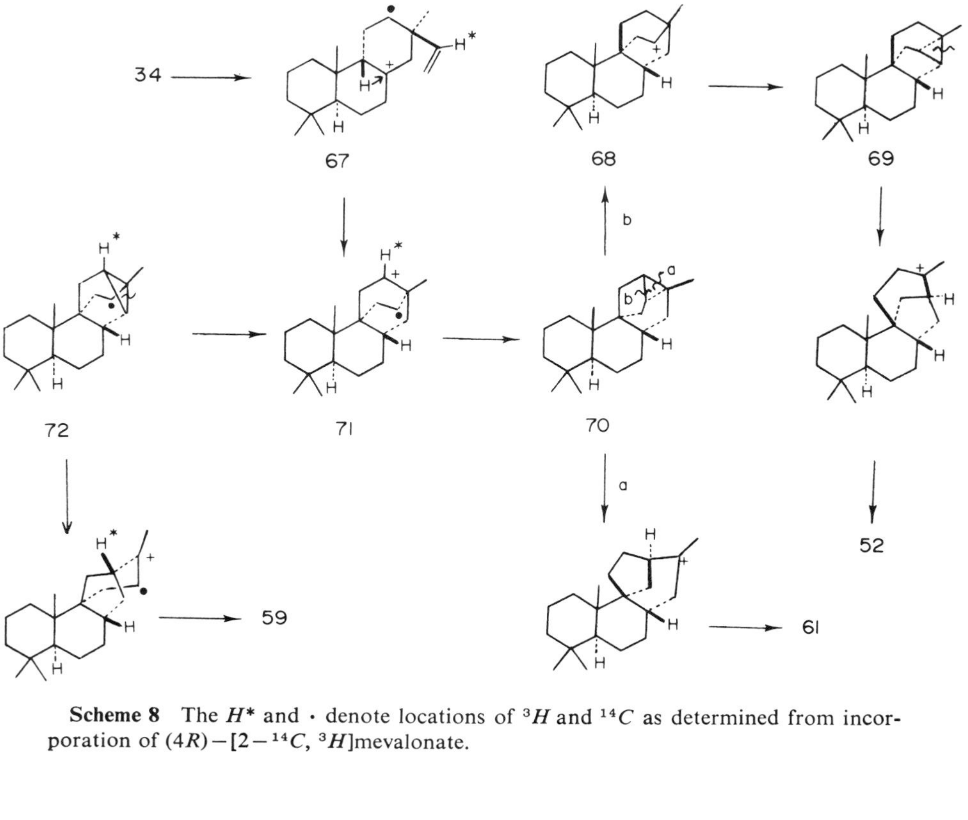

Scheme 8 The H^* and · denote locations of ^3H and ^{14}C as determined from incorporation of $(4R)-[2-^{14}C, \ ^3H]$mevalonate.

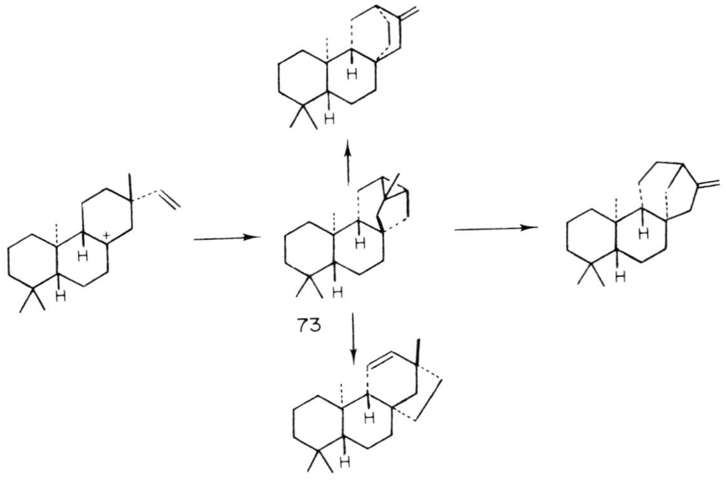

Scheme 9

rather than at C-8. An interesting feature of the set of intermediates proposed in Scheme 8 is the pseudosymmetry present in the tricyclo[3.2.1.0^{2,1}]octane moiety of pentacyclic intermediate **70**. This symmetry, which is absent in the trachylobane skeleton, permits interchange of ions **68** and **71** (via cleavage b) and is, of course, the reason why precursor **70** suffices for the stemodane and stemarane families.

The selection of pimarane configuration at C-13 for tricyclic precursor **67** is in conformity with label studies of Bu'Lock [52], although hydride migration (C-9 → C-8) cannot be concerted with bridging of the vinyl group in this case. In addition to allowing a hypothetical derivation of the three skeletal types from a common intermediate, Scheme 8 suggests that stemodane, stemarane, and aphidicolane diterpenes might be interconverted via a series of 1,3-eliminations and protonations analogous to those shown. The practical aspects of this idea remain to be explored.

CROTOFOLANE DITERPENES

The irritant and vesicant properties of the Euphorbiaceae are well known and have increasingly drawn the attention of natural-products chemists to this family [56]. The identification of esters of the diterpene phorbol (**74**) from *Croton tiglium*, a member of this family, as cocarcinogenic compounds has given additional impetus to the investigation of this group [57].

Isolation studies of a *Croton* species, *C. corylifolius*, collected in Jamaica,

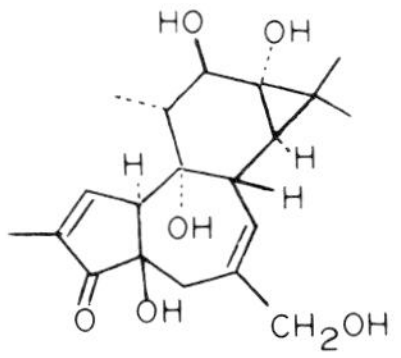

74

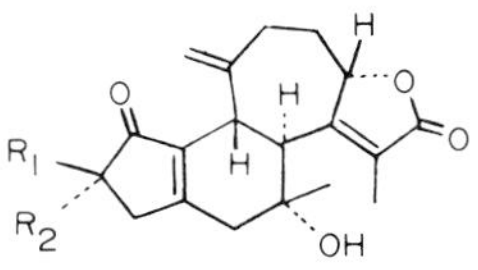

75, R_1 = OH, R_2 = CH_3
76, R_1 = CH_3, R_2 = OH

recently led to the structural elucidation of two diterpenes possessing a new skeleton. Crotofolin A (**75**), the structure and relative configuration of which were deduced from an X-ray analysis [58], possesses an angularly fused 5,6,7-tricarbocyclic (crotofolane) nucleus **83** [59] in which the isopropyl group is transposed by one position in the cycloheptane ring from a normal isoprenoid substitution pattern. Crotofolin B (**76**) was shown to be the C-2 epimer of **75** [60]. The absolute configuration of the crotofolins is presently unknown.

The crotofolane skeleton, like that of phorbol (**74**), appears to be a biogenetic derivative of casbene (**79**) [61], and a rationale for its origin from a geranylgeranyl precursor (**77**) is presented in Scheme 10. Casbene (**79**) most

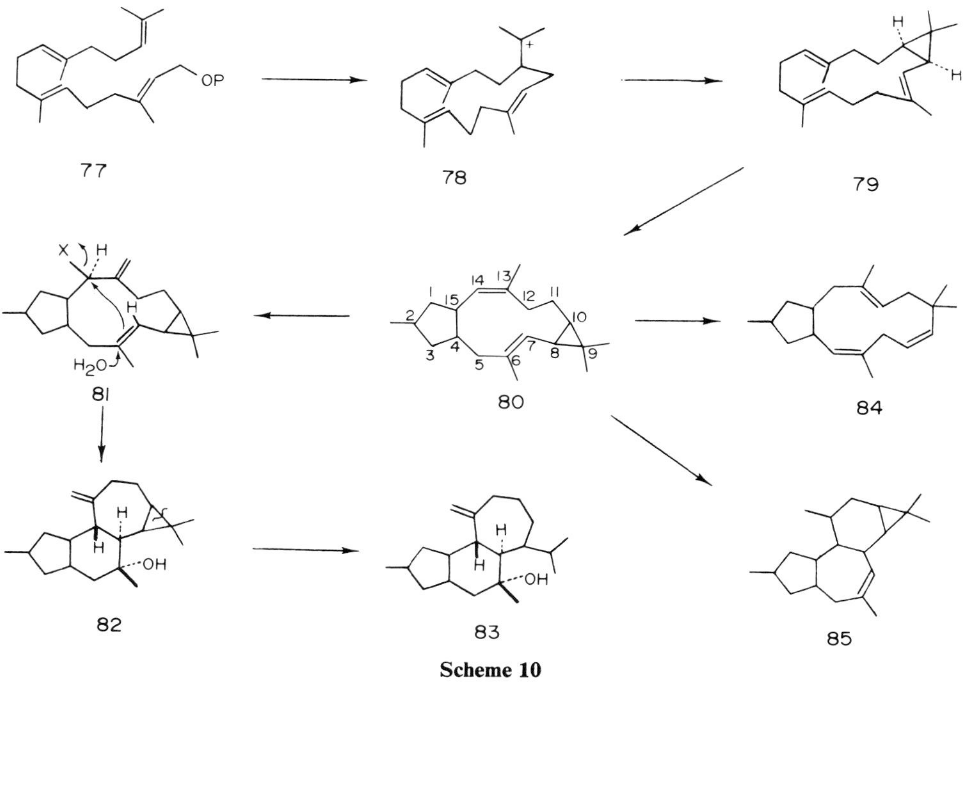

77   78   79

81   80   84

82   83   85

Scheme 10

plausibly comes from cation **78**, which is also the hypothetical progenitor of cembrene [62], and, in the Euphorbiaceae at least, a ring closure (C-4 to C-15) to give a cyclopentanoid prototype (**80**) seems to be a central feature of the secondary metabolism. The lathyrane skeleton (**80**) is found in structures such as bertyadionol (**86**) [63] as well as the lathyrol group [64], and a version (**84**) in which the 8,10 bond is removed is exemplified by jatrophone [65] and the kansuinines (e.g., **87**) [66]. A transannular cyclization (C-5 to C-11) of **80** gives rise to the tigliane framework **85** present in phorbol and related diterpenes such as daphnetoxin (**88**) [67]. The structure of ingenol (**89**) [68] can be derived from **85** by a shift (pinacol rearrangement?) of C-13 to C-15.

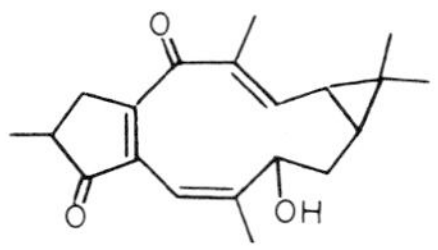

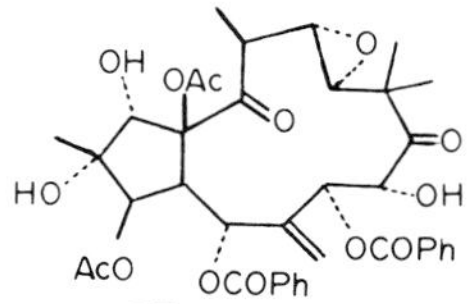

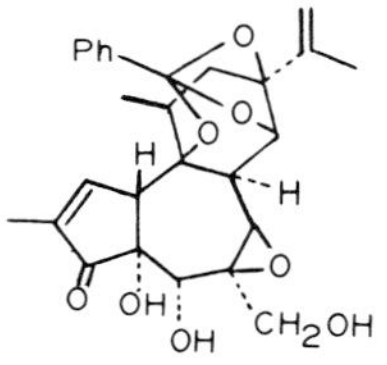

Formation of the crotofolins represents an alternate mode of transannular cyclization (C-7 to C-14) of the lathyrane structure **80**. An appropriately substituted derivative of this skeleton (e.g., **81**) can be well oriented for an internal displacement reaction at C-14 by the 6,7 double bond, with accompanying attack at the tertiary center (C-6) by water. Assuming an *anti* addition to a transoid $\Delta^{6,7}$ system, a tetracyclic intermediate (**82**) having the relative stereochemistry of the crotofolins at C-6, C-7, and C-14 is produced.

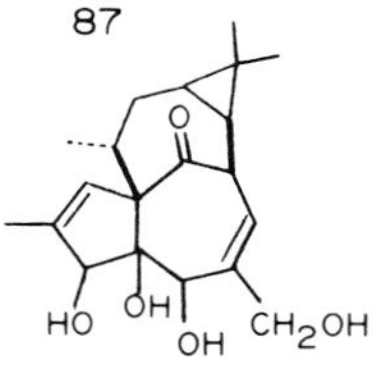

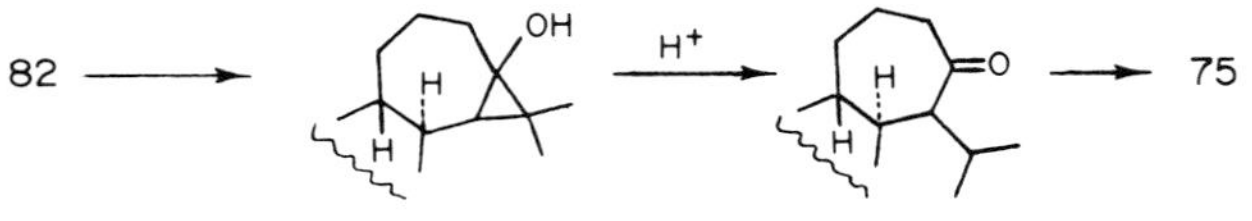

The residual skeletal transformation necessary to reach the crotofolane nucleus **83** involves cleavage of the cyclopropane at C-9—C-10. This process conceivably could be triggered by oxygenation of **82** at C-10 to give the hydroxycyclopropane analog **90** of phorbol (**74**). Very mild acidic treatment of phorbol is known to yield a product, crotophorbolone, with the structural elements shown in **91** [69]. Aside from the demonstration that ^{14}C-labeled geranylgeranyl pyrophosphate is efficiently incorporated into casbene [70], no biosynthetic evidence that might shed light on the set of pathways in Scheme 10 has been reported.

ACKNOWLEDGMENTS

We are grateful to the late Professor S. Morris Kupchan, University of Virginia, and Dr. Barry Hesp, Imperial Chemical Industries, for stimulating discussions and for making information available prior to publication. We are also indebted to Professor C. D. Adams, University of the West Indies, for assistance in the identification of plant species. Financial support for this work was provided by the National Science Foundation (to J.D.W.) and, during the early phase of this work carried out at Harvard University, by Hoffmann-LaRoche, Inc. One of us (J.D.W.) thanks the University of Virginia for a visiting appointment, during which this review was written.

REFERENCES

1. J. Simonsen and D. H. R. Barton, "The Terpenes," Vol. 3, p. 328. Cambridge Univ. Press, London and New York, 1952.
2. A. C. Oehlschlager and G. Ourisson, *in* "Terpenoids in Plants" (J. B. Pridham, ed.), p. 83. Academic Press, New York, 1967; J. R. Hanson, *in* "Chemistry of Terpenes and Terpenoids" (A. A. Newman, ed.), p. 155. Academic Press, New York, 1972.
2a. J. R. Hanson, *Progr. Phytochem.* **3**, 231 (1972).
2b. R. McCrindle and K. H. Overton, *Adv. Org. Chem.* **5**, 47 (1965).
2c. R. McCrindle and K. H. Overton, *in* "Chemistry of Carbon Compounds" (S. Coffey, ed.), Vol. 2, p. 369. Elsevier, Amsterdam, 1969; J. R. Hanson, "The Tetracyclic Diterpenes." Pergamon, Oxford, 1968; K. Nakanishi, T. Goto, S. Ito, S. Natori, and S. Nozoe, eds., "Natural Products Chemistry," Vol. 1, p. 185. Academic Press, New York, 1974.
3. For a source reference, see J. M. Watt and M. G. Breyer-Branwijk, "Medicinal and Poisonous Plants of Southern and Eastern Africa." Livingstone, Edinburgh, 1962.
4. G. R. Pettit, H. Klinger, N.-O. N. Jorgensen, and J. Occolowitz, *Phytochemistry* **5**, 301 (1966); see also J. Hartwell, *Lloydia* **32**, 247 (1969).
5. R. B. von Dreele, G. R. Pettit, R. H. Ode, R. E. Perdue, J. D. White, and P. S. Manchand, *J. Am. Chem. Soc.* **97**, 6236 (1975).
6. J. D. White and P. S. Manchand, *J. Am. Chem. Soc.* **92**, 5527 (1970).
7. J. D. White, P. S. Manchand, and W. B. Whalley, *Chem. Commun.* p. 1315 (1969).
8. J. D. White and P. S. Manchand, *J. Org. Chem.* **38**, 720 (1973).
9. E. R. Kaplan, K. Naidu, and D. E. A. Rivett, *J. Chem. Soc. C* p. 1656 (1970).
10. R. A. Appleton, J. W. B. Fulke, M. S. Henderson, and R. McCrindle, *J. Chem. Soc. C* p. 1943 (1967); D. M. S. Wheeler, M. M. Wheeler, M. Fetizon, and W. H. Castine, *Tetrahedron* **23**, 3909 (1967).
11. L. Mangoni, M. Adinolfi, G. Laonigro, and R. Caputo, *Tetrahedron* **28**, 611 (1972).
12. D. Burn and W. Rigby, *J. Chem. Soc.* p. 2964 (1957); C. Djerassi and W. Klyne, *ibid.* p. 4929 (1962).

13. P. S. Manchand, *Tetrahedron Lett.* p. 1907 (1973).
14. K. Purushothaman, S. Vasanth, and J. D. Connolly, *J. Chem. Soc., Perkin Trans. 1* p. 2661 (1974).
15. G. A. Eagle and D. E. A. Rivett, *J. Chem. Soc., Perkin Trans. 1* p. 1701 (1973).
16. G. Gafner, G. J. Kruger, and D. E. A. Rivett, *Chem. Commun.* p. 249 (1974).
17. O. S. Chizhov, A. V. Kessenikh, B. M. Zolotarev, and V. A. Petukhov, *Tetrahedron Lett.* p. 1361 (1961).
18. J. W. B. Fulke, M. S. Henderson, and R. McCrindle, *J. Chem. Soc. C* p. 807 (1968).
19. M. S. Henderson and R. McCrindle, *J. Chem. Soc. C* p. 2014 (1969).
20. T. Anthonsen, P. H. McCabe, R. McCrindle, and R. D. H. Murray, *Tetrahedron* **25**, 2233 (1969).
21. Y. Asaka, T. Kamikawa, and T. Kubota, *Chem. Lett.* p. 927 (1973).
21a. K. Hirotsu and A. Shimada, *Chem. Lett.* p. 1035 (1973).
22. M. J. Francis, P. K. Grant, K. S. Low, and R. T. Weavers, *Tetrahedron* **32**, 95 (1976).
23. G. A. Sim, T. A. Hamor, I. C. Paul, and J. M. Robertson, *Proc. Chem. Soc., London* p. 75 (1961); D. H. R. Barton, H. T. Cheung, A. D. Cross, L. M. Jackman, and M. Martin-Smith, *J. Chem. Soc.* p. 5061 (1961).
24. J. R. Hanson, *Terpenoids Steroids* **4**, 148 (1974).
25. G. Ferguson, W. C. March, R. McCrindle, and E. Nakamura, *Chem. Commun.* p. 399 (1975).
26. B. A. Burke, W. R. Chan, E. C. Prince, P. S. Manchand, N. Eickman, and J. Clardy, *Tetrahedron* **32**, 1881 (1976).
27. W. R. Chan, D. R. Taylor, and C. R. Willis, *J. Chem. Soc. C* p. 2781 (1968).
28. W. R. Chan, P. S. Manchand, J. O. Pezzanite, and J. Clardy, to be published.
29. H. Y. Hsu, V. P. Chen, and H. Kakisawa, *Phytochemistry* **10**, 2813 (1971); see also A. B. Anderson, R. McCrindle, and E. Nakamura, *Chem. Commun.* p. 453 (1974).
30. E. Fujita, I. Uchida, and T. Fujita, *J. Chem. Soc., Perkin Trans. 1* p. 1547 (1974).
31. W. B. T. Cruse, M. N. G. James, A. A. Al-Shamma, J. K. Beal, and R. W. Doskotch, *Chem. Commun.* p. 1278 (1971).
32. A. H. Wang, I. C. Paul, R. Zelnik, D. Lavie, and E. C. Levy, *J. Am. Chem. Soc.* **96**, 580 (1974).
33. S. M. Kupchan, W. A. Court, R. G. Dailey, C. J. Gilmore, and R. F. Bryan, *J. Am. Chem. Soc.* **94**, 7194 (1972).
34. M. Moir, P. Ruedi, and C. H. Eugster, *Helv. Chim. Acta* **56**, 2534 (1973).
35. D. Uemura and Y. Hirata, *Tetrahedron Lett.* p. 1387 (1972).
36. P. S. Manchand, J. D. White, J. Fayos, and J. Clardy, *J. Org. Chem.* **39**, 2306 (1974).
37. P. S. Manchand and J. F. Blount, *Tetrahedron Lett.* p. 2489 (1976).
38. S. M. Kupchan and R. M. Schubert, *Science* **185**, 791 (1974).
39. M. Moir, P. Ruedi, and C. H. Eugster, *Helv. Chim. Acta* **56**, 2539 (1973), and references cited.
40. S. M. Kupchan, R. J. Hemingway, and R. M. Smith, *J. Org. Chem.* **34**, 3898 (1969).
41. D. B. Borders and J. E. Lancaster, *J. Org. Chem.* **39**, 435 (1974).
42. C. S. Foote, S. Mazur, P. A. Burns, and D. Lerdal, *J. Am. Chem. Soc.* **95**, 586 (1973).
43. M. R. Demuth, P. E. Garrett, and J. D. White, *J. Am. Chem. Soc.* **98**, 634 (1976).
44. M. L. Bates, W. W. Reid, and J. D. White, *Chem. Commun.* p. 44 (1976).
45. C. D. Adams, "Flowering Plants of Jamaica," p. 662. Univ. of the West Indies Press, Mona, Jamaica, 1972.
46. P. S. Manchand, J. D. White, H. Wright, and J. Clardy, *J. Am. Chem. Soc.* **95**, 2705 (1973).

47. C. D. Hufford, R. O. Guerrero, and N. J. Doorenbos, *J. Pharm. Sci.* **65**, 778 (1976).
48. K. M. Brundret, W. Dalziel, B. Hesp, J. A. J. Jarvis, and S. Neidle, *Chem. Commun.* p. 1027 (1972).
49. W. Dalziel, B. Hesp, K. M. Stevenson, and J. A. J. Jarvis, *J. Chem. Soc., Perkin Trans. 1* p. 2841 (1973).
50. The numbering system used throughout this chapter corresponds to the usual convention applied to bridge structures, namely, the longer bridge (from bridgehead as starting point) in the larger ring takes precedence. Note that this is different from the numbering used by Dalziel *et al.* [49] for aphidicolin, which is based on the arbitrary system proposed by McCrindle and Overton (see reference 2c).
51. P. S. Manchand and J. F. Blount, *Chem. Commun.* p. 894 (1975).
52. M. R. Adams and J. D. Bu'Lock, *Chem. Commun.* p. 389 (1975).
53. E. Wenkert, *Chem. Ind. (London)* p. 282 (1955).
54. R. M. Coates and E. F. Bertrain, *J. Org. Chem.* **36**, 3722 (1971).
55. G. Hugel, L. Lods, J. M. Meller, D. W. Theobald, and G. Ourisson, *Bull. Soc. Chim. Fr.* p. 1974 (1963).
56. Y. Hirata, *Pure Appl. Chem.* **41**, 175 (1974); N. R. Farnsworth, R. N. Blomster, W. M. Messmer, J. C. King, G. J. Perinos, and T. D. Wilkes, *Lloydia* **31**, 1 (1969).
57. E. Hecker and R. Schmidt, *Fortschr. Chem. Org. Naturst.* **31**, 377 (1974).
58. W. R. Chan, E. C. Prince, P. S. Manchand, J. P. Springer, and J. Clardy, *J. Am. Chem. Soc.* **97**, 4437 (1975).
59. Cyathin A₃, a diterpene isolated from the fungus *Cyathus helenae*, also has an angularly fused 5,6,7-tricyclic ring system [W. A. Ayer and H. Taube, *Tetrahedron Lett.* p. 1917 (1972)]. Its substitution pattern, however, indicates a quite different biogenetic origin from that of the crotofolins.
60. W. R. Chan, E. C. Prince, P. S. Manchand, J. P. Springer, and J. Clardy, unpublished results.
61. W. E. Duncan, E. Ott, and E. E. Reid, *Ind. Eng. Chem.* **23**, 381 (1931). For a recent synthesis of casbene, see L. Crombie, G. Kneen, and G. Pattenden, *Chem. Commun.* p. 66 (1976). Casbene has tentatively been assigned an all-*trans* configuration.
62. H. Kobayashi and S. Akiyoshi, *Bull. Chem. Soc. Jpn.* **35**, 1044 (1962); **36**, 823 (1963); W. G. Dauben, W. E. Thiessen, and P. R. Resnick, *J. Org. Chem.* **30**, 1693 (1965). For a recent synthesis of cembrene, see W. G. Dauben, G. H. Beasley, M. D. Broadhurst, B. Muller, D. J. Peppard, P. Pesnelle, and S. Suter, *J. Am. Chem. Soc.* **97**, 4973 (1975).
63. E. L. Ghisalberti, P. R. Jeffries, T. G. Payne, and G. K. Worth, *Tetrahedron Lett.* p. 4599 (1970).
64. P. Narayanan, M. Rohrl, K. Zechmeister, D. W. Engel, W. Hoppe, E. Hecker, and W. Adolf, *Tetrahedron Lett.* p. 1325 (1971); M. Hergenhahn, S. Kusumoto, and E. Hecker, *Experientia* **30**, 1438 (1974).
65. S. M. Kupchan, C. W. Sigel, M. J. Matz, C. J. Gilmore, and R. F. Bryan, *J. Am. Chem. Soc.* **98**, 2295 (1976). This article corrects an earlier assignment [S. M. Kupchan, C. W. Sigel, M. J. Matz, J. A. Saenz Renauld, R. C. Haltiwanger, and R. F. Bryan, *J. Am. Chem. Soc.* **92**, 4476 (1970)] to jatrophone by revision of 5,6-double bond configuration to *cis*.
66. D. Uemura, Y. Hirata, Y. Chen, and H. Hsu, *Tetrahedron Lett.* p. 1697 (1975).
67. G. H. Stout, W. G. Balkenhol, M. Poling, and G. L. Hickernell, *J. Am. Chem. Soc.* **92**, 1070 (1970).

68. K. Zechmeister, F. Brandl, W. Hoppe, E. Hecker, H. J. Opferkuch, and W. Adolf, *Tetrahedron Lett.* p. 4075 (1970).
69. L. Crombie, M. L. Games, and D. J. Pointer, *J. Chem. Soc. C* p. 1347 (1968). A related phorbolone has been isolated from *Croton rhamnifolius* [K. L. Stuart and L. M. Barett, *Tetrahedron Lett.* p. 2399 (1969)].
70. B. R. Robinson and C. A. West, *Biochemistry* 9, 30 (1970).

Index